MW00992649

CMOS/CCD Sensors
and
Camera Systems

Gerald C. Holst
Terrence S. Lomheim

Copublished by

JCD Publishing
Winter Park, FL 32789

and

Bellingham, Washington USA

Library of Congress Cataloging-in-Publication Data

Holst, Gerald C.
CMOS/CCD sensors and camera systems/Gerald C. Holst, Terrence S. Lomheim.
p. cm.
Includes bibliographical references and index.
ISBN 9780819467300 and ISBN: 9780970774934
1. Charge coupled devices. 2. Metal oxide semiconductors, Complementary. 3. Information display systems. I. Lomheim, Terrence S. II. Title.

TK7871.99.C45H655 2007
621.3815'2--dc22 2007007184

Copublished by

JCD Publishing
2932 Cove Trail
Winter Park, FL 32789
Phone: 407/629-5370
Fax: 407/629-5370
JCDPublishing.com
ISBN: 9780970774934

SPIE
P.O. Box 10
Bellingham, WA 98227-0010
Phone: 360/676-3290
Fax: 360/647-1445
SPIE.org
SPIE Volume PM172
ISBN: 9780819467300

We dedicate this book to our wives, who believed we love our computers more than our families.

PREFACE

Charge-coupled devices (CCDs) were invented by Boyle and Smith in 1970. Since then, considerable literature has been written on CCD physics, fabrication, and operation. In the early 1990s, CMOS detectors became a contender. Each has its advantages and disadvantages. CMOS manufacturing capability can produce smaller-sized detectors, with 1 μm square detectors as a goal. These smaller detectors reduce manufacturing cost, but have a performance impact. The signal-to-noise ratio is relatively low. While even smaller detectors may be possible, the optical blur diameter ultimately limits spatial resolution.

Calling a sample a pixel does not seem bad. But, "pixel" is used to denote a detector, an element of a computer stored data array, and the primary element of a monitor. Unfortunately, there is no a priori relationship between the various device pixels. As indicated in Chapter 1, the various samples are called pixels, datels, and disels. This clarifies the relationship between what the detector array produces and what is presented on the monitor. The general assumption is that there is a one-to-one relationship between these elements – but there is no requirement that this be so. Chapter 2 provides the camera formula for both radiometric units and photometric units. Of prime importance is the relationship between the source spectral characteristics and detector spectral response.

CCD fundamentals are presented in Chapter 3. Devices may be described functionally according to their architecture (frame transfer, interline transfer, etc.) or by application. Certain architectures lend themselves to specific applications. For example, astronomical cameras typically use full-frame arrays whereas consumer video systems use interline transfer devices. Low-light-level devices include the electron multiplying CCD, electron bombardment CCD, and the intensified CCD (a CCD coupled to an image intensifier). Chapter 4 is the CMOS equivalent of Chapter 3. It lists the differences and similarities between CCD and CMOS sensors. Emphasis is placed on the difficulty and limitations of shrinking the detector size (less than 5 microns). Pixel structures (3T, 4T, 5T, and 6T) are compared with a view towards manufacturability.

Array parameters (well capacity, dark pixels, microelenses, and color filter arrays) apply both to CCD and CMOS arrays (Chapter 5). While Chapter 6 describes quantum efficiency, responsivity, and noise sources, the most important parameter is the signal-to-noise ratio. This is provided in radiometric units (scientific applications) and photometric units (consumer applications). The solid state *camera* requires a lens system and signal conditioning to create a viewable image. This forces array sizes to be consistent with standard displays (Chapter 7).

Image quality is linked to the system modulation transfer function (MTF). Chapter 8 provides an introduction to linear system theory, with all the MTFs provided in Chapter 9. A subsystem MTF cannot be analyzed in isolation. Its impact on the system MTF must be known. Sampling artifacts corrupt imagery (Chapter 10). Moiré patterns are created when viewing periodic targets (test targets, picket fences, and Venetian blinds). While moiré patterns are somewhat acceptable in monochrome imagery, in color imagery they are not. Undersampled monochrome imagery may have colored moiré patterns. These objectionable patterns are minimized by optical low-pass filters.

Chapter 11 discusses many different mathematical approaches to image quality. These formulas provide general design guidelines. However, they require imagery to prove a design is "good." Unfortunately, image quality depends on the camera design (as specified by its MTF and SNR) and the scene. Simulated imagery is provided for different scenes. While general content scenes (e.g., landscapes and Lena) will look "good," periodic targets accentuate sampling artifacts and may be considered "poor." Perhaps the starting point in system design should be the observer. The distance from the display significantly affects perceived image quality. Printers, monitors, and televisions are designed for an anticipated viewing distance.

When appropriate, we have relied on the works of others. This is evidenced by the extensive reference list at the end of each chapter. As a still-emerging technology, current CMOS design and performance parameters have not been published. We appreciate Jim Janesick's review and comments on Chapter 4, *Fundamentals of CMOS imagers*. His insight clarified many important concepts. We thank Chris Paul for the CMOS nonlinearity simulation data and Jerry Johnson for carefully reviewing Chapter 4.

Doug Marks, Pinnacle Communication Services, provided final editing and layout, and numerous items of artwork. We appreciated Doug's "instant" response to our requests for drawing modifications. Pat Carson, Aerospace Corporation, provided expert technical assistance in the preparation of our manuscript. Thank you, Pat.

April 2007

Gerald C. Holst
Terrence S. Lomheim

CMOS/CCD Sensors

and

Camera Systems

Table of contents

SYMBOL LIST

SYMBOL	DEFINITION	UNITS
α_{ABS}	spectrally averaged absorption coefficient	1/m
$\alpha_{ABS}(\lambda)$	spectral absorption coefficient	1/m
γ	gamma	numeric
ΔV	TDI velocity error	mm/s
$\Delta\lambda$	wavelength interval	μm
$\Delta\rho$	target-background reflectance difference	numeric
Δf_e	noise equivalent bandwidth	Hz
ΔL	luminance difference	lumen/m^2-sr
$\Delta n_{MINIMUM}$	minimum number of photoelectrons	numeric
Δn_{PE}	target-background photoelectron difference	numeric
Δt	time separation between two pulses	s
ε	charge transfer efficiency	numeric
η	quantum efficiency	numeric
$\eta_{CATHODE}$	photocathode quantum efficiency	numeric
η_{CCD}	CCD quantum efficiency	numeric
$\eta_{PHOTOCATHODE}$	intensifier photocathode quantum efficiency	numeric
η_{SCREEN}	intensifier screen quantum efficiency	numeric
λ	wavelength	μm
λ_{AVE}	average wavelength	μm
λ_{MAX}	maximum wavelength	μm
λ_{MIN}	minimum wavelength	μm
λ_O	selected wavelength	μm
λ_P	wavelength of peak response	μm
ρ	spectrally averaged reflectance	numeric
$\rho(\lambda)$	spectral reflectance	numeric
ρ_B	background reflectance	numeric
ρ_T	target reflectance	numeric
σ_{ATM}	atmospheric absorption	1/m
σ_{IIT}	image intensifier tube 1/e spot size	mm
σ_R	rms value of random motion	rms mm
σ_{SPOT}	display spot 1/e intensity size	mm
σ_{ATM}	atmospheric absorption	1/m
$\Phi_{CATHODE}$	flux incident onto to intensifier photocathode	W
Φ_V	luminous flux	lumens
A_D	detector area	m^2

A_I	area of source in image plane	m^2
A_O	lens area	m^2
A_{PIXEL}	projected CCD pixel on intensifier photocathode	m^2
A_S	source area	m^2
c	speed of light, $c = 3 \times 10^8$	m/s
c_1	first radiation constant $c_1 = 3.7418 \times 10^8$	W-$\mu m^4/m^2$
c_2	second radiation constant $c_2 = 1.4388 \times 10^4$	μm-K
c_3	third radiation constant $c_3 = 1.88365 \times 10^{27}$	photons-μm^3/s-m^2
C	sense node capacitance	farad
C_{NODE}	sense node integrated capacitance	farad
C_O	target's inherent contrast	numeric
C_{RO}	target's contrast at entrance aperture	numeric
d	detector width	mm
d_{CC}	detector pitch	mm
d_{CCH}	horizontal detector pitch	mm
d_{CCV}	vertical detector pitch	mm
d_{CIRCLE}	circular detector diameter	mm
d_{ERROR}	target displacement in TDI systems	mm
d_H	horizontal detector size	mm
d_O	selected target dimension	mm
d_{OLPF}	distance between spots created by birefringent crystal	mm
d_{FPW}	flat panel horizontal element size	mm
d_T	target detail	mm
d_V	vertical detector size	mm
D	observer to display distance	m
D_O	optical diameter	mm
D_W	flat panel horizontal size	mm
DAS_H	horizontal detector angular subtense	mrad
DAS_V	vertical detector angular subtense	mrad
E_e	spectral radiant incidance	W/(m^2-μm)
E_G	detector band gap	eV
E_T	impurity band gap	eV
E_v	luminous incidence	Candela (cd)
f_{BOOST}	boost frequency	Hz
f_c	cutoff frequency of an ideal circuit	Hz
f_{CLOCK}	pixel clock rate	Hz
f_{e3dB}	one-half power frequency	Hz
f_e	electrical frequency	Hz
f_{EYE}	spatial frequency at eye	cycles/deg
f_N	Nyquist frequency	cycles/mm
f_o	selected spatial frequency	cycles/mm
f_{PEAK}	peak frequency of eye MTF	cycles/deg
f_S	sampling frequency	cycles/mm

f_v	electrical frequency in the video domain	Hz
f_{v3dB}	one-half power frequency (video domain)	Hz
f_{vN}	Nyquist frequency in the video domain	Hz
f_{vS}	sampling frequency in the video domain	Hz
f_x	horizontal spatial frequency	cycles/mm
f_y	vertical spatial frequency	cycles/mm
fl	focal length	m
F	focal ratio (f-number)	numeric
F_{MAX}	maximum frame rate	Hz
F_R	frame rate	Hz
g	EMCCD unit cell gain	numeric
G	on-chip amplifier gain	numeric
G_1	off-chip amplifier gain	numeric
G_{MCP}	microchannel gain	numeric
h	Planck's constant, $h = 6.626 \times 10^{-34}$	J-s
h_c	target critical dimension	m
H	target height	m
$H_{MONITOR}$	monitor height	m
$HFOV$	horizontal field of view	mrad
J_D	dark current density	A/cm^2
k	Boltzmann's constant, $k = 1.38 \times 10^{-23}$	J/K
k_1	a constant	numeric
k_{HIGH}	Super CCD gain multiplier	numeric
k_{LOW}	Super CCD gain multiplier	numeric
K_M	luminous efficacy, $K_M = 683$ (photopic)	lumens/W
k_{MCP}	microchannel excess noise	numeric
L_B	background luminance	lumen/m^2-sr
L_D	depletion length	μm
L_{DIFF}	diffusion length	μm
$L_{DISPLAY}$	display brightness	lumen/m^2
L_e	spectral radiant sterance	W/(m^2-μm-sr)
L_q	spectral photon sterance	photons/(s-m^2-μm-sr)
L_T	target luminance	lumen/m^2-sr
Lv	luminance sterance	nit
m	index	numeric
M	EMCCD gain	numeric
M_e	spectral radiant exitance	W/(m^2-μm)
M_{OPTICS}	optical magnification	numeric
M_P	spectral power	W/μm
M_q	spectral photon exitance	photons/(s-m^2-μm)
M_v	luminous exitance	lumens/m^2

n	index	numeric
$n_{CATHODE}$	number of electrons created by a photocathode	numeric
n_{DARK}	number of dark electrons	numeric
$n_{DETECTOR}$	number of photons incident onto detector	numeric
n_e	number of electrons	numeric
n_{EYE}	number of photons incident onto eye	numeric
n_{HIGH}	number of photoelectrons created by a Supper CCD	numeric
n_{IMAGE}	number of photons incident onto image plane	numeric
n_{LENS}	number of photons incident onto lens	numeric
n_{LOW}	number of photoelectrons created by a Supper CCD	numeric
n_{MCP}	number of photons incident on microchannel plate	numeric
n_{PE}	number of photoelectrons	numeric
$n_{PHOTOCATHODE}$	number of photoelectrons created by a photocathode	numeric
$n_{PE\text{-}B}$	background photoelectrons	numeric
$n_{PE\text{-}T}$	target photoelectrons	numeric
$n_{PHOTON\text{-}CCD}$	number on photons incident onto CCD	numeric
n_{READ}	pixels between active array and sense node	numeric
n_{SCREEN}	photons incident onto intensifier screen	numeric
n_{WINDOW}	number of photons incident onto intensifier	numeric
N	index	numeric
$N_{DETECTORS}$	total number of detectors	numeric
N_H	number of horizontal detectors	numeric
N_{LINE}	number of raster lines	numeric
N_T	equivalent number of cycles on target	cycles
N_{TDI}	number of TDI elements	numeric
N_{TRANS}	number of charge transfers	numeric
N_{TV}	display resolution	TVL/PH
N_V	number of vertical detectors	numeric
N_{WELL}	charge well capacity	numeric
P	display line spacing or pixel spacing	mm/line
PAS_H	horizontal pixel angular subtense	mrad
PAS_V	vertical pixel angular subtense	mrad
q	electron charge, $q = 1.6 \times 10^{-19}$	coul
R	range to target	m
R_1	distance from lens to source	m
R_2	distance from lens to detector	m
R_{AVE}	spectral averaged responsivity	V/(μJ-cm^{-2}) or DN/(μJ-cm^{-2})
R_e	average spectral responsivity	A/W
$R_e(\lambda)$	spectral responsivity	A/W
R_P	peak responsivity	A/W
$R_{PHOTOMETRIC}$	responsivity	V/lux
R_q	spectrally averaged quantum efficiency	numeric
$R_q(\lambda)$	spectral quantum efficiency	numeric
R_R	raster resolution	lines/mm
R_{SYS}	system angular resolution	mrad
R_{TVL}	horizontal display resolution	TVL/PH

R_V	responsivity	A/lumen
$R_{VERTICAL}$	vertical display resolution	lines
S	display spot size (FWHM intensity)	mm
t_{ARRAY}	time to clock out the full array	s
t_{CLOCK}	time between pixels	s
t_{H-LINE}	time to read one pixel line	s
t_{INT}	integration time	s
t_{LINE}	video active line time	s
T	absolute temperature	Kelvin
$T/\#$	T-number	numeric
T_{ASPECT}	target aspect ratio	numeric
T_{ATM}	spectrally averaged atmospheric transmittance	numeric
$T_{ATM}(\lambda)$	spectral atmospheric transmittance	numeric
T_{fo}	fiber optic bundle transmittance	numeric
T_{illum}	absolute temperature of illuminating source	Kelvin
$T_{IR-FILTER}$	spectrally averaged transmittance of IR filter	numeric
T_{OPTICS}	spectrally averaged optical transmittance	numeric
$T_{OPTICS}(\lambda)$	spectral optical transmittance	numeric
$T_{RELAY\ LENS}$	relay lens transmittance	numeric
T_{WINDOW}	transmittance of intensifier window	numeric
u	horizontal spatial frequency object space)	cycles/mrad
u_d	horizontal spatial frequency (display space)	cycles/mm
u_i	horizontal spatial frequency (image space)	cycles/mm
u_{iC}	horizontal optical cutoff (image space)	cycles/mm
u_{iD}	horizontal detector cutoff (image space)	cycles/mm
u_{iN}	horizontal Nyquist frequency (image space)	cycles/mm
u_P	spatial frequency (printer space)	cycles/mm
u_r	radial spatial frequency (image space)	cycles/mm
u_{RES}	system resolution (image space)	cycles/mm
U_{FPN}	fixed pattern noise	numeric
U_{PRNU}	photoresponse nonuniformity	numeric
v	vertical spatial frequency	cycles/mm
v_d	vertical spatial frequency (display space)	cycles/mm
v_i	vertical spatial frequency (image space)	cycles/mm
v_{iD}	vertical detector cutoff (image space)	cycles/mm
v_{iN}	vertical Nyquist frequency (image space)	cycles/mm
$V(\lambda)$, $V'(\lambda)$	photopic, scotopic eye response	numeric
V_{CAMERA}	camera output voltage	V
V_{GRID}	voltage on CRT grid	V
V_{LSB}	voltage of the least significant bit	V
V_{MAX}	maximum signal	numeric
V_{MIN}	minimum signal	numeric
V_{NOISE}	noise voltage after on-chip amplifier	V rms
V_{OUT}	voltage after on-chip amplifier	V

V_{RESET}	reset voltage after on-chip amplifier	V
V_{SCENE}	video voltage before gamma correction	V
V_{SIGNAL}	signal voltage after on-chip amplifier	V
V_{VIDEO}	video voltage after gamma correction	V
$VFOV$	vertical field of view	mrad
W	target width	m
$W_{MONITOR}$	monitor width	m
x	horizontal distance	m
y	vertical distance	m
$\langle n_1 \rangle$	noise source 1	rms electrons
$\langle n_{ADC} \rangle$	quantization noise	rms electrons
$\langle n_{CCD\text{-}DARK} \rangle$	dark current CCD noise in an ICCD	rms electrons
$\langle n_{CCD\text{-}PHOTON} \rangle$	noise before CCD in an ICCD	rms electrons
$\langle n_{CCD} \rangle$	CCD noise in an ICCD	rms electrons
$\langle n_{DARK} \rangle$	dark current shot noise	rms electrons
$\langle n_{FLOOR} \rangle$	noise floor	rms electrons
$\langle n_{FPN} \rangle$	fixed pattern noise	rms electrons
$\langle n_{MCP} \rangle$	microchannel noise	rms electrons
$\langle n_m \rangle$	noise source m	rms electrons
$\langle n_{OFF\text{-}CHIP} \rangle$	off-chip amplifier noise	rms electrons
$\langle n_{ON\text{-}CHIP} \rangle$	on-chip amplifier noise	rms electrons
$\langle n_{PATTERN} \rangle$	pattern noise	rms electrons
$\langle n_{PC\text{-}DARK} \rangle$	intensifier dark current shot noise	rms electrons
$\langle n_{PC\text{-}SHOT} \rangle$	intensifier photon shot noise	rms electrons
$\langle n_{PE} \rangle$	photon shot noise	rms electrons
$\langle n_{PRNU} \rangle$	photoresponse nonuniformity noise	rms electrons
$\langle n_{RESET} \rangle$	reset noise	rms electrons
$\langle n_{SCREEN} \rangle$	intensifier screen noise	rms electrons
$\langle n_{SHOT} \rangle$	shot noise	rms electrons
$\langle n_{SYS} \rangle$	system noise	rms electrons
$\langle V_{NOISE} \rangle$	noise voltage	rms V

1
INTRODUCTION

Charge-coupled devices (CCDs) were invented by Boyle and Smith[1,2] in 1970. Since then, considerable literature[3-9] has been written on CCD physics, fabrication, and operation. However, the array does not create an image by itself. It requires an optical system to image the scene onto the array's photosensitive area. The array requires a bias and clock signals. Its output is a series of analog pulses that represent the scene intensity at a series of discrete locations.

Devices may be described functionally according to their architecture (frame transfer, interline transfer, etc.) or by application. Certain architectures lend themselves to specific applications. For example, astronomical cameras typically use full frame arrays whereas consumer video systems use interline transfer devices.

The heart of the solid state camera is the solid state array. It provides the conversion of light intensity into measurable voltage signals. With appropriate timing signals, the temporal voltage signal represents spatial light intensities. When the array output is amplified and formatted into a standard video format, a solid state camera is created. Because CCDs were the first solid state detectors, cameras are popularly called CCD cameras even though they may contain charge injection devices (CIDs) or complementary metal-oxide-semiconductor (CMOS) devices as detectors. These are solid state cameras – more commonly called digital cameras.

The array specifications, while the first place to start an analysis, are only part of the overall system performance. The system image quality depends on all the components. Array specifications, capabilities, and limitations are the basis for the camera specifications. Camera manufacturers cannot change these. A well-designed camera will not introduce additional noise nor adversely affect the image quality provided by the array.

A camera is of no use by itself; its value is only known when an image is evaluated. The camera output may be directly displayed on a monitor, stored on videotape or disk for later viewing, or processed by a computer. The computer may be part of a machine vision system, be used to enhance the imagery, or be used to create hard copies of the imagery. If interpretation of image quality is performed by an observer, then the observer becomes a critical component of the imaging system. Consideration of human visual system attributes should probably be the starting point of the camera design. But some machine vision systems may not create a "user friendly" image. These systems are designed from a traditional approach: resolution, high signal-to-noise ratio, and ease of operation.

Effective camera design requires an orderly integration of diverse technologies and languages associated with radiation physics, optics, solid state sensors, electronic circuitry, and image processing algorithms. Each field is complex and is a separate discipline.

Electro-optical imaging system analysis is a mathematical construct that provides an optimum design through appropriate tradeoff analyses. A comprehensive model includes the target, background, the properties of the intervening atmosphere, the optical system, detector, electronics, display, and the human interpretation of the displayed information. While any of these components can be studied in detail separately, the electro-optical imaging system cannot.

1.1. SOLID STATE DETECTORS

CCD refers to a semiconductor architecture in which charge is read out of storage areas. The CCD architecture has three basic functions: a) charge collection, b) charge transfer, and c) conversion of the charge into a measurable voltage. The basic building block of the CCD is the metal-oxide-semiconductor (MOS) capacitor. The capacitor is called a gate. By manipulating the gate voltages, charge can be either stored or transferred. Charge generation in most devices occurs under a MOS capacitor (also called a photogate). For some devices (notably interline transfer devices), photodiodes create the charge. After charge generation, the transfer occurs in the MOS capacitors for all devices.

Because most sensors operating in the visible region use CCD-type architectures to read the signal, they are popularly called CCD cameras. For these devices, charge generation is often considered as the initial function of the CCD. More explicitly, these cameras should be called solid state cameras with a CCD readout. As film-based cameras are replaced by digital still cameras, they will simply be referred to as a camera with "digital" understood.

CCDs and detectors can be integrated either monolithically or as hybrids. Monolithic arrays combine the detector and CCD structure on a single chip. The most common detectors are sensitive in the visible region of the spectrum. They use silicon photogates or photodiodes and are monolithic devices. CCDs have been successfully used for infrared detectors such as Schottky barrier devices that are sensitive to radiation in the 1.2 μm to 5 μm spectral region.

Hybrid arrays avoid some pitfalls associated with growing different materials on a single chip and provide a convenient bridge between well-developed but otherwise incompatible technologies. HgCdTe (sensitive to 8 to 12 μm radiation) is bump bonded to a CCD readout using indium as the contact and, as such, is a hybrid array.

With charge-injection devices (CIDs), the pixels consist of two MOS capacitors whose gates are separately connected to rows and columns. Usually the column capacitors are used to integrate charge while the row capacitors sense the charge after integration. With the CID architecture, each pixel is addressable; i.e., it is a matrix-addressable device.

CID readout is accomplished by transferring the integrated charge from the column capacitors to the row capacitors. After this nondestructive signal readout, the charge moves back to the columns for more integration or is injected (discarded) back into the silicon substrate. By suspending charge injection, the user initiates "multiple frame integration" (time lapse exposure) and can view the image on a display as the optimum exposure develops. Integration may proceed for a few milliseconds up to several hours. With individual capacitors on each sensing pixel, blooming is not transported so overloads cannot propagate.

In the 1990s, active pixels[10-15] were introduced. These devices are fabricated with complementary-metal-oxide-semiconductor (CMOS) technology. The advantage is that one or more active transistors can be integrated into the pixel. As such, they become fully addressable (can read selected pixels) and can perform on-chip image processing. CMOS technology is appealing to those applications requiring low power consumption and minimal support circuitry (e.g., mobile phones).

1.2. IMAGING SYSTEM APPLICATIONS

There are four broad applications: general imagery (which includes professional television broadcast, consumer camcorder systems and digital still cameras), machine vision, scientific, and military. Trying to appeal to all four applications, manufacturers use words such as low noise, high frame rate, high resolution, reduced aliasing, and high sensitivity. These words are simply adjectives with no specific meaning. They only become meaningful when compared (i.e., camera A has low noise compared to camera B).

Table 1-1 lists several design categories. While the requirements vary by category, a camera may be used for multiple applications. For example, a consumer video camera often is adequate for many scientific experiments. A specific device may not have all the features listed or may have additional features not listed. The separation between professional broadcast, consumer video, machine vision, scientific, and military devices becomes fuzzy as technology advances.

Color cameras are used for professional television, camcorder, and film replacement systems (digital still cameras). With machine vision systems, color cameras are used to verify the color consistency of objects such as printed labels or paint mixture colors. While color may not be the primary concern, it may be necessary for image analysis when color is the only information to distinguish boundaries. While consumers demand color camera systems, this is not true for other applications. Depending on the application, monochrome (black and white) cameras may be adequate. A monochrome camera has higher sensitivity and therefore is the camera of choice in low-light conditions.

Image enhancement helps an observer extract data. Some images belong to a small precious data set (e.g., remote sensing imagery from interplanetary probes). The images must be processed repeatedly to extract every piece of information. Some images are part of a data stream that is examined once (e.g., real-time video) and others have become popular and are used routinely as standards. These include the three-bar or four-bar test patterns, Lena, and the African mandrill.

The camera cannot perfectly reproduce the scene. The array spatially samples the image, and noise is injected by the array electronics. Spatial sampling creates ambiguity in target edge location and produces moiré patterns when viewing periodic targets. While this is a concern to scientific and military applications, it typically is of little consequence to the average professional television broadcast and consumer markets.

To some extent the goals listed in Table 1-1 are incompatible, thereby dictating design compromises. The demand for machine vision systems is increasing dramatically. Smaller target detail can be discerned with magnifying optics. However, this reduces the field-of-view. For a fixed field of view, higher resolution (more pixels) cameras are required to discern finer detail. When selecting an imaging system, the environment, camera, data storage, and final image format must be considered (Table 1-2).

Table 1-1
DESIGN GOALS

DESIGN CATEGORY	GENERAL IMAGERY	MACHINE VISION	SCIENTIFIC	MILITARY
Image processing algorithms	Gamma correction Extended dynamic range	Application specific	Menu-driven multiple options	Application specific
Image processing time	Real time	Application-specific with emphasis on high speed operation	Real-time not usually required	Real time
Resolution	Matched to video format (e.g., EIA 170)	For a fixed field of view, increased resolution is desired	High resolution	High resolution
Dynamic range	8 bits/color	Up to 10 bits/color or black and white	Up to 16 bits	10 or 12 bits
Sensitivity (high signal-to-noise ratio)	High-contrast targets (noise not necessarily a dominant design factor)	Application-specific (not necessarily an issue because lighting can be controlled)	Low-noise operation	Low-noise operation

Table 1-2
SYSTEM DESIGN CONSIDERATIONS

ENVIRONMENT	CAMERA	TRANSMISSION and STORAGE	DISPLAY
Target size Target reflectance Distance to target Atmospheric transmittance Lighting conditions	Noise Frame rate Sensitivity Detector size Array format Dynamic range Color capability	Data rate Type of storage Storage capacity Video compression	Hard copy Soft copy Resolution

Finally, the most convincing evidence of system performance is image quality. Every time an image is transferred to another device, that device's tonal transfer and modulation transfer function affect the displayed image. Analog video recorders degrade resolution. Imagery will look different on different displays. Similarly, hard copies produced by different printers may appear different, and there is no guarantee that the hard copy will look the same as the soft copy in every respect. Hard copies should only be considered as representative of system capability. Even with its over 100-year history, wet film developing and printing must be controlled with extreme care to create "identical" prints.

1.2.1. GENERAL IMAGERY

Cameras for the professional broadcast television and consumer camcorder markets are designed to operate in real time with an output that is consistent with a standard broadcast format. The resolution, in terms of array size, is matched to the bandwidth recommended in the standard. An array that provides an output of 768 horizontal by 484 vertical pixels creates a satisfactory image for conventional television. Eight bits (256 intensity levels or gray levels) provide an acceptable image in the broadcast and camcorder industry. The largest consumer markets are the camcorder and digital still camera markets.

Because solid state cameras have largely replaced image vacuum tubes, the terminology associated with these tubes is also used with solid state cameras. For example, compared to image vacuum tubes, solid state cameras have no image burn-in, no residual imaging, and usually are not affected by microphonics.

1.2.2. MACHINE VISION

In its simplest version, a machine vision system consists of a light source, camera, and computer software that rapidly analyzes digitized images with respect to location, size, flaws, and other preprogrammed data. Unlike other types of image analysis, a machine vision system also includes a mechanism that immediately reacts to images that do not conform to the parameters stored in the computer. For example, defective parts are taken off a production conveyor line.

Machine vision functions include location, inspection, gauging, counting, identification, recognition, and motion tracking. These systems do not necessarily need to operate at a standard frame rate. For industrial inspection, linear arrays operating in the time delay and integration (TDI) mode can be used to measure objects moving at a high speed on a conveyor belt.

While a multitude of cues are used for target detection, recognition, or identification, machine vision systems cannot replace the human eye. The eye processes intensity differences over 11 orders of magnitude, color differences, and textual cues. The solid state camera, on the other hand, can process limited data much faster than the human. Many operations can be performed faster, cheaper, and more accurately by machines than by humans. Machine vision systems can operate 24 hours a day without fatigue. They operate consistently, whereas variability exists among human inspectors. Furthermore, these cameras can operate in harsh environments that may be unfriendly to humans (e.g., extreme heat, cold, or ionizing radiation).

1.2.3. SCIENTIFIC APPLICATIONS

For scientific applications, low noise, high responsivity, large dynamic range, and high resolution are dominant considerations. To exploit a large dynamic range, scientific cameras may digitize the signal into 12, 14, or 16 bits. Array linearity and analog-to-digital converter linearity are important. Resolution is specified by the number of detector elements, and scientific arrays may have 5000×5000 detector elements.[14] Theoretically the array can be any size, but manufacturing considerations may ultimately limit the array size.

Low noise means low dark current and low readout noise. The dark current can be minimized by cooling the CCD. Long integration times can increase the signal value so that the readout noise is small compared to the photon shot noise. Although low-light-level cameras have many applications, they tend to be used for scientific applications. There is no industry-wide definition of a "low light level" imaging system. To some, it is simply a solid state camera that can provide a usable image when the lighting conditions are less than 1 lux. To others, it refers to an intensified camera and is sometimes called a low-light-level television (LLLTV) system. An image intensifier amplifies a low-light-level image so that it can be sensed by a solid state camera. The image-intensifier/CCD camera combination is called an intensified CCD or ICCD. The image intensifier provides tremendous amplification but also introduces additional noise.

1.2.4. MILITARY APPLICATIONS

The military is interested in detecting, recognizing, and identifying targets at long distances. This requires high-resolution, low-noise sensors. Target detection is a perceptible act. A human determines if the target is present. The military uses the minimum resolvable contrast (MRC) as a figure of merit.

Solid state cameras are popular because of their ruggedness and small size. They can easily be mounted on remotely piloted vehicles. They are replacing wet-film systems for mapping and photo interpretation.

1.3. CONFIGURATIONS

Imaging systems for the four broad application categories may operate in a variety of configurations. The precise setup depends on the specific requirements. Figure 1-1 is representative of a closed-circuit television system where the camera output is continuously displayed. The overall image quality is determined by the camera capability (which is based on the array specifications), the bandwidth of the video format (e.g., EIA 170, NTSC, PAL, or SECAM), the display performance, and the observer. EIA 170 was formerly called RS 170, and the NTSC standard is also called RS 170A.

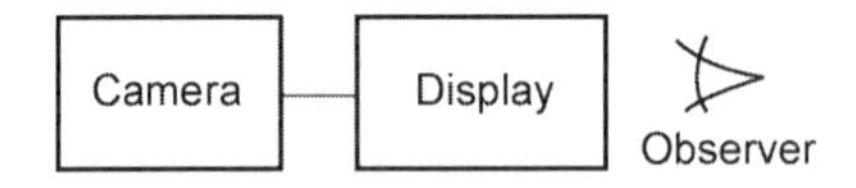

Figure 1-1. A closed circuit television system. Image quality is determined by an observer.

Figure 1-2 illustrates a generic transmission system. The transmitter and receiver must have sufficient electronic bandwidth to provide the desired image quality. For remote sensing, the data may be compressed before the link.

Compression may alter the image and the effects of compression will be seen on selected imagery. But compression effects are not objectionable in most imagery.

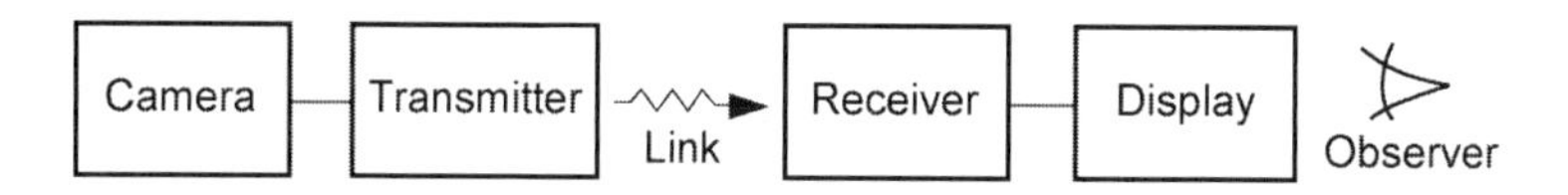

Figure 1-2. Generic remote transmission system. The transmitter and receiver must have adequate electronic bandwidth to support the image quality created by the solid state camera. Conventional television uses analog transmission. It is slowly being replaced with digital transmission.

For remote applications, the image may be recorded on a separate video recorder[15,16] (Figure 1-3) or with a camcorder. The most popular player is the video cassette recorder (VCR). The recorder further modifies the image by reducing the signal-to-noise ratio (SNR) and image sharpness. As the desire for portability increases, cameras and recorders will shrink in size. Flash memories are replacing VCRs with future systems being all-digital.

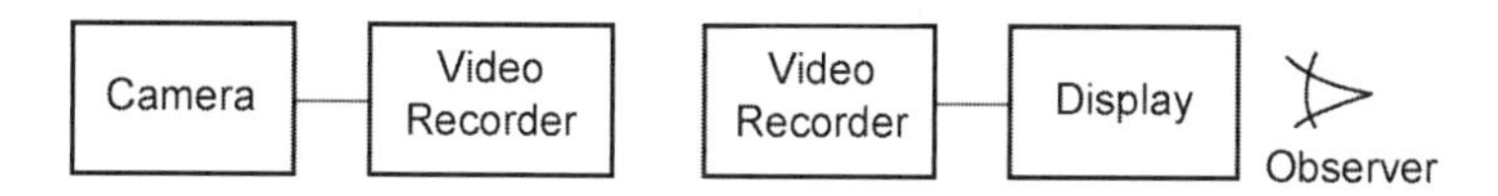

Figure 1-3. Imagery can be stored on analog video tape. However, the recorder circuitry may degrade the image quality. Digital storage has replaced the analog tape.

For scientific applications, the camera output is digitized and then processed by a computer (Figure 1-4). After processing, the image may be presented on a monitor (soft copy), printed (hard copy), or stored. The digital image can also be transported to another computer via the Internet, local area network, CD, or memory stick. For remote applications, the digital data may be stored on a digital recorder (Figure 1-5).

Perhaps the most compelling reason for adopting digital technology is the fact that the quality of digital signals remains intact through copying and reproduction unless the signals are deliberately altered. Digital signal "transmission" was first introduced into tape recorders. Because a bit is either present or not, multiple-generation copies retain high image quality.

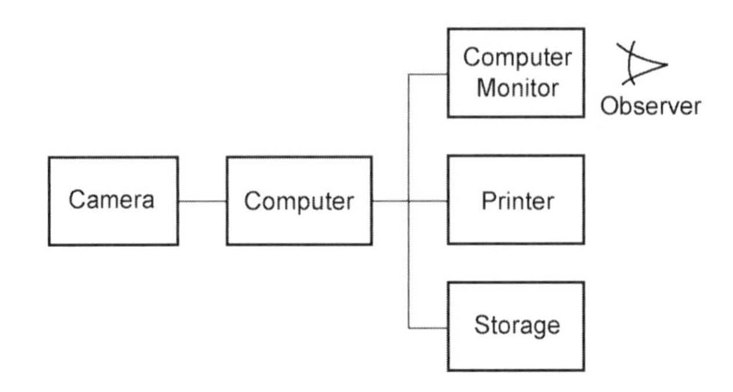

Figure 1-4. Most imagery today is enhanced through image processing. The camera may view a scene directly or may scan a document. The computer output hard copy may be used in newspapers, advertisements, or reports.

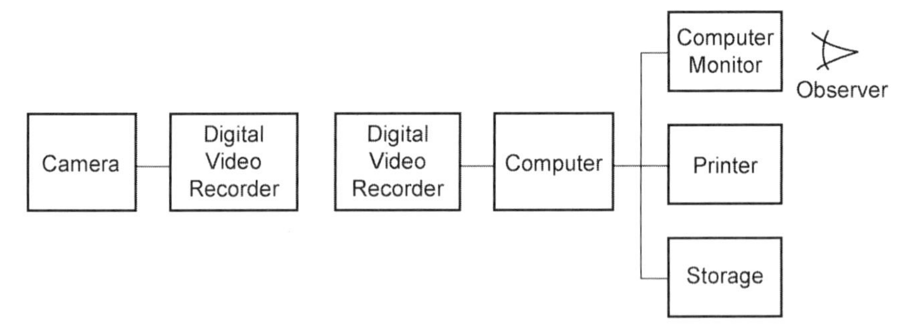

Figure 1-5. Digital systems can provide very high-quality imagery. Electronic bandwidth limitations may impose data compression requirements. Data compression may alter the image, but this alteration may not be obvious on general imagery.

In comparison, analog recorders, such as in video home systems (VHSs), provide very poor quality after just a few generations. The first digital recorder used a format called D-1, which became known as the CCIR 601 component digital standard. Digital recording formats now include[16] D-1, D-2, D-3, D-4, D-5, and D-6. Error-correcting codes can be used to replace missing bits.

New digital recorders[17,18] combine advanced technologies in electronics, video compression, and mechanical transport design. Real-time digital video systems operate at data rates that are faster than many computers. While errors in the imagery are never considered desirable, the eye is very tolerant of defects. In comparison, a computer is considered worthless if the error rate is high.

When it comes to displaying images (either hard copy or soft copy), the dynamic range of the camera digitizer is often greater than the display device. A camera may offer 12, 14, or 16 bits whereas the display may only be 8 bits. Here, a look-up table is employed to match the camera output to the display. This may be a simple linear relationship or specific expansion of a part of the camera's gray scale. False color can also be introduced to enhance the displayed image. False color is often useful for the human observer, but serves no function in machine vision systems. The greatest challenge is matching the camera's dynamic range to the scene dynamic range and the conversion of this range into the limited display dynamic range.

While a machine vision system does not require a monitor (Figure 1-6), a monitor is often used during system setup and for diagnostic evaluation. That is, a computer algorithm compares an image to a stored standard. If the target does not compare favorably to the standard, then the process is changed. For example, this may mean sending a rejected item for rework or changing the light intensity.

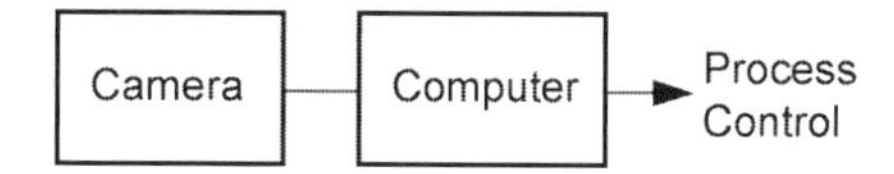

Figure 1-6. Machine vision systems do not require a monitor. The computer output controls the manufacturing process. Monitors are used for setup and diagnostic evaluation.

1.4. IMAGE QUALITY

Image quality is a subjective impression of ranking imagery from poor to excellent. It is a somewhat learned skill. It is a perceptual ability, accomplished by the brain, affected by and incorporating other sensory systems, emotions, learning, and memory. The relationships are many and not well understood. Perception varies between individuals and over time. There exist large variations in an observer's judgment as to the correct rank ordering from best to worst and therefore image quality cannot be placed on an absolute scale. Visual psychophysical investigations have not measured all the properties relevant to imaging systems.

Many formulas exist for predicting image quality. Each is appropriate under a particular set of viewing conditions. These expressions are typically obtained from empirical data in which multiple observers view many images with a known amount of degradation. The observers rank order the imagery from worst to best and then an equation is derived that relates the ranking scale to the amount of degradation.

Early metrics were created for film-based cameras. Image quality was related to the camera lens and film modulation transfer functions (MTFs). With the advent of television, image quality centered on the perception of raster lines and the minimum SNR required for good imagery. Here, it was assumed that the studio camera provided a perfect image and only the receiver affected the image quality.

The system MTF is the major component of system analysis. It describes how sinusoidal patterns propagate through the system. Because any target can be decomposed into a Fourier series, the MTF approach indicates how imagery will appear on the display.

Many tests have provided insight into image metrics that are related to image quality. Most metrics are related to the system MTF, resolution, or the signal-to-noise ratio. In general, images with higher MTFs and less noise are judged as having better image quality. There is no single ideal MTF shape that provides the best image quality.

Digital processing is used for image enhancement and analysis. Because the pixels are numerical values in a regular array, mathematical transforms can be applied to the array. The transform can be applied to a single pixel, group of pixels, or the entire image. Many image processing designers think of images as an array of numbers that can be manipulated with little regard to who is the final interpreter.

With a solid state camera system, the lens, array architecture, array noise, and display characteristics all affect system performance. Only an end-to-end assessment will determine the overall image quality. There is no advantage to using a high-quality camera if the display cannot produce a faithful image. Often, the display is the limiting factor in terms of image quality and resolution. No matter how good the camera is, if the display resolution is poor, then the overall system resolution is poor. Only if the display's contrast and spatial resolution are better than the camera will the camera's image quality be preserved.

For consumer applications, system resolution is important and SNR is secondary. Scientific applications may place equal importance on resolution and MTF. The military, interested in target detection, couples the system MTF and system noise to create a perceived contrast. When the perceived contrast is above a threshold value, the target is just detected. The contrast at the entrance aperture depends on the range and atmospheric transmittance.

1.5. PIXELS, DATELS, DISELS, AND RESELS

The overall system may contain several independent sampling systems. The array spatially samples the scene, the computer may have its own digitizer, and the monitor may have a limited resolution. A monitor "pixel" may or may not represent a "pixel" in camera space. The designer and user must understand the differences among the sampling lattices.

Sampling theory describes the requirements that lead to the reconstruction of a digitized signal. The original analog signal is digitized, processed, and then returned to the analog domain. For electrical signals, each digitized value is simply called a sample and the digitization rate is the sampling rate.

Electronic imaging systems are more complex, and several sampling lattices are present. The detector output represents a sampling of the scene. The detector output voltage is digitized and placed in a memory. After image processing, the data are sent to a display medium. Although the display medium provides an analog signal, it is typically digitally controlled.

Each device has its own minimum sample size or primary element. Calling a sample a pixel or a pel (picture element) does not seem bad. Unfortunately, there is no a priori relationship between the various device pixels. The various digital samples in the processing path are called pixels, datels, and disels (sometimes called dixels) (Table 1-3 and Figure 1-7). There is no standard definition for a resel. When a conversion takes place between the analog and digital domains, the resel may be different from the digital sample size. The analog signal rather than the digital sample may limit the system resolution. In oversampled systems, the resel consists of many samples.

Table 1-3
THE "-ELS"

ELEMENT	DESCRIPTION
Pixel or pel (picture element)	A sample created by a detector.
Datel (data element)	Each datum is a datel. Datels reside in a computer memory.
Disel (display element)	The smallest element (sample) that a display medium can access.
Resel (resolution element)	The smallest signal supported by an analog system

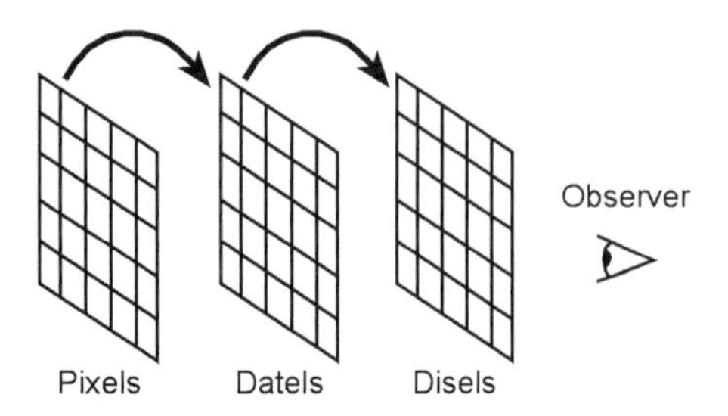

Figure 1-7. Each array is mapped onto the next. The number of elements in each array may be different. Not every array exists in every system.

For staring arrays, the total number of pixels is equal to the number of detectors. The detector's spatial response is determined by the detector's size (e.g., photosensitive area). If the extent is d_H in the horizontal direction and the optics focal length is fl, then the horizontal detector-angular-subtense (DAS_H) is

$$DAS_H = \frac{d_H}{fl} \tag{1-1}$$

Similarly, in the vertical direction,

$$DAS_V = \frac{d_V}{fl} \tag{1-2}$$

Staring arrays are often specified by the detector center-to-center spacing (pitch). The horizontal pixel-angular-subtense (PAS_H) is

$$PAS_H = \frac{d_{CCH}}{fl} \tag{1-3}$$

where d_{CCH} is the horizontal pitch. Similarly, in the vertical direction

$$PAS_V = \frac{d_{CCV}}{fl} \tag{1-4}$$

The fill factor is the ratio of areas:

$$Fill\ factor = \frac{d_H\ d_V}{d_{CCH}\ d_{CCV}} \tag{1-5}$$

With a staring array that has a 100% fill factor, the PAS is equal to the DAS (e.g., $d_H = d_{CCH}$ and $d_V = d_{CCV}$).

If the Airy disk (a resel) is much larger than the PAS, then the optical resel determines the system resolution. If the electronic imaging system output is in a digital format, then the number of datels (samples) equals the number of pixels. If the camera's analog output is digitized, then the number of datels is linked to the frame grabber's digitization rate. This number can be much greater than the number of pixels. This higher number does not create more resolution in terms of the "-els."

Image processing algorithms operate on datels. A datel may not represent a pixel or a resel. Because there must be two samples to define a frequency, the Nyquist frequency associated with pixels, datels, and disels may be different. If there are more datels than pixels, the Nyquist frequency associated with an image processing algorithm is higher than that for the detector array. Most image processing books illustrate datels and call them pixels. The image processing specialist must understand the differences between the sampling lattices and take into account who will be the final data interpreter.

After image processing, the datels are outputted to a display medium. For monitors, each datel is often mapped, one-to-one, onto each disel. Monitors are often specified by the number of addressable pixels (defined as disels in this text). Consider a printer that provides 600 dpi. If the number of pixels is 640×480 and there is one-to-one mapping, the image size would be approximately 1-inch square; there is a one-to-many mapping with printers.

Finally, the system designer must be aware of which subsystem limits the overall system resolution. In some respects, the starting point for system design should begin with the final interpreter of the data. The minimum "-el" should be just discernible by the interpreter to ensure maximum transfer of information. However, the observer may not find this image aesthetically pleasing.

1.6. REFERENCES

1. W. S. Boyle and G. E. Smith, "Charge Coupled Semiconductor Devices," *Bell Systems Technical Journal*, Vol. 49, pp. 587-593 (1970).
2. G. F. Amelio, M. F. Tompsett, and G. E. Smith, "Experimental Verification of the Charge Coupled Concept," *Bell Systems Technical Journal*, Vol. 49, pp. 593-600 (1970).
3. M. J. Howes and D. V. Morgan, eds., *Charge-Coupled Devices and Systems,* John Wiley and Sons, NY (1979).
4. C. H. Sequin and M. F. Tompsett, *Charge Transfer Devices,* Academic Press, NY (1975).
5. E. S. Yang, *Microelectronic Devices*, McGraw-Hill, NY (1988).
6. E. L. Dereniak and D. G. Crowe, *Optical Radiation Detectors*, John Wiley and Sons, NY (1984).

7. A. J. P. Theuwissen, *Solid-State Imaging with Charge-Coupled Devices*, Kluwer Academic Publishers, Dordrecht, The Netherlands (1995).
8. R. Janesick, *Scientific Charge-Coupled Devices,* SPIE Press Bellingham, WA (2001).
9. Numerous articles can be found in the Proceedings of the SPIE conferences: *Sensors, Cameras, and Systems for Scientific/Industrial Applications* and *Semiconductor Photodetectors* SPIE Press, Bellingham, WA.
10. S. K. Mendis, S. E. Kemeny, and E. R. Fossum, "A 128 × 128 CMOS Active Pixel Sensor for Highly Integrated Imaging Systems," *IEEE IEDM Technical Digest*, pp. 583-586 (1993).
11. E. R. Fossum, "Active pixel sensors: Are CCDs dinosaurs?" in *Charge-Coupled Devices and Solid State Optical Sensors* III; Morley M. Blouke; ed., SPIE Proceedings Vol. 1900, 2-12, (1993).
12. E. R. Fossum, "CMOS Image Sensors: Electronic Camera-on-a-chip," *IEEE Transactions on Electron Devices*, Vol. 44(10), pp. 1689-1698 (1997).
13. M. Schanz, W. Brockherde, R. Hauschild, B. Hosticka, and M. Schwarz, "Smart CMOS Image Sensor Arrays," *IEEE Transactions on Electron Devices*, Vol. 44(10), pp. 1699-1704 (1997).
14. S. G. Chamberlain, S. R. Kamasz, F. Ma, W. D. Washkurak, M. Farrier, and P.T. Jenkins, "A 26.3 Million Pixel CCD Image Sensor," in *IEEE Proceedings of the International Conference on Electron Devices*, pp. 151-155, Washington, D.C. December 10, 1995.
15. Z. Q. You and T. H. Edgar, *Video Recorders: Principles and Operation*, Prentice Hall, New York, NY (1992).
16. M. Hobbs, *Video Cameras and Camcorders*, Prentice Hall, New York, NY (1989).
17. J. Hamalainen, "Video Recording Goes Digital," IEEE Spectrum, Vol. 32(4), pp. 76-79 (1995).
18. S. Winkler, *Digital Video Quality: Vision Models and Metrics*, John Wiley & Sons (2005).

2
RADIOMETRY AND PHOTOMETRY

The radiometric/photometric relationship between the scene and camera is of primary importance. It determines the output signal. The second radiometric/photometric consideration is the match between the display's spectral output and the eye's spectral response. Display manufacturers have considered this in detail, which reduces the burden on the system designer. However, the displayed image typically is not a precise reproduction of the scene. While the displayed image usually represents the scene in spatial detail, the intensity and color rendition may be different.

Scenes in the visible and near-infrared are illuminated by the sun, moon, starlight, night glow, or artificial sources. Since both the target and its background are illuminated by the same source, targets are detected when differences in reflectance exist. The camera's output voltage depends on the relationship between the scene spectral content and the spectral response of the camera.

Historically, cameras were designed to operate in the visible region only. For those systems, it was reasonable to specify responsivity in photometric units (e.g., volts/lux). However, when the spectral response extends past the visible region as with silicon photodetectors, the use of photometric units can be confusing and even misleading.

Radiometry describes the energy or power transfer from a source to a detector. When normalized to the eye's response, photometric units are used. Radiometric and photometric quantities are differentiated by subscripts: e is used for radiometric units, q for photons, and v is used for photometric units.

The symbols used in this book are summarized in the *Symbol List*, which appears after the *Table of Contents*.

2.1. RADIATIVE TRANSFER

Spectral radiant sterance, L_e, is the basic quantity from which all other radiometric quantities can be derived. It contains both the areal and solid angle concept[1] that is necessary to calculate the radiant flux incident onto a system. It is the amount of radiant flux, (watts), per unit wavelength (micrometers) radiated into a cone of incremental solid angle (steradians) from a source whose area is measured in meters (Figure 2-1)

$$L_e(\lambda) = \frac{\partial^2 \Phi(\lambda)}{\partial A_S \partial \Omega} \qquad \frac{\mathbf{W}}{\mathbf{m^2 - \mu m - sr}} \tag{2-1}$$

Similarly, L_q is the spectral photon sterance expressed in photons/(s-m^2-μm-sr). The variables L_e and L_q are invariant for an optical system that has no absorption or reflections. That is, L_e and L_q remain constant as the radiation transverses through the optical system. Table 2-1 provides the standard radiometric units. For Lambertian sources, the radiance is emitted into a hemisphere. Integrating Equation 2-1 provides

$$M_e(\lambda,T) = \pi\, L_e(\lambda,T) \qquad \frac{\mathbf{w}}{\mathbf{m^2 - \mu m}} \tag{2-2}$$

and the spectral photon exitance is

$$M_q(\lambda,T) = \pi\, L_q(\lambda,T) \qquad \frac{\mathbf{photons}}{\mathbf{s - m^2 - \mu m}} \tag{2-3}$$

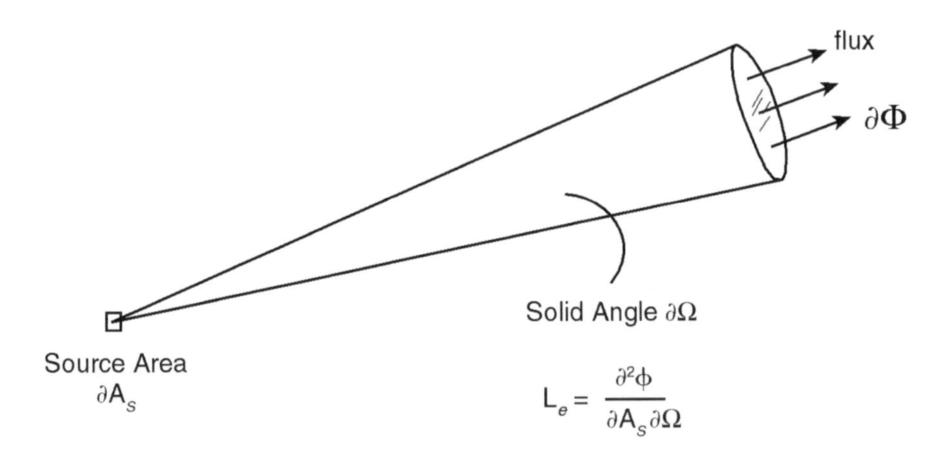

Figure 2-1. Radiant sterance.

Table 2-1
STANDARD SPECTRAL RADIOMETRIC UNITS

DEFINITION	SYMBOL	UNITS
Radiant energy	$\mathbf{Q}_e$	Joule (J)
Radiant flux	$\mathbf{\Phi}_e$	Watts (W)
Spectral radiant intensity	$\boldsymbol{I}_e$	W/(m^2-μm-sr)
Spectral radiant exitance (from a source) (also radiant emittance)	$\boldsymbol{M}_e$	W/(m^2-μm)
Spectral radiant incidance (onto a target) (also irradiance)	$\boldsymbol{E}_e$	W/(m^2-μm)
Spectral radiant sterance (also radiance)	$\boldsymbol{L}_e$	W/(m^2-μm-sr)

2.2. PLANCK'S BLACKBODY LAW

The spectral radiant exitance of an ideal blackbody source whose absolute temperature is $\boldsymbol{T}$ (Kelvin), can be described by Planck's blackbody radiation law

$$M_e(\lambda,T)=\frac{c_1}{\lambda^5}\left(\frac{1}{e^{(c_2/\lambda T)}-1}\right)\quad\frac{\mathbf{W}}{\mathbf{m}^2-\boldsymbol{\mu}\mathbf{m}} \tag{2-4}$$

where the first radiation constant is c_1 = 3.7418×10^8 W-μm^4/m^2, the second radiation constant is c_2 = 1.4388×10^4 μm-K, and λ is the wavelength expressed in micrometers. The value $\boldsymbol{T}$ is also called the color temperature. Figure 2-2 illustrates Planck's spectral radiant exitance in logarithmic coordinates. Since a photodetector responds linearly to the available power, linear coordinates may provide an easier representation to interpret (Figure 2-3). Each curve has a maximum at λ_{PEAK}. Wien's displacement law provides $\lambda_{\mathrm{PEAK}}T$ = 2898 μm-K. A source must have an absolute temperature above about 700 K to be perceived by the human eye.

The spectral photon exitance is simply the spectral radiant exitance divided by the energy of one photon ($\boldsymbol{hc/\lambda}$):

$$M_q(\lambda,T)=\frac{c_3}{\lambda^4}\left(\frac{1}{e^{(c_2/\lambda T)}-1}\right)\quad\frac{\textbf{photons}}{\mathbf{s-m^2-\boldsymbol{\mu}m}} \tag{2-5}$$

where $\boldsymbol{h}$ is Planck's constant ($\boldsymbol{h} = 6.626 \times 10^{-34}$ J-s), $\boldsymbol{c}$ is the speed of light ($\boldsymbol{c} = 3 \times 10^{8}$ m/s), and the third radiation constant is $c_3 = 1.88365\times10^{27}$ photons-µm^3/s-m^2. Figures 2-4 and 2-5 provide the spectral photon exitance in logarithmic and linear coordinates, respectively, for sources typically used for CCD calibration. These curves have a maximum at $\boldsymbol{\lambda}_{\text{PEAK}}\boldsymbol{T} = 3670$ µm-K.

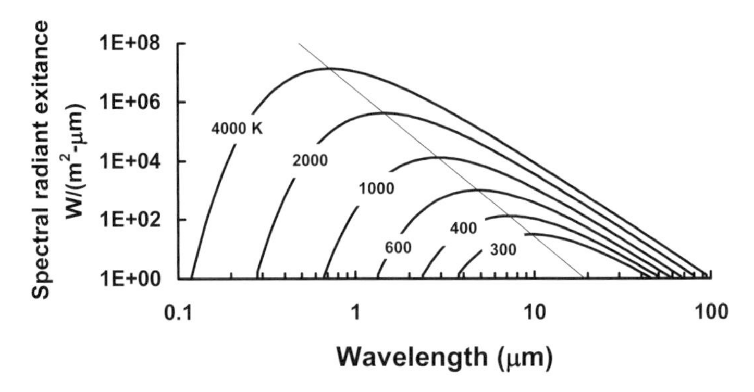

Figure 2-2. Planck's spectral radiant exitance plotted in logarithmic coordinates. The light line is Wien's law.

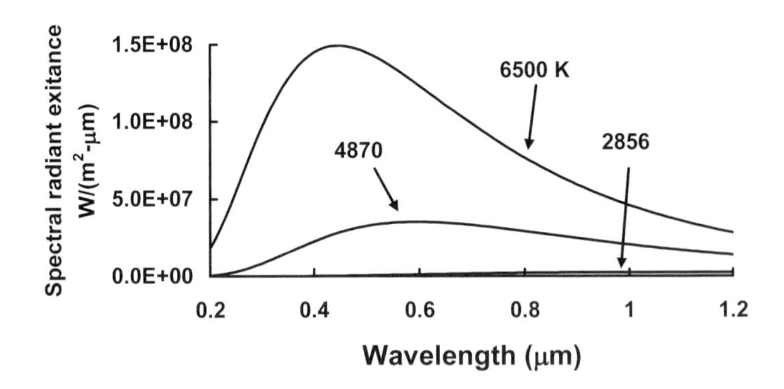

Figure 2-3. Planck's spectral radiant exitance plotted in linear coordinates for typical sources used for calibrating CCD arrays. The maximum values occur at 1.01, 0.60, and 0.45 µm for color temperatures of 2856, 4870, and 6500 K, respectively.

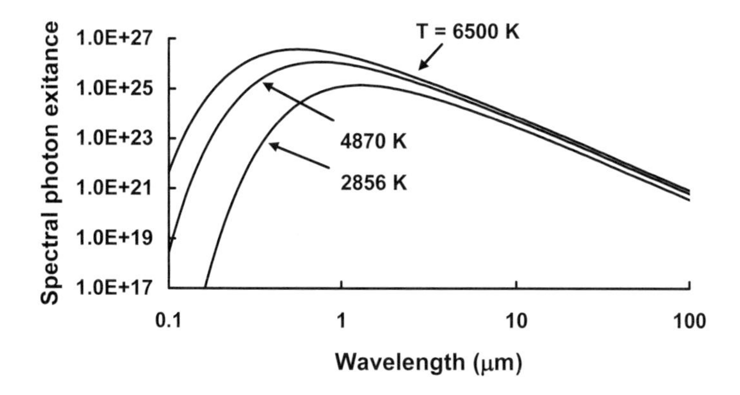

Figure 2-4. Planck's spectral photon exitance for $\boldsymbol{T}$ = 2856, 4870, and 6500 K. The maximum values occur at 1.29, 0.75, and 0.56 μm, respectively. The units are photons/(s-m^2-μm).

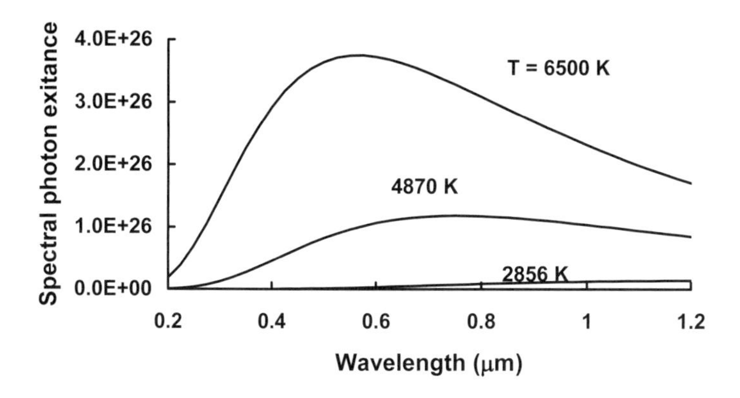

Figure 2-5. Planck's spectral photon exitance plotted in linear coordinates. The units are photons/(s-m^2-μm).

2.3. PHOTOMETRY

Photometry describes the radiative transfer from a source to a detector where the units of radiation have been normalized to the spectral sensitivity of the eye. Photometry applies to all systems that are sensitive to visible radiation. The luminous flux emitted by a source is

$$\Phi_v = K_M \int_{0.38}^{0.72} V(\lambda) M_P(\lambda, T)\, d\lambda \quad \text{lumens} \tag{2-6}$$

The variable $M_P(\lambda,T)$ has units of watts/μm and K_M is the luminous efficacy for photopic vision. It is 683 lumens/W at the peak of the photopic curve ($\lambda = 0.55$ μm) and 1746 lumens/W for the scotopic region at $\lambda = 0.505$ μm. Although both photopic and scotopic units are available, usually only the photopic units are used (Table 2-2 and Figure 2-6). Displays are bright and printed imagery is viewed under "average" lighting conditions. In both cases, the eye's cones are operating (photopic response).

Table 2-2
PHOTOPIC and SCOTOPIC EYE RESPONSE

Wavelength nm	Photopic V(λ)	Scotopic V'(λ)	Wavelength nm	Photopic V(λ)	Scotopic V'(λ)
380		0.00059	570	0.952	0.2076
390	0.00012	0.00221	580	0.870	0.1212
400	0.0004	0.00929	590	0.757	0.0655
410	0.0012	0.03484	600	0.631	0.03315
420	0.0040	0.0966	610	0.503	0.01593
430	0.0116	0.1998	620	0.381	0.00737
440	0.023	0.3281	630	0.265	0.00335
450	0.038	0.455	640	0.175	0.00150
460	0.060	0.567	650	0.107	0.00067
470	0.091	0.676	660	0.061	0.00031
480	0.139	0.793	670	0.032	
490	0.208	0.904	680	0.017	
500	0.323	0.982	690	0.0082	
510	0.503	0.997	700	0.0041	
520	0.710	0.935	710	0.0021	
530	0.862	0.811	720	0.00105	
540	0.954	0.650	730	0.00052	
550	0.995	0.481	740	0.00025	
560	0.995	0.3288	750	0.00012	

The photopic response can be approximated[2] by

$$V(\lambda) \approx 1.019 e^{-285.4(\lambda - 0.559)^2} \tag{2-7}$$

and the scopotic response by[2]

$$V'(\lambda) \approx 0.992 e^{-321(\lambda - 0.503)^2} \tag{2-8}$$

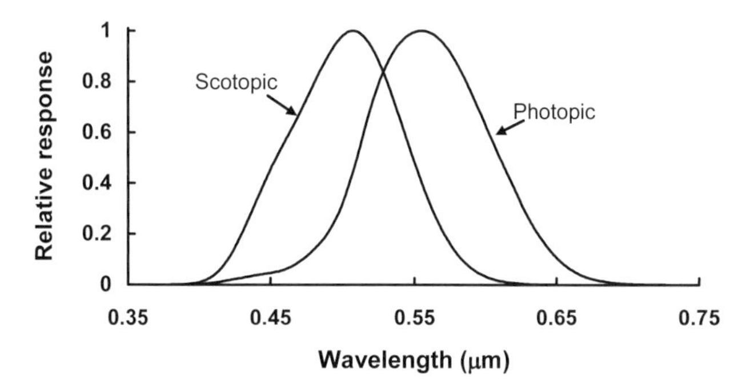

Figure 2-6. Photopic and scotopic eye responses.

2.3.1. UNITS

Unfortunately, there is overabundance of terminology being used in the field of photometry (Table 2-3). The SI units are recommended. Luminance emittance and illuminance are sometimes used as alternatives to luminous exitance and luminous incidance, respectively. Brightness and luminance may be used in place of luminous sterance. Luminance sterance is also called the normalized intensity or nit. For Lambertian sources, the luminance is emitted into a hemisphere whose solid angle is 2π. The luminous exitance of a Lambertian surface is simply the luminance exitance given in Table 2-3 divided by π. Figure 2-7 illustrates the geometric relationship between the SI, CGS, and English luminous incidance units. The numeric relationship is provided in Table 2-4.

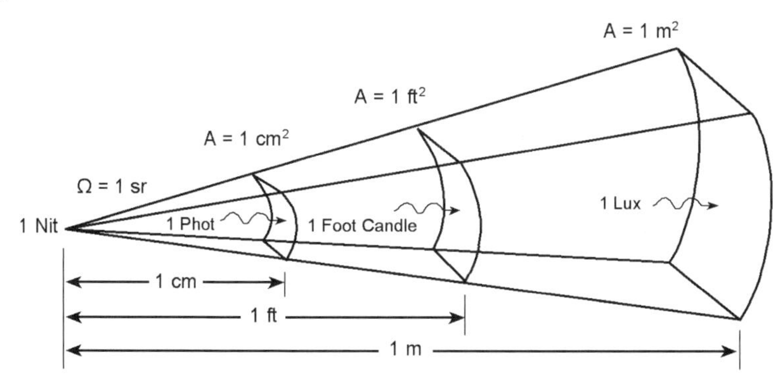

Figure 2-7. Geometric relationship between the SI, CGS, and English luminous incidance units. The solid angle is one steradian. (Not to scale).

Table 2-3
STANDARD PHOTOMETRIC UNITS
(The SI units are recommended)

DEFINITION	SYMBOL	SI and MKS UNITS	CGS UNITS	ENGLISH UNITS
Luminous energy	$\boldsymbol{Q_v}$	Talbot (T)	Talbot (T)	Talbot (T)
Luminous flux	$\boldsymbol{\Phi v}$	Lumen (lm)	Lumen (lm)	Lumen (lm)
Luminous intensity	$\boldsymbol{I_v}$	Candela (cd) lm/sr	Candela (cd) lm/sr	Candela (cd) lm/sr
Luminous exitance (from a source) (also luminous emittance)	$\boldsymbol{M_v}$	Lux (lx) lumen/m^2	Phot (ph) lumen/cm^2	Footcandle (fc) limen/ft^2
Luminous incidance (onto a target) (also illuminance)	$\boldsymbol{E_v}$	Lux (lx) lumen/m^2	Phot (ph) lumen/cm^2	Footcandle (fc) limen/ft^2
Luminous sterance (also brightness and luminance)	$\boldsymbol{L_v}$	Nit Lumen/m^2-sr or cd/m^2	Stilb (sb) cd/cm^2	candela/ft^2

Table 2-4
CONVERSION BETWEEN SI, CGS, and ENGLISH UNITS

	PHOT	FOOTCANDLE	LUX
1 Phot =	1	929	1×104
1 Footcandle =	1.076×10^{-3}	1	10.764
1 Lux =	1×10^{-4}	0.0929	1

2.3.2. TYPICAL ILLUMINATION LEVELS

Natural lighting levels can vary by over nine orders of magnitude (Table 2-5). The minimum level is limited by night glow and the maximum level is provided by the sun. At very low light levels (less than 5×10^{-3} lux), the eye's rods operate (scotopic response). For light levels above 5×10^{-2} lux, the eye's cones (photopic response) respond. Between these two values, both rods and cones are operating and the eye's response is somewhat between the two values. This composite response is called the mesopic response. Table 2-6 provides typical artificial lighting levels.

The eye automatically adapts to the ambient lighting conditions to provide an optimized image. Cameras operate over a limited region. They may need neutral density filters if the light level is too high. Cameras may have an automatic iris and shutter speed to optimize the image on the detector array. If the light level is too low, an intensified camera may be necessary.

Table 2-5
NATURAL ILLUMINANCE LEVELS

SKY CONDITION	EYE RESPONSE	AVERAGE LUMINOUS INCIDANCE (lux)
Direct sun	Photopic	10^5
Full daylight	Photopic	10^4
Overcast sky	Photopic	10^3
Very dark day	Photopic	10^2
Twilight	Photopic	10
Deep twilight	Photopic	1.0
Full moon	Photopic	10^{-1}
Quarter moon	Mesopic	10^{-2}
Moonless, clear night (starlight)	Scotopic	10^{-3}
Moonless, overcast (night glow)	Scotopic	10^{-4}

Table 2-6
TYPICAL ARTIFICIAL ILLUMINANCE LEVELS

LOCATION	AVERAGE LUMINOUS INCIDANCE (lux)
Hospital operating theater	10^5
TV studio	10^3
Shop windows	10^3
Drafting office	500
Business office	250
Good street lighting	20
Poor street lighting	10^{-1}

2.4. SOURCES

As shown in Figure 2-2, the peak wavelength shifts toward the blue end of the spectrum as the color temperature increases. This puts more energy in the visible region. There is approximately 70% more luminous flux available from a lamp operating at 3200 K than a light operating at 3000 K based on blackbody curves. This is the basis for using a higher-color temperature lamp for higher illumination levels. However, the lifetime of tungsten halogen bulbs decreases dramatically with increasing color temperature. It is far better to use more lamps to increase the luminous flux than to increase the color temperature and sacrifice lifetime.

2.4.1. CALIBRATION SOURCES

The CIE (Commission Internationale de l'Eclairage, or International Commission on Illumination) recommended four illuminants that could be used for calibration of cameras sensitive in the visible region (Table 2-7). Commercially available sources simulate these illuminants only over the visible range (0.38 to 0.75 μm). The source intensity is not specified; only the effective color temperature is specified. Illuminants A and D65 are used routinely, whereas illuminants B and C are of historical interest. The apparent solar color temperature depends upon latitude, time of day, and a variety of environmental conditions (e.g., aerosols, pollutants, and cloud cover). The standardization of illuminant C at 6770 K was arbitrary. It was considered representative of "average" daylight.

Illuminant A is a tungsten filament whose output follows Planck's blackbody radiation law. The other illuminants are created by placing specific filters in front of illuminant A. These filters have spectral characteristics that, when combined with illuminant A, provide the relative outputs illustrated in Figure 2-8. On a relative basis, the curves can be approximated by blackbodies whose color temperatures are 4870, 6770, and 6500 K for illuminant B, C, D65, respectively.

Table 2-7
CIE RECOMMENDED ILLUMINANTS

CIE ILLUMINANT	EFFECTIVE COLOR TEMPERATURE	DESCRIPTION
A	2856 K	Light from an incandescent source
B	4870 K	Average noon sunlight
C	6770 K	Average daylight (sun + sky)
D65 or D6500	6500 K	Daylight with a corrected color temperature

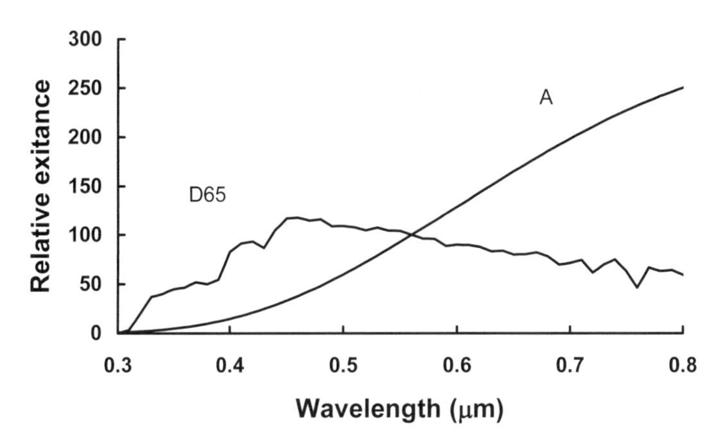

Figure 2-8. Relative spectral exitance of the standard illuminants A and D65 (D6500). Illuminant C spectral output is very similar to D65.

2.4.2. REAL SOURCES

A tungsten filament bulb emits light that is closely matched to an ideal blackbody radiator. This is not so for discharge lamps such as fluorescent lamps. For these lamps, a blackbody curve that approximates the output is fit to the spectral radiant exitance (Table 2-8). This approximation is for convenience and should not be used for scientific calculations. The actual spectral photon sterance must be used for non-ideal sources[3]. Although many discharge lamps seem uniform white in color, they have peaks in the emission spectra (Figures 2-9 and 2-10).

Table 2-8
APPROXIMATE COLOR TEMPERATURE

SOURCE	APPROXIMATE COLOR TEMPERATURE
Northern sky light	7500 K
Average daylight	6500 K
Xenon (arc or flash)	6000 K
Cool fluorescent lamps	4300 K
Studio tungsten lamps	3200 K
Warm fluorescent lamps	3000 K
Floodlights	3000 K
Domestic tungsten lamps	2800 - 2900 K
Sunlight at sunset	2000 K
Candle flame	1800 K

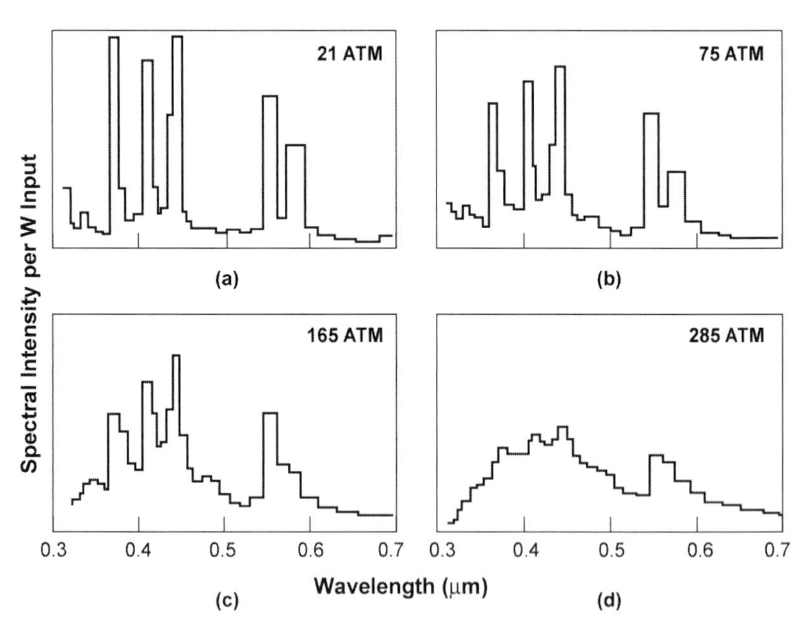

Figure 2-9. Relative output of a mercury arc lamp as a function of pressure.[4] (a) 21 atm, (b) 75 atm, (c) 165 atm, and (d) 285 atm.

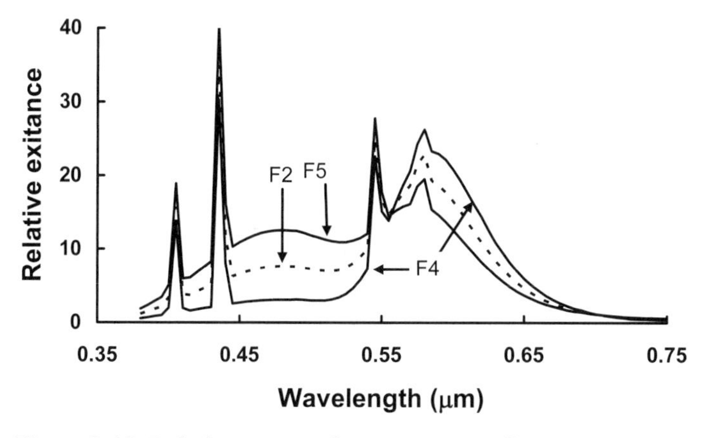

Figure 2-10. Relative output of some common fluorescent tubes,

Usually there is no relationship between the color temperature and contrast of a scene and what is seen on a display. This is not the fault of the camera and display manufacturers; they strive for compatibility. The observer, who usually has no knowledge of the original scene, will adjust the display for maximum visibility and aesthetics.

The video signal is simply a voltage with no color temperature associated with it. The display gain and offset can be adjusted such that a white target appears as if it was illuminated by a source whose color temperature is between 3200 K and 10,000 K. Most displays are preset to either 6500 K or 9300 K (Figure 2-11). As the color temperature increases, whites appear to change from a yellow tinge to a blue tinge. The perceived color depends on the adapting illumination (e.g., room lighting) Setting the color temperature to 9300 K provides aesthetically pleasing imagery and this setting is unrelated to the actual scene color temperature. Some displays allow tuning from 3200 to 10,000 K. The displayed color temperature is independent of the scene color temperature. A scene illuminated with a 2854 K source (yellow tinge) can appear whitish when displayed at 9300 K.

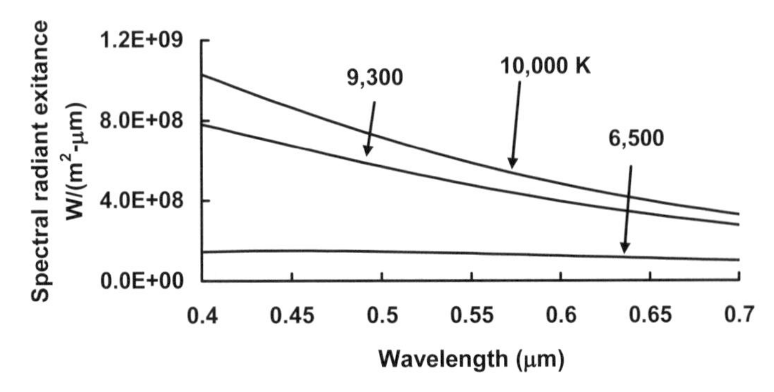

Figure 2-11. Display color temperature is typically preset at either 6500 or 9300 K.

2.5. POINT SOURCES and EXTENDED SOURCES

An array may be calibrated by a point source or an extremely large source (extended source). The on-axis spectral radiant incidance (Figure 2-12) from a diffuse source is

$$E_e(\lambda) = \pi L_e(\lambda) \frac{r^2}{r^2 + R^2} \tag{2-9}$$

When $r \ll R$, the source appears as a point source and the irradiance decreases by R^2. When $r \gg R$, the source appears infinitely large and $E_e(\lambda) \approx \pi L_e(\lambda)$. The equations are identical for photon flux and photometric calculations.

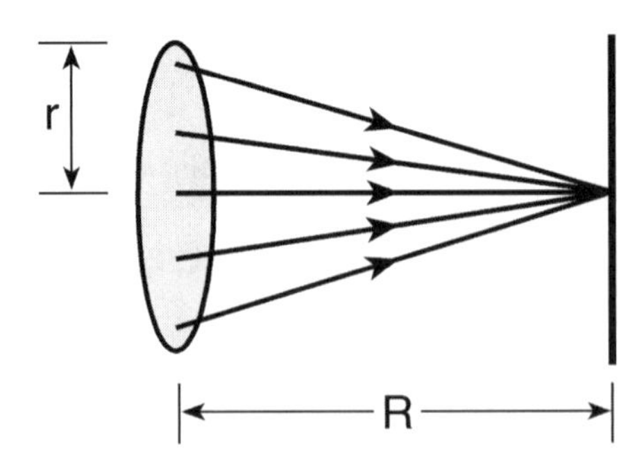

Figure 2-12. Spectral radiant incidance from an ideal diffuse source.

2.6. CAMERA FORMULA

If an imaging system is at a distance R_1 from a source (Figure 2-13), the number of photons incident onto the optical system of area A_o is during time t_{INT} is

$$n_{LEXS} = L_q \frac{A_o}{R_1^2} A_S T_{ATM} t_{INT} \tag{2-10}$$

where the small angle approximation was used (valid when $R_1^2 \gg A_o$). Here it is assumed that the quantities have no spectral features (spectrally flat). T_{ATM} is the intervening atmospheric transmittance.

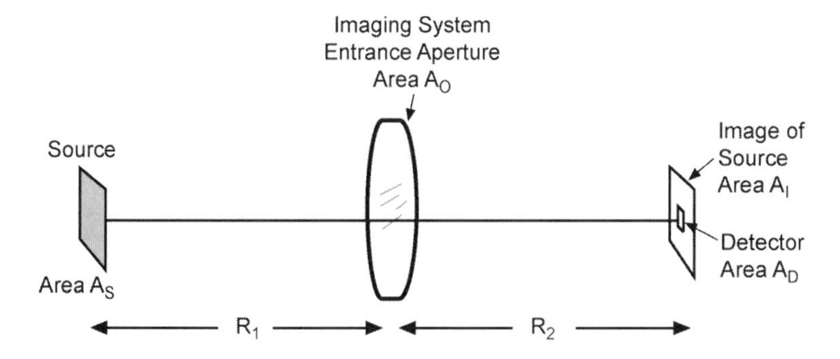

Figure 2-13. An imaging system directly viewing a source.

The number of on-axis photons reaching the image plane is

$$n_{IMAGE} = L_q \frac{A_o}{R_1^2} A_S T_{OPTICS} T_{ATM} t_{INT} \tag{2-11}$$

The variable T_{OPTICS} is the system's optical transmittance, which includes all the intervening filters: IR filter, color filter array, and the microlens (all discussed in later chapters).

When the image size is much larger than the detector area ($A_i \gg A_D$), the source is said to be resolved or the system is viewing an extended source. Equivalently, the detector is flood-illuminated. In most solid state camera applications, the source is resolved. The value A_D is the effective sensitive detector area. Microlenses (discussed in Section 5.5. *Microlenses*) increase the effective detector area.

The number of photons incident onto the detector is simply the ratio of the areas:

$$n_{DETECTOR} = n_{IMAGE} \frac{A_D}{A_i} \tag{2-12}$$

Using the small angle approximation for paraxial rays ($R_1^2 \gg A_S$ and $R_2^2 \gg A_i$),

$$\frac{A_s}{R_1^2} = \frac{A_i}{R_2^2} \tag{2-13}$$

We ignore the sign convention found in optics texts. We take R_1 and R_2 as positive quantities.

The number of photons reaching the detector becomes

$$n_{DETECTOR} = \frac{L_q A_o A_D}{fl^2 (1 + M_{OPTICS})^2} \tau_{OPTICS} T_{ATM} t_{INT} \tag{2-14}$$

The optical magnification is $M_{OPTICS} = R_2/R_1$. Here, R_1 and R_2 are related to the system's effective focal length, fl, by

$$\frac{1}{R_1} + \frac{1}{R_2} = \frac{1}{fl} \tag{2-15}$$

Assuming a circular aperture ($A_o = \pi D_o^2/4$) and defining the focal ratio (also called the f-number) as $F = fl/D$

$$n_{DETECTOR} = \frac{\pi}{4} \frac{L_q A_D}{F^2 (1 + M_{OPTICS})^2} T_{OPTICS} T_{ATM} t_{INT} \tag{2-16}$$

While valid for paraxial rays (1st order approximation), when $F < 3$, the equation must be modified (see Appendix). An off-axis image will have reduced incidance compared to an on-axis image by cosine$^4\theta$. As required, cosine$^4\theta$ can be added to all the equations.

Many of the variables are a function of wavelength. The number of photoelectrons generated in a solid state detector is

$$n_{PE} = \int_{\lambda_1}^{\lambda_2} R_q(\lambda) n_{DETECTOR}(\lambda)\, d\lambda \tag{2-17}$$

or

$$n_{PE} = \int_{\lambda_1}^{\lambda_2} R_q(\lambda) \frac{\pi}{4} \frac{L_q(\lambda,T) A_D t_{INT}}{F^2(1+M_{OPTICS})^2} T_{OPTICS}(\lambda) T_{ATM}(\lambda)\, d\lambda \qquad (2\text{-}18)$$

The detector's quantum efficiency [often expressed as $\eta(\lambda)$] is $R_q(\lambda)$ and has units of electrons per photon. The integration time is t_{INT}. As the source moves to infinity, M_{OPTICS} approaches zero.

In photography, shutter speeds (exposure times) vary approximately by a factor of two (e.g., 1/30, 1/60, 1/125, 1/250, etc.). Thus, changing the shutter speed by one setting changes n_{PE} approximately by a factor of 2. f-stops have been standardized to 1, 1.4, 2, 2.8, 4, 5.6, 8..... The ratio of adjacent f-stops is $\sqrt{2}$. Changing the lens speed by one f-stop changes the focal ratio (f-number) by a factor of $\sqrt{2}$. Here, also, the n_{PE} changes by a factor of 2.

In an actual application, what is of interest is the signal difference produced by a target and its immediate background. Here, both the target and the background are assumed to be illuminated by the same source (artificial lighting, sun, moon, night glow, star light). Let $\Delta\rho = \rho_T(\lambda) - \rho_B(\lambda)$ where $\rho_T(\lambda)$ and $\rho_B(\lambda)$ are the spectral reflectances of the target and background, respectively. Although the number of electrons is used for solid state array calculations, the camera output is a voltage

$$\Delta V_{CAMERA} = G_{CAMERA} \Delta n_{PE} \qquad (2\text{-}19)$$

Where

$$\Delta n_{PE} = \int_{\lambda_1}^{\lambda_2} R_q(\lambda) \frac{\pi}{4} \frac{\Delta\rho\, E_{q-SCENE}(\lambda,T) A_D t_{INT}}{F^2(1+M_{OPTICS})^2} \tau_{OPTICS}(\lambda) T_{ATM}(\lambda)\, d\lambda \qquad (2\text{-}20)$$

The value G_{CAMERA} contains both the array output conversion gain (units of volts/electron) and the subsequent amplifier gain. It would appear that by simply decreasing the f-number, n_{PE} or Δn_{PE} would increase. Often, lower focal ratio lens systems have more optical elements and therefore may have lower transmittance. The value τ_{OPTICS}/F^2 must increase to increase n_{PE} or Δn_{PE}. Lens systems may be described by the T-number

$$T/\# = \frac{F}{\sqrt{T_{OPTICS}}} \qquad (2\text{-}21)$$

If T_{OPTICS} is a function of wavelength, then T/# will also be a function of wavelength. Here T_{OPTICS}/F^2 in Equations 2-16, 2-18, or 2-20 is replaced by $1/(T/\#)^2$.

For back-of-the-envelope calculations, the atmospheric transmittance is assumed to have no spectral features $T_{ATM}(\lambda) \approx T_{ATM}$ (discussed in Section 12.1, *Atmospheric Transmittance*). For detailed calculations, the exact spectral transmittance[5] must be used.

Example 2-1
VISUAL THRESHOLD

An object is heated to incandescence. What is the approximate color temperature for it to be just perceived?

After adapting to the dark at least 60 min, the eye can perceive a few photons per second (absolute threshold). Here, the pupil dilates to about 8 mm (A_o = 5×10^{-5} m^2). When the eye is dark adapted, only the rods are functioning and the eye's spectral response is approximately 0.38 μm to 0.66 μm. The number of photons per second reaching the eye from an extended Lambertian source is

$$n_{EYE} = \int_{0.38}^{0.66} M_q(\lambda,T) A_o \, d\lambda \approx M_q(\lambda_o,T) A_o \Delta\lambda \qquad (2\text{-}22)$$

where $\Delta\lambda$ = 0.28 μm. The peak scotopic eye response occurs at λ_o = 0.515 μm. The observer can just perceive this object about 600 K when he is in a completely blackened room. When an object is just glowing, it is said to be heated to incandescence.

As the ambient illumination increases, the pupil constricts (A_o decreases). Furthermore, the eye's detection capability depends on the ambient lighting. As the light level increases, the target flux must also increase to be perceptible.

The eye's ability to just discern intensity differences depends on the intensity of the surrounding illuminance. At 0.01 lux the object must provide 10 times more flux than the background to be discerned by the eye. If the surround is at 700 K, the target must be at 750 K to be perceptible. At 1000 lux, the object must provide 1000 times more flux than the surround. If the surround color temperature is 1000 K, then the target must be heated to 1350 K to be perceptible. Photometric units are not linearly related to color temperature.

These back-of-the-envelope calculations do not include the eye's spectral response and therefore only illustrate required intensity differences. The eye changes its ability to perceive differences with the surround illumination.

2.7. NORMALIZATION

"Normalization...is the process of reducing measurement results as nearly as possible to a common scale"[6]. Normalization is essential to ensure that appropriate comparisons are made. Figure 2-14 illustrates the relationship between the spectral response of a system to two different sources. The output of a system depends on the spectral features of the input and the spectral response of the imaging system. Simply stated, the camera output depends on the source used.

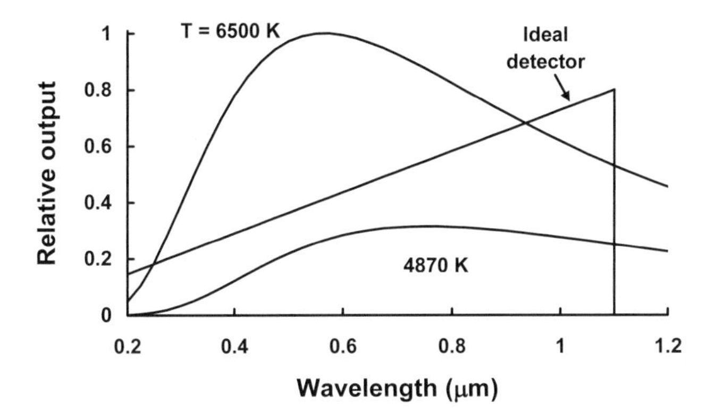

Figure 2-14. Sources with different spectral outputs can produce different system outputs. The 6500 K source provides more in-band radiant flux than the 4870 K source. The system output will be higher when viewing the 6500 K source.

Variations in output can also occur if "identical" systems have different spectral responses. Equation 2-18 is integrated over the wavelength interval of interest. Since arrays may have different spectral responses, an imaging system whose spectral response is 0.4 to 0.7 μm may have a different responsivity than a system that operates 0.38 to 0.75 μm although both systems are labeled as visible systems. The spectral response of solid state arrays varies from manufacturer to manufacturer (discussed in Section 6.1. *Quantum efficiency*). Systems can be made to appear as equivalent or one can be made to provide better performance by simply selecting an appropriate source.

While the sensor may be calibrated with a standard illuminant, it may not be used with the same illumination. Calibration with a standard illuminant is useful for comparing camera responsivities under controlled conditions. However, if the source characteristics in a particular application are significantly different from the standard illuminant, the selection of one sensor over another must be done with care. For example, street lighting from an incandescent bulb is different from that of mercury-arc lights (compare Figure 2-8 with Figure 2-9). The only way to determine the relative responsivities is to perform the calculation indicated in Equation 2-18.

The effect of ambient lighting is particularly noticeable when using low-light-level televisions at nighttime. The spectral output of the sky (no moon) and moon (clear night) are significantly different. Figure 2-15 illustrates the natural night sky luminous incidance (irradiance). Since an abundance of photons exists in the near infrared (0.7 to 1.1 μm), most night vision systems (e.g., image intensifiers, starlight scopes, snooper scopes, etc.) are sensitive in this region. Full moon and quarter moon irradiances are provided in Figures 2-16 and 2-17. As the moon's zenith angle changes, the air mass increases and spectral irradiance changes. Figure 2-18 illustrates direct daylight.

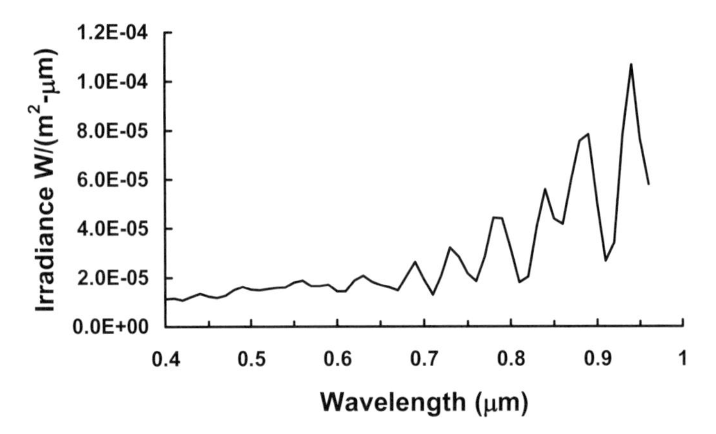

Figure 2-15. Starlight irradiance[7] at ground level looking straight upwards (air mass = **1**). As the viewing angle changes, the air mass increases and the irradiance changes. For 50% starlight, overcast starlight, and deep overcast starlight, divide the irradiance by 2, 10, and 100, respectively.

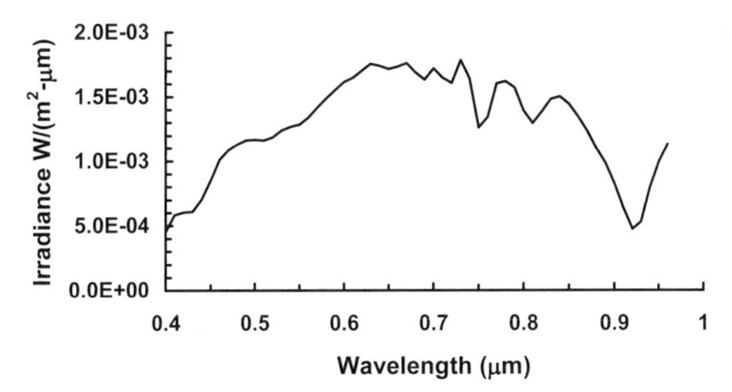

Figure 2-16. Moonlight irradiance[7] at ground level looking straight upwards (air mass = **1**).

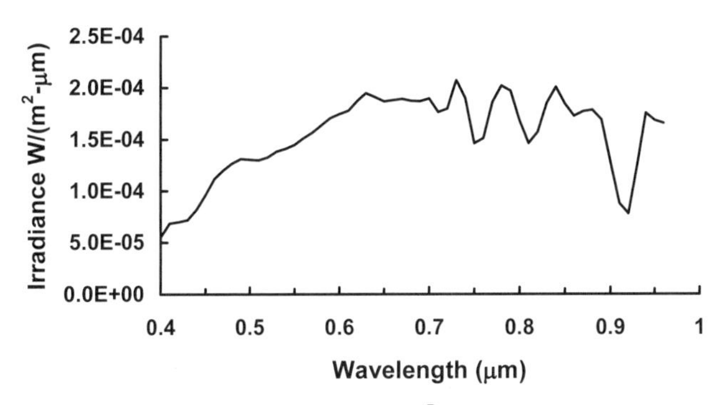

Figure 2-17. Quarter moonlight irradiance[7] at ground level (air mass = **1**).

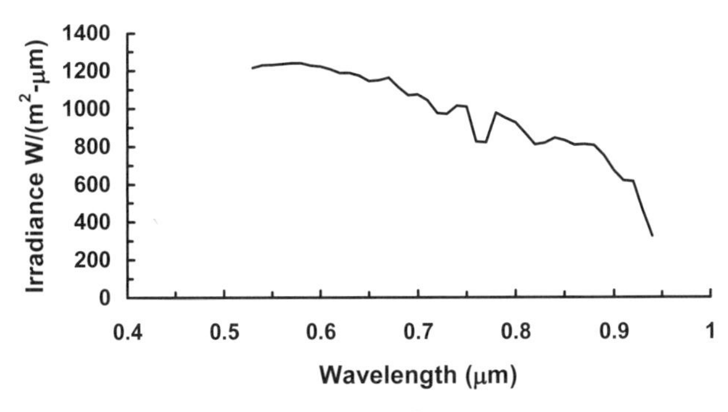

Figure 2-18. Direct daylight irradiance[7] at ground level (air mass = **1**).

2.8. NORMALIZATION ISSUES

A camera's output depends on both the system spectral response and the color temperature of the illuminating source (Equation 2-18). When the camera spectral response is within the eye's response, then photometric units are reasonable. When the camera's response is outside the eye's response curve, photons contribute to signal but the photometry remains constant. Even though two sources may provide the same luminous incidance, the camera output can be quite different. Sometimes an infrared filter is used to restrict the wavelength response. When the filter is added, the average responsivity becomes

$$R_V = \frac{\int_{\lambda_1}^{\lambda_2} T_{IR-FILTER}(\lambda) R_e(\lambda) M_e(\lambda,T)\, d\lambda}{683 \int_{0.38}^{0.75} V(\lambda) M_p(\lambda,T)\, d\lambda} \quad \frac{\text{A}}{\text{lumen}} \tag{2-23}$$

This average type responsivity is useful for comparing the performance of cameras with similar spectral responses. It is appropriate if the source employed during actual use is similar to the one used for the calibration.

For ideal photon detectors, the spectral responsivity (expressed in A/W) is

$$\begin{aligned} R_e(\lambda) &= \frac{\lambda}{\lambda_P} R_P \quad \textit{when } \lambda \le \lambda_P \\ &= 0 \quad \textit{elsewhere} \end{aligned} \tag{2-24}$$

where R_P is the peak responsivity and λ_P is the wavelength at which R_P occurs. Since R_q is the quantum efficiency (η)

$$R_e(\lambda) = \frac{q\lambda}{hc} R_q \tag{2-25}$$

When the wavelength is expressed in micrometers, $R_e(\lambda) = (\lambda/1.24)R_q$ and $R_P = R_q/E_G$. For silicon, the band gap is $E_G = 1.12$ eV resulting in a cutoff wavelength of $\lambda_P \approx 1.24/E_G \approx 1.1$ μm. The quantity q is the electron charge ($q = 1.6\times10^{-19}$ coul).

The ideal silicon detector response is used to estimate the relative outputs of two systems: one with an ideal IR filter (transmittance is unity from 0.38 to 0.70 μm and zero elsewhere) and one without. The relative outputs are normalized to that expected when illuminated with a CIE A illuminant (Table 2-9). The sources are considered ideal blackbodies. As the source temperature increases, the output increases but not linearly with color temperature.

Table 2-9
RELATIVE OUTPUT
Assuming ideal blackbodies and ideal Si spectral response.
A specific sensor may deviate significantly from these values.

SOURCE TEMPERATURE	RELATIVE OUTPUT (0.38 to 1.1 μm)	RELATIVE OUTPUT (0.38 to 0.70 μm)
2856	1.00	0.196
3200	1.96	0.482
4870	15.7	6.95
6500	48.9	27.4
6770	56.5	32.4

Only photons within the visible region affect the number of available lumens. For the system without the IR filter, the detector is sensing photons whose wavelengths are greater than 0.7 μm even though the number of lumens is not affected by these photons (Figure 2-19). As the color temperature increases, the flux available to the observer (lumens) increases faster than the output voltage. As a result, responsivity, expressed in A/lumens, decreases with increasing color temperature (Table 2-10). This decrease affects the wide spectral response systems to a greater extent. Specifying the output in A/lumens is of little value for an electro-optical sensor whose response extends beyond the eye's response. Radiometric units are advised here.

The values in Tables 2-9 and 2-10 should be considered as illustrative. Solid state detectors do not follow the ideal response (discussed in Section 6.1. *Quantum efficiency*). The IR filter cutoff also varies by manufacturer. Only detailed spectral response evaluation permits correct comparisons among detectors. Since there is considerable variation in spectral response and type of IR filter used, it is extremely difficult to compare systems based on responsivity values only.

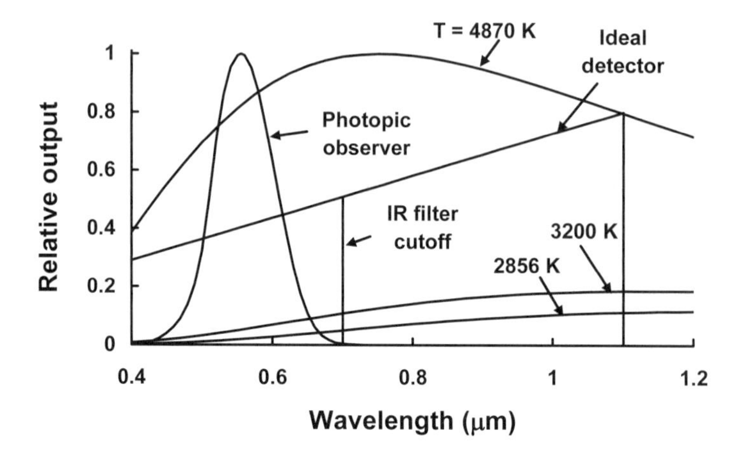

Figure 2-19. Spectral relationship between source characteristic, detector response, and the photopic observer. As the color temperature increases, the energy available to the observer increases rapidly. Therefore, the number of lumens increases dramatically.

Table 2-10
RELATIVE RESPONSIVITY (A/lumen)
A specific sensor may deviate significantly from these values

SOURCE TEMPERATURE	RELATIVE RESPONSIVITY (0.38 to 1.1 μm)	RELATIVE RESPONSIVITY (0.38 to 0.70 μm)
2856	1.00	0.196
3200	0.754	0.185
4870	0.387	0.171
6500	0.314	0.176
6770	0.308	0.177

System output does not infer anything about the source other than that an equivalent blackbody of a certain color temperature would provide the same output. This is true no matter what output units are used (volts, amps, or any other arbitrary unit). These units, by themselves, are not very meaningful for system-to-system comparison. For example, ΔV_{CAMERA} (and equivalently the camera responsivity) can be increased by increasing the system gain. As such, it is dangerous to compare system response based on only a few numbers.

Example 2-2
SOURCE SELECTION

When viewing a scene, the signal-to-noise ratio is too small. The engineer can either change the light source to one with a higher color temperature or add more lamps. What are the differences?

With an ideal blackbody, the color temperature completely specifies the luminous exitance. Increasing the color temperature (by increasing the voltage on the bulb) will increase the luminous flux. For real sources, the color temperature is used only to denote the relative spectral content. A 50W bulb whose output approximates a 3200 K source will have a lower photometric output than a 1000W bulb operating at 2856 K. Adding more lamps does not change the relative spectral content but increases the luminous flux.

Example 2-3
RELATIVE VERSUS ABSOLUTE OUTPUT

Should a detector be calibrated with a 5W or 1000W bulb?

If the relative spectral output of the two bulbs is the same (e.g., CIE A), the responsivity is independent of the source flux. The bulb intensity should be sufficiently high to produce a good signal-to-noise ratio. It should not be so high that the detector saturates.

Arrays may be specified by their radiometric responsivity ($V/(J/m^2)$) or by their photometric responsivity (V/lux). The relationship between the two depends on the spectral content of the source and the spectral response of the array. In principle, the source can be standardized (e.g., selection of the CIE A illuminant), but the array spectral response varies among manufacturers and may vary within one manufacturing environment if different processes are used. Therefore, there is no simple (universal) relationship between radiometric and photometric responsivities.

The correct analysis is to perform a radiometric calculation using the source spectral output (which may not be an ideal blackbody) and using the spectral response of the camera system. Using standard sources for quality control on a production line is valid since the spectral response of the detectors should not change very much from unit to unit. The camera spectral response must include the optical spectral transmittance of all filters present.

For machine vision systems, the infrared blocking filter is not typically used. However, since the human observer is not familiar with infrared images, the imagery may appear slightly different compared to that obtained with the IR blocking filter.

Low-light-level televisions can have a variety of spectral responses (discussed in Section 6.1. *Quantum efficiency*). The methodology presented in this chapter can be used to calculate responsivity and output voltage. The results given in Tables 2-9 and 2-10 do not apply to intensified systems.

2.9. REFERENCES

1. C. L. Wyatt, *Radiometric System Design*, Chapter 3, Macmillan Publishing Co., New York, NY (1987).
2. J. M. Palmer, "Radiometry and Photometry" in *Handbook of Optics*, Vol. III, p 7.11, McGraw Hill, 2001.
3. D. Kryskowski and G. H. Suits, "Natural Sources," in *Sources of Radiation*, G. J. Zissis, ed., pp. 151-209, Volume 1, *The Infrared and Electro-Optical Systems Handbook*, J. S. Accetta and D. L. Shumaker, eds., copublished by Environmental Research Institute of Michigan, Ann Arbor, MI, and SPIE Press, Bellingham, WA (1993).
4. E. B. Noel, "Radiation from High Pressure Mercury Arcs, "*Illumination Engineering*, Vol. 36, pg. 243 (1941).
5. The ONTAR Corporation, 129 University Road, Brookline, MA 02146-4532 and JCD Publishing, 2932 Cove Trail, Winter Park, FL 32789-1159, www.JCDPublishing.com offer a variety of atmospheric transmittance codes.
6. F. E. Nicodemus, "Normalization in Radiometry," *Applied Optics*, Vol. 12(12), pp. 2960-2973 (1973).
7. R. Vollmerhausen and T. Maurer, "Night Illumination in the Visible, NIR, and SWIR Spectral Bands," in *Infrared Imaging System: Design, Analysis, Modeling, and Testing XIV*, SPIE Proceeding Vol. 5076, pp. 60-69 (2003).

3
CCD FUNDAMENTALS

CCD refers to a semiconductor architecture in which charge is transferred through storage areas. The CCD architecture has three basic functions: a) charge collection, b) charge transfer, and c) the conversion of charge into a measurable voltage. A CCD can be attached to a variety of detectors. The basic building block of the CCD is the metal-insulator-semiconductor (MIS) capacitor. It is also called a gate. The most important MIS is the metal-oxide-semiconductor (MOS). Because the oxide of silicon is an insulator, it is a natural choice.

For monolithic devices, charge generation is often considered as the initial function of the CCD. With silicon photodetectors, each absorbed photon creates an electron-hole pair. Either the electrons or holes can be stored and transferred. For frame transfer devices, charge generation occurs under a MOS capacitor (also called a photogate). For some devices (notably interline transfer devices) photodiodes create the charge. The charge created at a pixel site is in proportion to the incident light level present. The aggregate effect of all the pixels is to produce a spatially sampled representation of the continuous scene.

Because most sensors operating in the visible region use a CCD architecture to move a charge packet, they are popularly called CCD arrays. However, the charge injection device (CID) does not use a CCD for charge transfer. Rather, two overlapping silicon MOS capacitors share the same row and column electrode. Pixel readout occurs by sensing the charge transfer between the capacitors at the pixel site.

Active pixels are fabricated with complementary-metal-oxide-semiconductor (CMOS) technology. The advantage is that one or more active transistors can be integrated into the pixel. As such, they become fully addressable (can read selected pixels) and can perform on-chip image processing. To denote the number of transistors in the unit cell, they are called 1T, 2T, etc. CMOS fundamentals are provided in Chapter 4. Section 4.1 compares CCDs and CMOS array and highlights the material in this chapter.

Considerable literature[1-10] has been written on the physics, fabrication, and operation of CCDs. The charge transfer physics is essentially the same for all CCD arrays. However, the number of phases and number and location of the serial shift readout registers vary by manufacturer. The description that follows should be considered as illustrative. A particular design may vary significantly from the simplified diagrams shown in this chapter. A specific device may not have all the features listed, or may have additional features not listed.

Although CCD arrays are common, the fabrication is quite complex. Theuwissen[11] provides an excellent step-by-step procedure for fabricating CCDs. He provides a 29-step procedure supplemented with detailed diagrams. Actual CCDs may require more than 150 different operations. The complexity depends on the array architecture.

Devices may be described functionally according to their architecture (frame transfer, interline transfer, etc.) or by application. To minimize cost, array complexity, and electronic processing, the architecture is typically designed for a specific application. For example, astronomical cameras typically use full-frame arrays whereas video systems generally use interline transfer devices. The separation between general imagery, machine vision, scientific, and military devices becomes fuzzy as technology advances.

The symbols used in this book are summarized in the *Symbol List*, which appears after the *Table of Contents*.

3.1. PHOTODETECTION

When an absorbed photon creates an electron-hole pair, the photodetection process has occurred. To be a useful detector in a solid state array, the photogenerated carrier must be collected in a storage site. The absorption coefficient is wavelength-specific and decreases with increasing wavelength. Some of the incident photons will be reflected at the array's surface. The remainder may pass through layers of electrodes and insulators before reaching the photoactive material. Thus, the absorption characteristics of the overlying layers, thickness of the photoactive material, and the location of the storage site determine the spectral quantum efficiency (discussed in Section 6.1. Quantum efficiency).

After photodetection, the stored charge is read out. The array architecture often is optimized for a specific application. General imagery systems are designed for interlace operation consistent with a standard video (e.g., EIA 170). Research devices may operate in the progressive scan mode. This reduces some image distortion when the object is moving. For scientific applications the output may be in the slow scan mode, which simply means that the data read out rate is lower than a standard frame rate (e.g., EIA 170).

3.1.1. PHOTOGATE

The photogate is a photoactive MOS capacitor where the photogenerated carriers are stored in a depletion region. The storage capacity depends on substrate doping, gate voltage, and oxide thickness. Very large capacities are possible. Because photogates have overlaying electrodes, the quantum efficiency in the blue region of the spectrum is reduced. Back-sided illumination overcomes this limitation.

3.1.2. PHOTODIODE

With photodiodes, a depletion region is formed between the n- and p-type regions. The photogenerated carriers are usually stored at the junction until readout. The width of the junction limits the storage capability. The junction is initially reverse biased by a MOS gate. Depending upon the substrate, either holes or electrons can be stored. For this text, it is assumed to be electrons. As the electrons, n_e, are collected, the voltage drops by

$$\Delta V = \frac{n_e\, q}{C} \tag{3-1}$$

After readout, the voltage is reset to its original value. If the number of electrons is large, the voltage is reduced to zero and saturation occurs. Excess photoelectrons will diffuse to neighboring wells and create blooming. Photodiodes do not have an overlaying structure and therefore have higher quantum efficiency than comparable photogates.

With a photodiode, a transfer gate is required to move the stored electrons into the CCD. Devices with photodiodes are often labeled as CCPD for charge-coupled photodiodes. While transfer gates are essential for CCPDs, they may also exist in devices with photogates.

The hole accumulation diode (HAD) is a pinned photodiode that offers the best features of the photogate and photodiode. It offers the high blue response of the photodiode and large well capacity of the photogate.

3.2. CCD ARRAY OPERATION

Applying a positive voltage to the CCD gate causes the mobile positive holes in the p-type silicon to migrate toward the ground electrode because like charges repel. This region, which is void of positive charge, is the depletion region (Figure 3-1). If a photon whose energy is greater than the energy gap is absorbed in the depletion region, it produces an electron-hole pair. The electron stays within the depletion region whereas the hole moves to the ground electrode. The amount of negative charge (electrons) that can be collected is proportional to the applied voltage, oxide thickness, and gate electrode area. The total number of electrons that can be stored is called the well capacity.

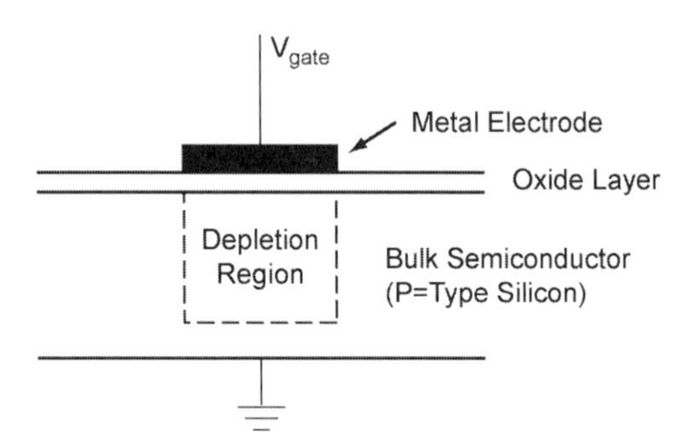

Figure 3-1. Metal-oxide-semiconductor (MOS) gate for p-type silicon. Although the depletion region (charge well) is shown to have an abrupt transition, its actual shape is gradual. That is, the well is formed by a two-dimensional voltage gradient.

The CCD register consists of a series of gates. Manipulation of the gate voltage in a systematic and sequential manner transfers the electrons from one gate to the next in a conveyor-belt-like fashion. For charge transfer, the depletion regions must overlap. The depletion regions are actually gradients, and the gradients must overlap for charge transfer to occur. Efficient overlap occurs when the gate electrodes overlap (Figure 3-2). The multilevel polysilicon structure minimizes the gap between adjacent electrodes. Without this structure, charge transfer would be poor. For clarity, throughout the remainder of the text, the structure is illustrated as adjacent, non-overlapping gates.

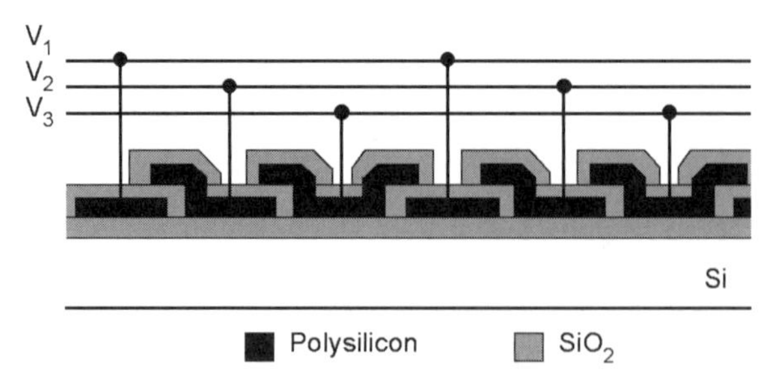

Figure 3-2. Three-phase CCD. The gates must overlap for efficient charge transfer.

Each gate has its own control voltage that is varied as a function of time. The voltage is called the clock or clocking signal. When the gate voltage is low, it acts as a barrier; when the voltage is high, charge can be stored (Figure 3-3). The process is repeated many times until the charge is transferred through the shift register.

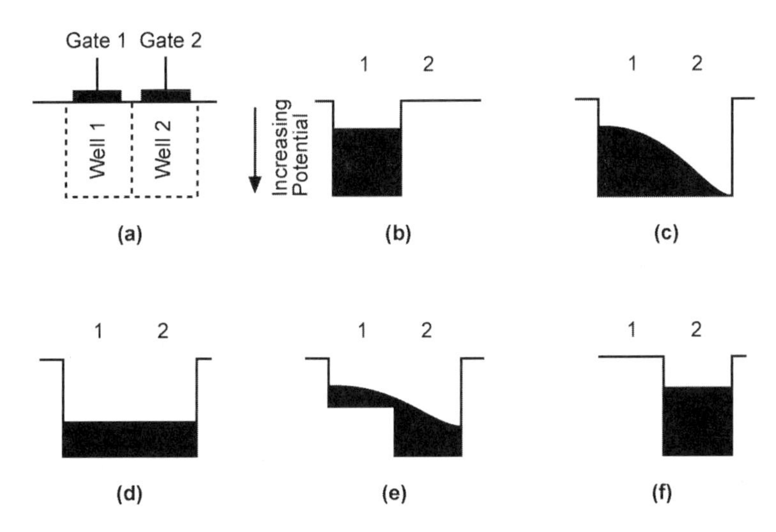

Figure 3-3. Charge transfer between two wells. (a) Adjacent wells. (b) Charge in **well 1**. (c) After a voltage is applied to **gate 2**, electrons flow into **well 2** in a waterfall manner. (d) Equilibration of charge. (e) Reduction of **gate 1** voltage causes the electrons in **well 1** to transfer into **well 2**. (f) All electrons have been transferred to **well 2**.

The CCD array is a series of column registers (Figure 3-4). The charge is kept within rows or columns by channel stops or channel blocks, and the depletion regions overlap in one direction only. At the end of each column is a horizontal register of pixels. This register collects a line at a time and then transports the charge packets in a serial fashion to an output amplifier. The entire horizontal serial register must be clocked out to the sense node before the next line enters the serial register. Therefore, separate vertical and horizontal clocks are required for all CCD arrays. This process creates a serial data stream that represents the two-dimensional image.

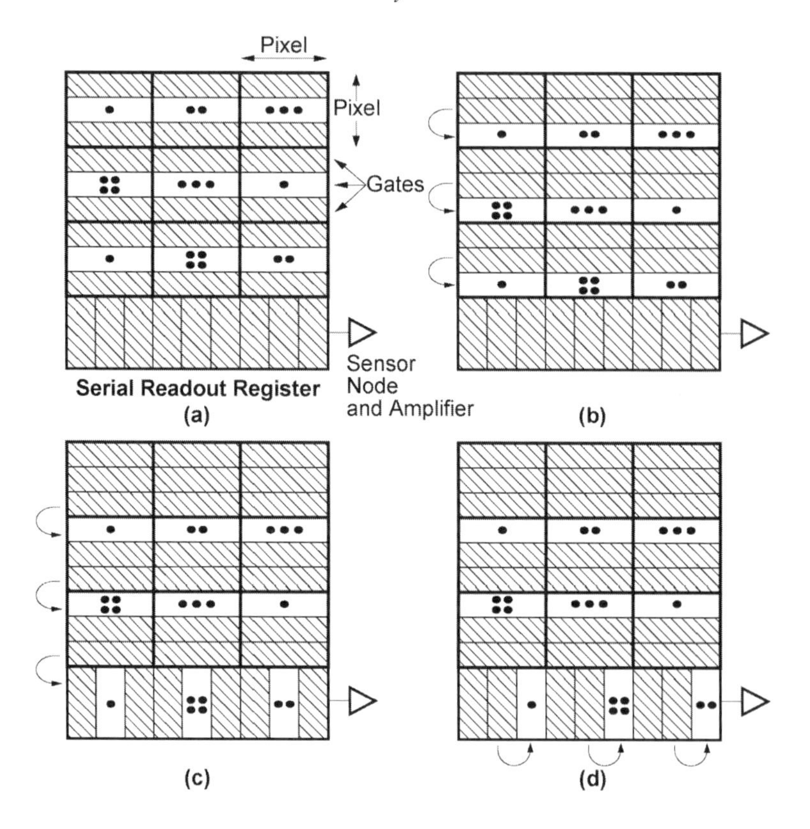

Figure 3-4. Representative CCD operation of a 3×3 array with three gates per pixel. Nearly all the photoelectrons generated within a pixel are stored in the well beneath that pixel. (a) The CCD is illuminated and an electronic image is created. (b) The columns are shifted down in parallel, one gate at a time. (c) Once in the serial row register, the pixels are shifted right to the output node. (d) The entire serial register must be clocked out before the next packet can be transferred into the serial readout register.

Although any number of transfer sites (gates) per detector area can be used, it generally varies from two to four. With a four-phase device, the charge is stored under two or three wells depending upon the clock cycle. Figure 3-5 represents the steady-state condition after the gates have switched and the charge has equilibrated within the wells. It takes a finite time for this to happen. Figure 3-6 illustrates the voltage levels that created the wells shown in Figure 3-5. The time that the charge has equilibrated is also shown. The clock rate is

$f_{CLOCK} = 1/t_{CLOCK}$ and is the same for all phases. The output video is valid once per clock pulse (selected as T_2) and occurs after the charge has equilibrated (e.g., Figure 3-3b or Figure 3-3f). Only one master clock is required to drive the array. Its phase must be varied for the gates to operate sequentially. An anti-parallel clocking scheme is also possible. Here V_3 is the inverse of V_1 and V_4 is the inverse of V_2. That is, when V_1 is high, V_3 is low. The anti-parallel clock system is easier to implement because phasing is not required. With the anti-parallel scheme, charge is always stored in two wells. For equal pixel sizes, four-phase devices offer the largest well capacity compared to the two- or three-phase systems. With the four-phase system, 50% of the pixel area is available for the well.

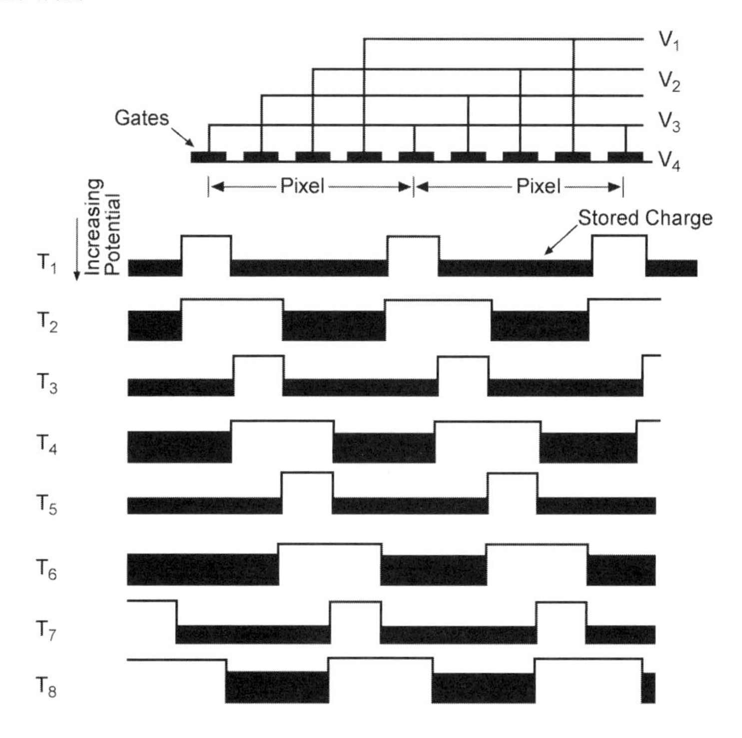

Figure 3-5. Charge transfer in a four-phase device after equilibration. This represents one column. Rows go into the paper. Charge is moved through storage sites by changing the gate voltages as illustrated in Figure 3-6. A four-phase array requires four separate clock signals to move a charge packet to the next pixel. A four-phase device requires eight steps to move the charge from one pixel to the next. (After Reference 12).

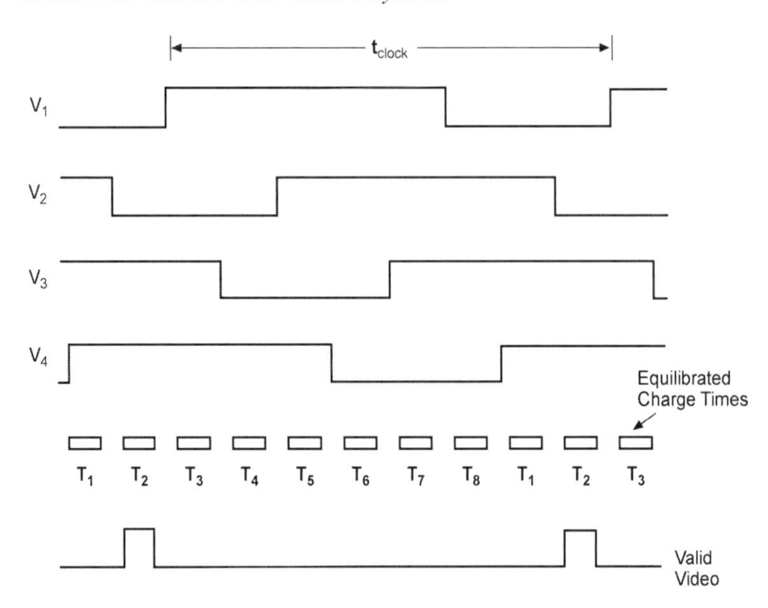

Figure 3-6. Voltage levels for a four-phase system. As the voltage is applied, the charge packet moves to that well. By sequentially varying the gate voltage, the charge moves off the horizontal shift register and onto the sense capacitor. The clock signals are identical for all four phases but offset in time (phase). The valid video is available once per clock pulse (selected as T2). (After Reference 12).

The three-phase system (Figure 3-7 and Figure 3-8) stores the charge under one or two gates. Only 33% of the pixel area is available for the well capacity. With equal potential wells, a minimum number of three phases are required to clock out charge packets efficiently.

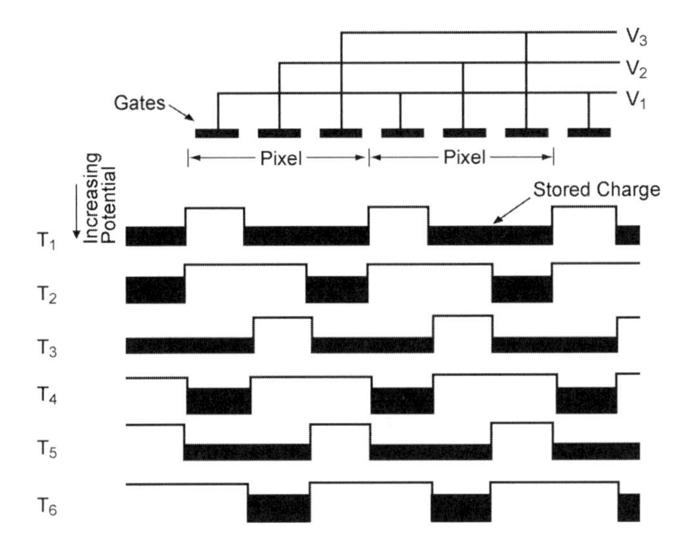

Figure 3-7. Charge transfer in a three-phase device. This represents one column. Rows go into the paper. Six steps are required to move the charge one pixel. (After Reference 12).

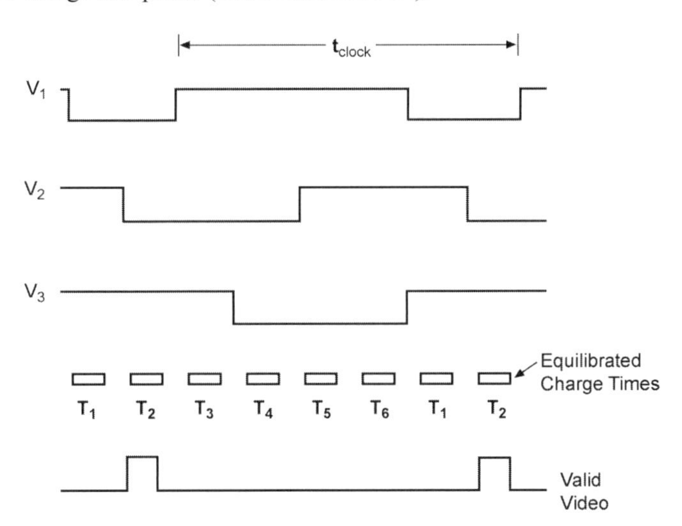

Figure 3-8. Voltage levels for a three-phase system. The clock signals are identical for all three phases but offset in time. T2 was selected as the valid video time. Valid video exists only after the charge has equilibrated. (After Reference 12).

The well potential depends on the oxide thickness underneath the gate. In Figure 3-9, the alternating gates have different oxide thicknesses and therefore will create different well potentials. With different potentials it is possible to create a two-phase device. Charge will preferentially move to the right-hand side of the pixel, where the oxide layer is thinner. Assume initially that the wells controlled by V_2 are empty. This will occur after a few clock pulses. At time T_1, both clocks are low. When V_2 is raised, the potential increases as shown at time T_2. Because the effective potential increases across the gates, the charge cascades down to the highest potential. Then V_2 is dropped to zero and the charge is contained under the V_2 gate at time T_3. This process is repeated until the charge is clocked off the array. The voltage timing is shown in Figure 3-10. While Figure 3-9 illustrates variations in the oxide layer, the potential wells can also be manipulated though ion implantation. As with the four-phase device, the two-phase device can also be operated in the anti-parallel mode: V_2 is a mirror image of V_1 That is, when V_1 is high, V_2 is low.

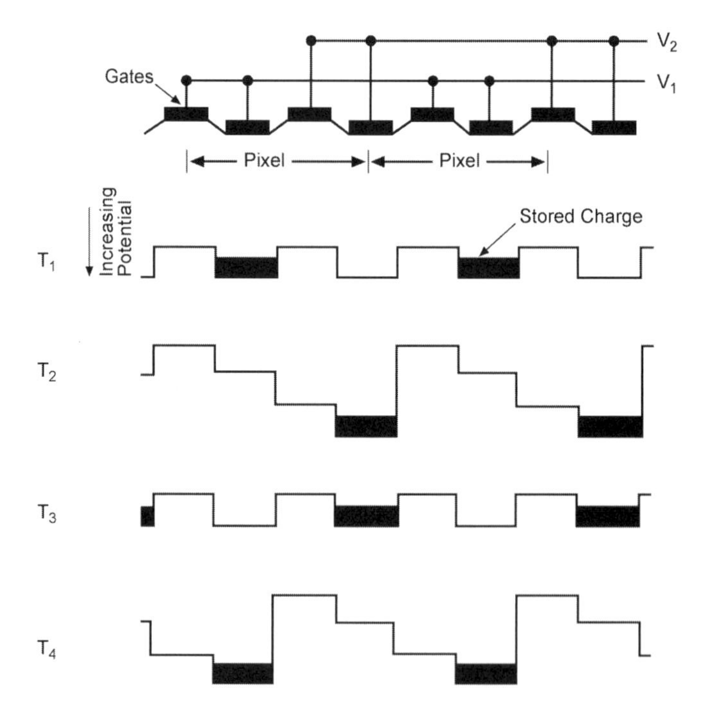

Figure 3-9. Charge transfer in a two-phase device. Well potential depends on the oxide thickness. (After Reference 12).

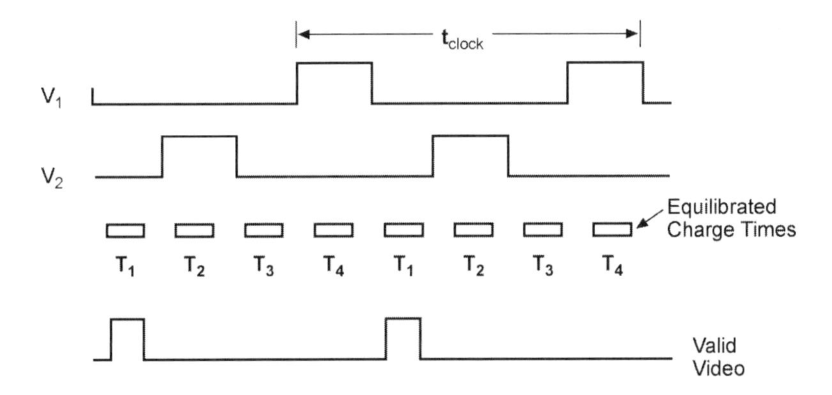

Figure 3-10. Voltages for a 2-phase system. (After Reference 12).

The virtual phase device requires only one clock (Figure 3-11). Additional charge wells are created by internally biased gates. Charge is stored either beneath the active gate or the virtual gate. When V_1 is low, the charge cascades down to the highest potential (which is beneath the virtual well). When V_1 is applied at T_2, the active gate potential increases and the charge moves to the active gate well.

The final operating step is to convert the charge packet to a measurable voltage. This is accomplished by a floating diode or floating diffusion. The diode, acting as a capacitor, creates a voltage proportional to the number of electrons [discussed in Section 3.4, *Charge conversion (output structure)*]. With many arrays, it is possible to shift more than one row of charge into the serial register. Similarly, it is possible to shift more than one serial register element into a summing gate just before the output node. This is called charge grouping, binning, super pixeling, or charge aggregation. Binning increases signal output and dynamic range at the expense of spatial resolution. Because it increases the signal-to-noise ratio, binning is useful for low-light applications where resolution is less important. Serial registers and the output node require larger capacity charge wells for binning operation. If the output capacitor is not reset after every pixel, then it can accumulate charge.

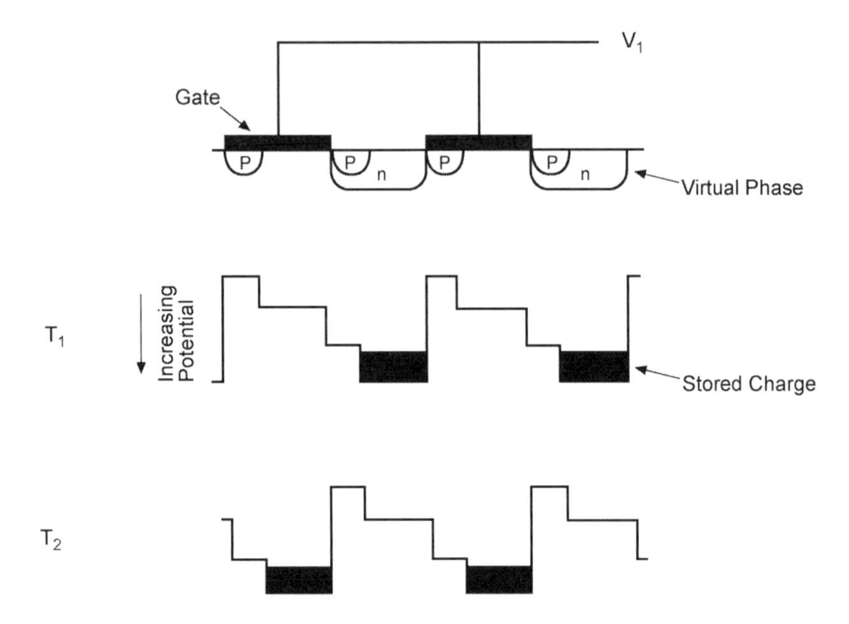

Figure 3-11. Charge transfer in a device with a virtual phase. The virtual well is created by p- and n-material implants. These ions create a fixed bias and therefore a well with fixed potential. By changing V_1, the active gate potential can be lower than the virtual well (T_1) or higher (T_2). This represents one column. Rows go into the paper.

3.3. CCD ARRAY ARCHITECTURE

Array architecture is driven by the application. Full-frame and frame transfer devices tend to be used for scientific applications. Interline transfer devices are used in consumer camcorders, digital still cameras, and professional television systems. Linear arrays, progressive scan, and time delay and integration (TDI) are used for industrial applications. Despite an ever increasing demand for color cameras, black-and-white cameras are widely used for many scientific and industrial applications. Tutorials on the operation of different architectures can be found at a variety of websites. See, for example, Princeton Instruments Reference Library at http://www.piacton.com/library/tutorials/detectors/.

3.3.1. LINEAR ARRAYS

The simplest arrangement is the linear array or single line of detectors. While these could be photogates, photodiodes are used more often because they have higher quantum efficiency. Linear arrays are used in applications where either the camera or object is moving in a direction perpendicular to the row of sensors. They are used where rigid control is maintained on object motion such as in document scanning.

Located next to each sensor are the transfer gate and then the CCD shift register (Figure 3-12a). The shift register is also light sensitive and is covered with a metal light shield. Overall pixel size is limited by the photodiode size. For example, with a three-phase system, the pixel width is three times the width of a single gate. For a fixed gate size, the active pixel width can be reduced with a bilinear readout (Figure 3-12b). Because resolution is inversely related to the detector-to-detector spacing (pixel pitch), the bilinear readout has twice the resolution. For fixed pixel size, the bilinear readout increases the effective charge transfer efficiency because the number of transfers is reduced by a factor of two (discussed in Section 9.6. *Charge Transfer Efficiency*).

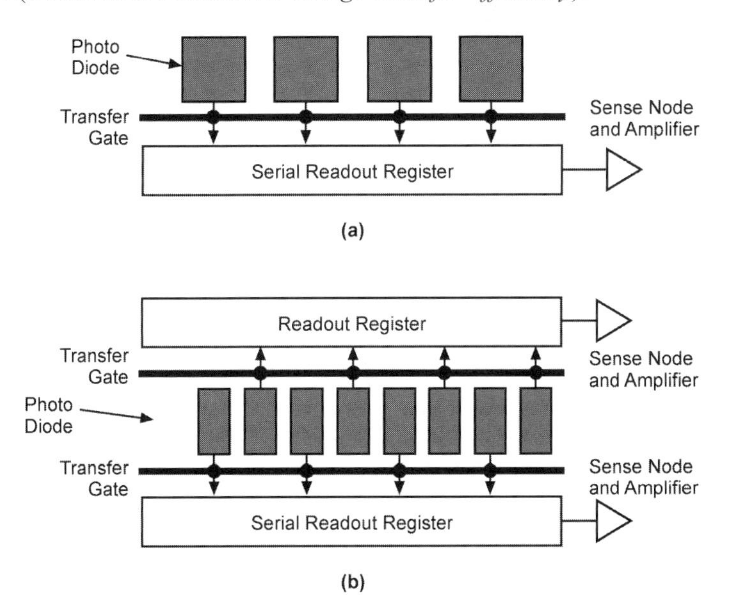

Figure 3-12. Linear array. (a) Simple structure and (b) bilinear readout. The transfer gate clock pulse allows charge to move from the photodiode into the CCD transfer register.

3.3.2. FULL FRAME ARRAYS

Figure 3-13 illustrates a full frame transfer (FFT) array. After integration, the image pixels are read out line-by-line through a serial register that then clocks its contents onto the output sense node (see Figure 3-4). All charge must be clocked out of the serial register before the next line can be transferred. In full-frame arrays, the number of pixels is often based on powers of two (e.g., 512×512, 1024×1024) to simplify memory mapping. Scientific arrays have square pixels, and this simplifies image processing algorithms. Full-frame arrays approach 100% fill factor and are the choice for applications that require the highest possible quantum efficiency.

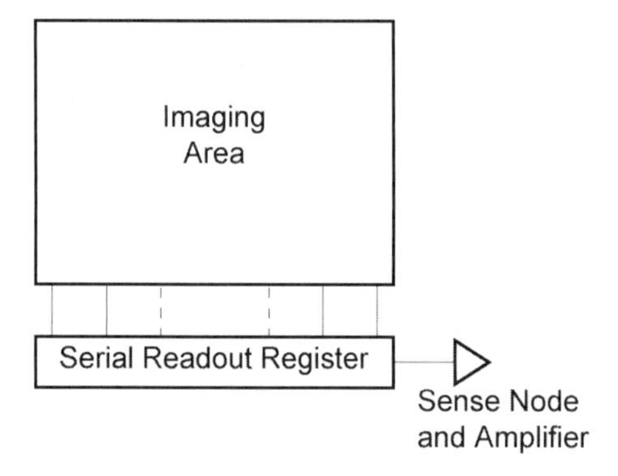

Figure 3-13. Full frame architecture. Because photogates are used, transfer gates are not essential.

During read out, the photosites are continually illuminated resulting in a smeared image. The smear will be in the direction of the charge transport in the imaging part of the array. A mechanical or external electronic shutter can be used to shield the array during readout to avoid smear. When using strobe lights to illuminate the image, no shutter is necessary if the transfer is between strobe flashes. If the image integration time is much longer than the readout time, then the smear may be considered insignificant. This situation often occurs with astronomical observations.

Data rates are limited by the amplifier bandwidth and, if present, the conversion capability of the analog-to-digital converter. To increase the effective readout rate, the array can be divided into subarrays that are read out simultaneously. In Figure 3-14, the array is divided into four subarrays. Because they are all read out simultaneously, the effective clock rate increases by a factor of four. Software then reconstructs the original image.

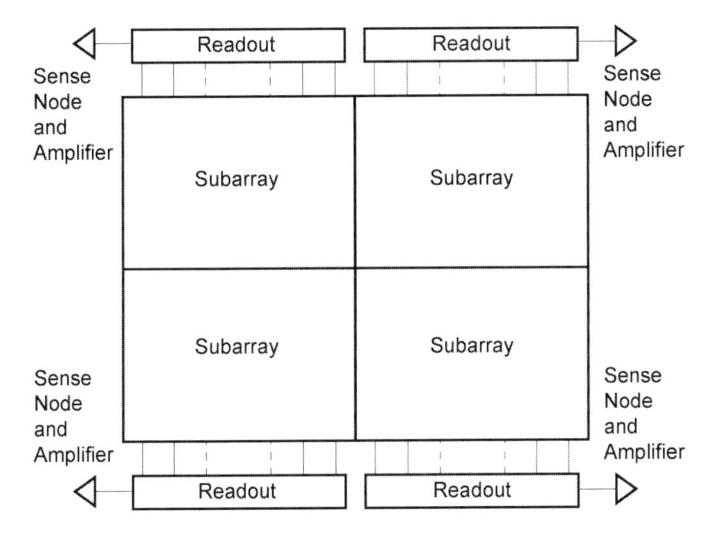

Figure 3-14. A large array divided into four subarrays. Each subarray is read out simultaneously to increase the effective data rate. Very large arrays may have up to 32 parallel outputs. The subimages are reassembled in electronics external to the device to create the full image.

3.3.3. FRAME TRANSFER

A frame transfer (FT) imager consists of two almost identical arrays, one devoted to image pixels and one for storage (Figure 3-15). The storage cells are identical in structure to the light sensitive cells but are covered with a metal light shield to prevent any light exposure. After the integration cycle, charge is transferred quickly from the light-sensitive pixels to the storage cells. Transfer time to the shielded area depends on the array size but is typically less than 500 μs. Smear is limited to the time it takes to transfer the image to the storage area. This is considerably shorter than that required by a full-frame device.

The array may have a few dummy charge wells between the active and storage areas. These wells collect any anticipated light leakage and their data are ignored. Some arrays do not have the integral light shield. These arrays can either be operated in the full-frame mode (e.g., 512×512) or in the frame transfer mode (e.g., 256×512). Here, it becomes the user's responsibility to design the light shield for the frame transfer mode.

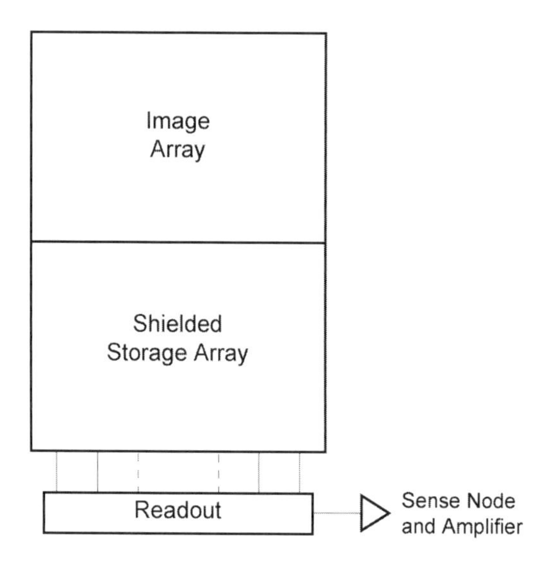

Figure 3-15. Basic architecture for a full-frame CCD. Frame transfer devices can also have a split architecture similar to that shown in Figure 3-14.

Example 3-1
SHIELD DESIGN

A camera manufacturer buys a full-frame transfer array (512×512) but wants to use it in a frame transfer mode (256×512). An engineer designs an opaque shield that is placed 1 mm above the array. He is using f/5 optics. What is the usable size of the array if the pixels are 20 μm square and the shield edge is centered on the array?

When placed near the focal plane, the shield edge will create a shadow that is approximately the distance from the focal plane (***d*** = 1 mm) divided by the focal ratio (***F*** = 5). The full shadow is 200 μm or 10 pixels. One-half of the shadow falls on the active pixels and one-half illuminates the pixels under the shield. This results in a usable array size of 251×512. An alignment error of 0.5 mm adds an additional 25 pixels. If the alignment error covers active pixels, the active area is reduced to 226×512. If the error covers shielded pixels, the storage area is reduced to 226×512. Either way, the usable array is 226×512.

Example 3-2
USEFUL ARRAY SIZE

A consumer wants to operate the camera designed in Example 3-1 under relatively low illumination levels. He changes the lens to one with focal ratio of 1.25. Will he be satisfied with the imagery?

The lower focal ratio number increases the shadow to 40 pixels. Assuming an alignment error of 0.5 mm, the minimum usable area size has dropped to 211×512 with a "soft edge" of 40 pixels. Acceptability of this image size depends on the application.

3.3.4. INTERLINE TRANSFER

The interline transfer (IL) is the most widely used architecture in volume applications. The IL array consists of photodiodes separated by vertical transfer registers covered by an opaque metal shield (Figure 3-16). Although photogates could be used, photodiodes offer higher quantum efficiency (discussed in Section 6.1. *Quantum efficiency*). After integration, the charge generated by the photodiodes is transferred to the vertical CCD registers in about 1 μs and smear is minimized. The main advantage of interline transfer is that the transfer from the active sensors to the shielded storage is quick. There is no need to shutter the incoming light for average conditions. However, bright lights can still smear. A digital camera may use a mechanical shutter. An electronic shutter open time can be variable if an antibloom drain is present (discussed in Section 3.6. *Antibloom drain*). The shields act like a Venetian blind that obscures half the information available in the scene. The area fill factor may be as low as 20%. Because the detector area is only 20% of the pixel area, the output voltage is only 20% of a detector that would completely fill the pixel area. Microlenses increase the optical fill factor (discussed in Section 5.5. *Microlenses*).

The combination of the capacitances associated with the photodiode and transfer gate provide a time constant that limits the speed of transfer to the vertical CCD register. As a result, not all the charge is read out and some is left behind. This residual charge is added to the next frame and results in image lag. That is, with photodiodes, the remaining charge is drained in successive readouts, causing an after-image. With hole-accumulation devices wells can be completely emptied and no image lag occurs.

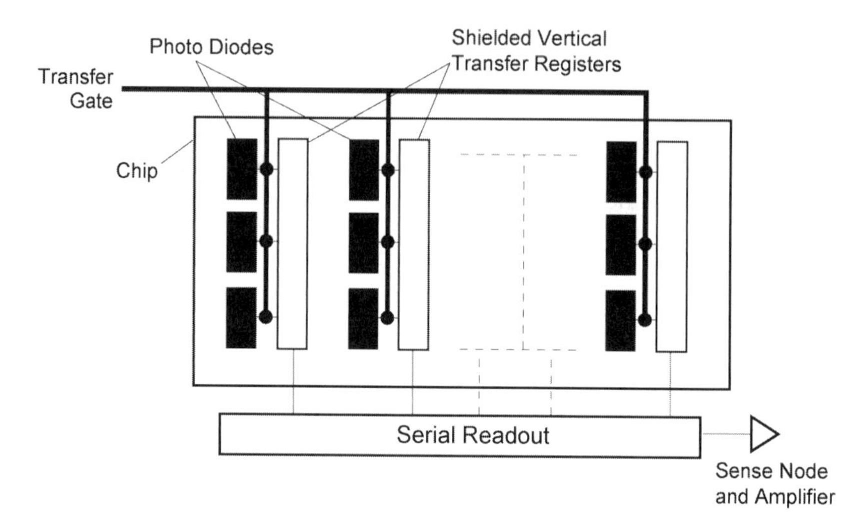

Figure 3-16. Interline transfer architecture. The charge is rapidly transferred to the interline transfer registers via the transfer gate. The registers may have three or four gates. Interline transfer devices can also have a split architecture similar to that shown in Figure 3-14.

A fraction of the light can leak into the vertical registers. This effect is most pronounced when viewing an ambient scene that has a bright light in it. For professional television applications, the frame interline transfer (FIT) array was developed to achieve even lower smear values (Figure 3-17).

A variety of transfer register architectures are available. A register can operate with virtual, two, three, or four phases. The number of phases selected depends on the application. Because interline devices are most often found in general imagery products, most transfer register designs are based on standard video timing (discussed in Section 7.5. *Video formats*). Figure 3-18 illustrates a four-phase transfer register that stores charge under two gates. With 2:1 interlace, both fields are collected simultaneously but are read out alternately. This is called frame integration. With EIA 170 (formerly called RS 170), each field is read every 1/60 s. Because the fields alternate, the maximum integration time is 1/30 s for each field (see Section 3.3.5, Figure 3-21).

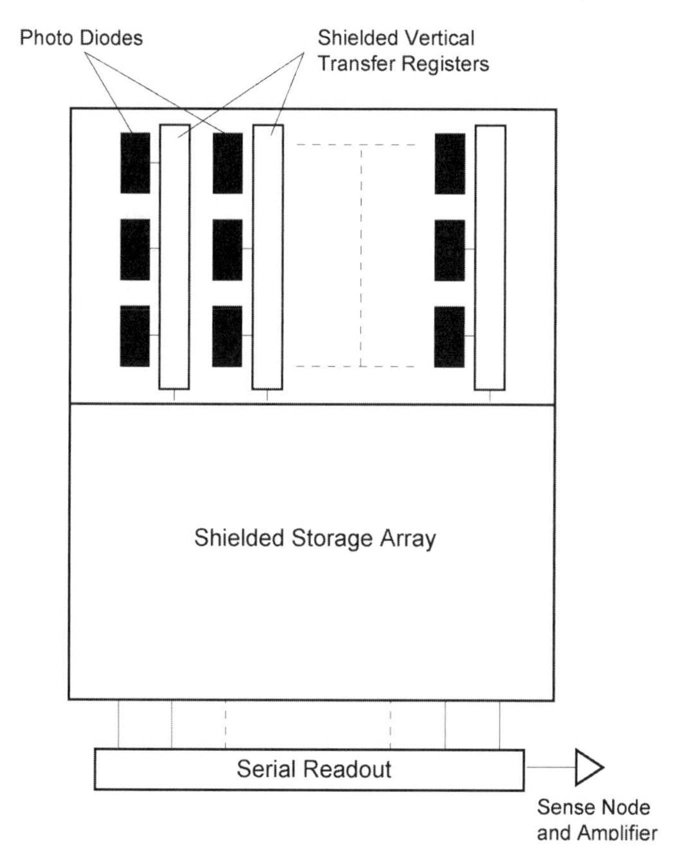

Figure 3-17. Frame interline transfer architecture. Both the vertical transfer registers and storage area are covered with an opaque mask to prevent light exposure. The transfer gates are not shown. FITs are used in broadcast cameras.

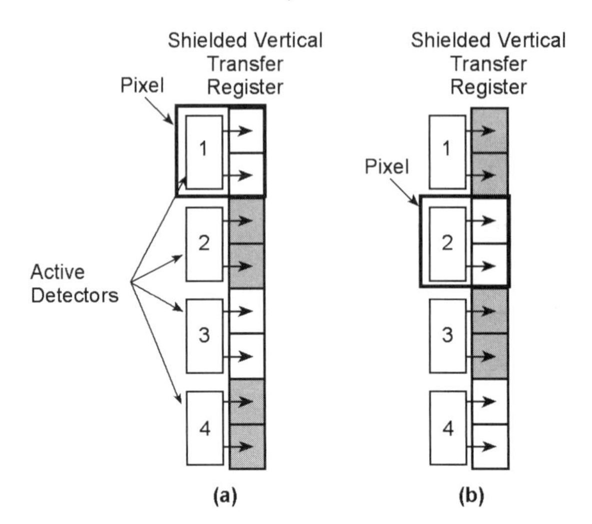

Figure 3-18. Detailed layout of the 2:1 interlaced array. (a) The odd field is clocked into the vertical transfer register and (b) the even field is transferred. The vertical transfer register has four gates and charge is stored under two wells. The pixel is defined by the detector center-to-center spacing and it includes the shielded vertical register area. The transfer gate is not shown. (After Reference 13).

Pseudo-interlacing (sometimes called field integration) is shown in Figure 3-19. Changing the gate voltage shifts the image centroid by one-half pixel in the vertical direction. This creates 50% overlap between the two fields. The pixels have twice the vertical extent of standard interline transfer devices and therefore have twice the sensitivity. An array that appears to have 240 elements in the vertical direction is clocked so that it creates 480 lines. However, this reduces the vertical MTF. With some devices, the pseudo-interlace device can also operate in a standard interlace mode.

While any architecture can be used to read the charge, the final video format generally limits the selection. Although the charge is collected in rows and columns, the camera reformats the data into a serial stream consistent with the monitor requirements. This serial stream is related to a fictitious "scan" pattern. Figure 3-20 illustrates the video lines and field of view of a standard-format imaging system.

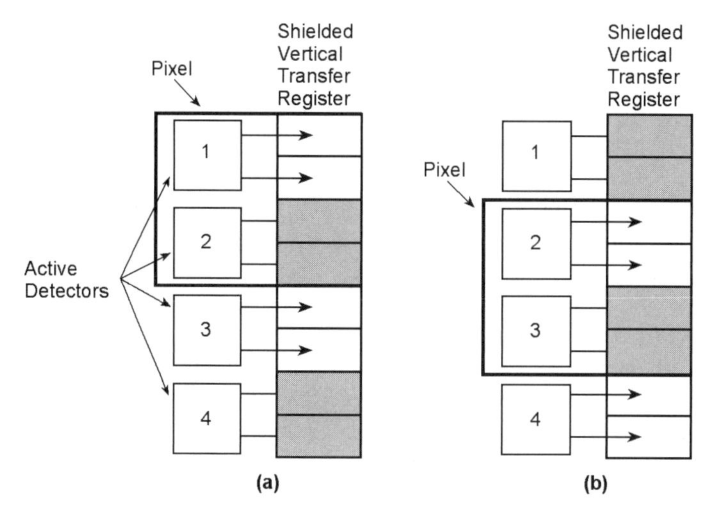

Figure 3-19. Pseudo-interlace. By collecting charge from alternating active detector sites, the pixel centroid is shifted by one-half pixel. (a) Odd field and (b) even field. The detector numbers are related to the "scan" pattern shown in Figure 3-20b. The transfer gate is not shown. (After Reference 13).

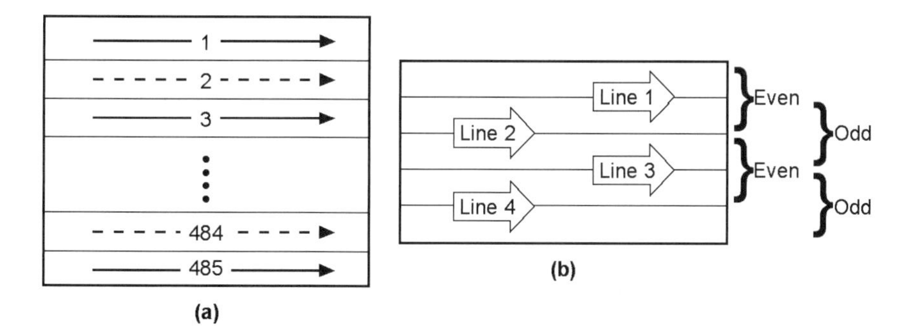

Figure 3-20. Fictitious "scan" patterns. The scan patterns help to visualize how the scene is dissected by the array architecture. (a) Interlace "scanning" achieved with the array shown in Figure 3-18. The solid lines are the odd field and the dashed lines are the even field. (b) Pseudo-interlace "scanning" provided by the array shown in Figure 3-19.

3.3.5. PROGRESSIVE SCAN

CCD arrays do not scan the image, but it is convenient to represent the output as if it were scanned. Progressive scan simply means the noninterlaced or sequential line-by-line scanning of the image. This is important to machine vision because it supplies accurate timing and has a simple format. Any application that requires digitization and a computer interface will probably perform better with progressively scanned imagery.

The primary advantage of progressive scan is that the entire image is captured at one instant of time. In contrast, interlaced systems collect fields sequentially in time (Figure 3-21). Any vertical image movement from field-to-field smears the interlaced image in a complex fashion. Horizontal movement serrates vertical lines. With excessive movement, the two fields will have images that are displaced one from the other (Figure 3-22). Slight movement (usually unavoidable) causes image edges to dance (called twitter). This is most noticeable with small alphanumeric characters.

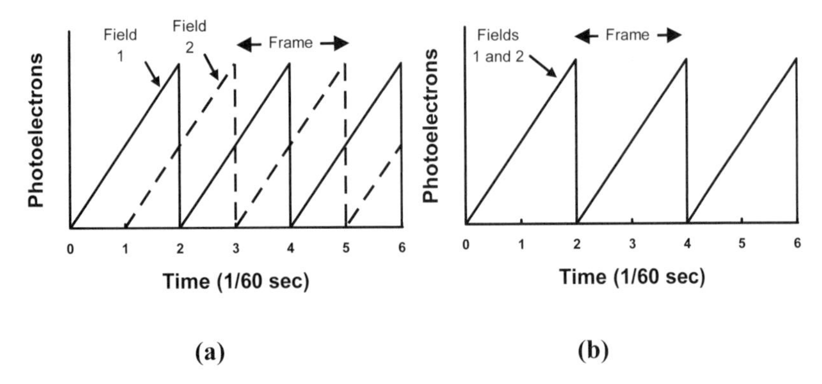

Figure 3-21. (a) Interlace scan. Each field integrates for 1/30 sec (staggered). The read out is every 1/60 sec. (b) Progressive scan. Both fields integrate at the same time. After readout, the charge well is reset to zero.

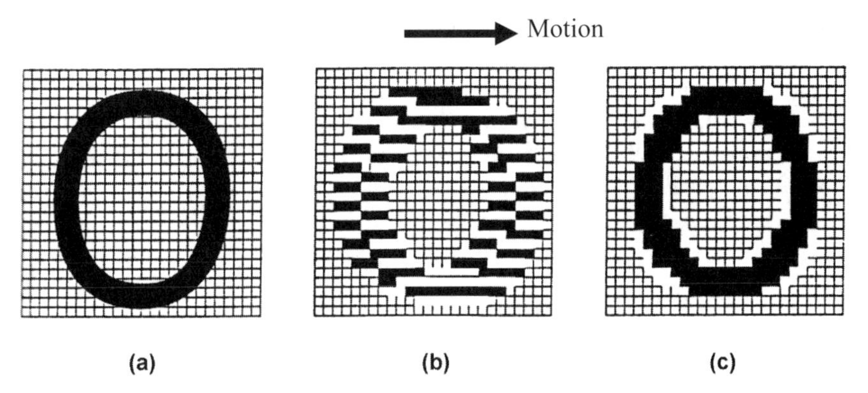

Figure 3-22. Effects of movement. (a) Target, (b) serration that occurs with an interlace system, and (c) output of a progressively scanned array. An image can still be smeared with progressive scanning, but it won't be serrated.

If a strobe light is used to stop motion, the image will appear only in the field that was active during the light pulse. Because of image motion effects, only one field of data can be used for image processing. This reduces the vertical resolution by 50% and will increase vertical image aliasing.

Because an entire frame is captured, progressive scan devices do not suffer from same image motion effects seen with interlaced systems. Thus, progressive scan cameras are said to have improved vertical resolution over interlaced systems. Full-frame and frame transfer devices inherently provide progressive scan readout. Although early interline transfer architecture was designed for interlace scanning, some current devices also provide progressive scan output.

Electronics can convert progressively scanned data into a standard video format. This allows the progressively scanned camera output to be viewed on any standard interlaced display. For scientific and industrial applications, the progressive output is captured by a frame grabber. Once in the computer, algorithms automatically reformat the data into the format required by the monitor. The camera pixel layout and array timing do not need to fit a specified format for these applications.

3.3.6. TIME DELAY AND INTEGRATION

Time delay and integration (TDI) is analogous to taking multiple exposures of the same object and adding them[14]. The addition takes place automatically in the charge well, and the array timing produces the multiple images. Figure 3-23 illustrates a typical TDI application.

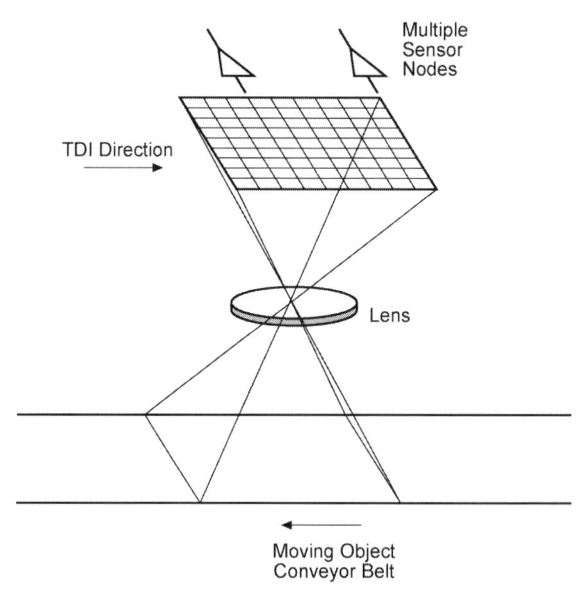

Figure 3-23. Typical TDI operation. The pixel clock rate must be matched to the image velocity.

A simple camera lens inverts the image so the image moves in the opposite direction from the object. As the image is swept across the array, the charge packets are clocked at the same rate. The relative motion between the image and the target can be achieved in many ways. In airborne reconnaissance, the forward motion of the aircraft provides the motion with respect to the ground. With objects on a conveyor belt, the camera is stationary and the objects move. In a document flatbed scanner, the document is stationary but either the camera moves or a scan mirror moves.

Figure 3-24 illustrates four detectors operating in TDI mode. The charge increases linearly with the number of detectors in TDI. Theoretically, this results in an SNR improvement by the square root of the number of TDI elements. The well capacity limits the maximum number of TDI elements that can be used.

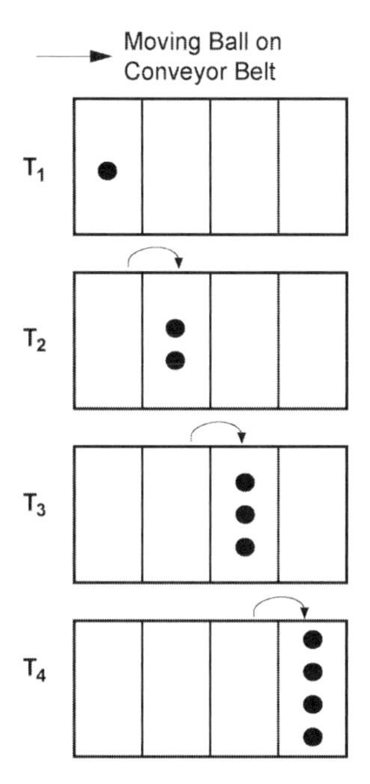

Figure 3-24. TDI concept. At time T_1, the image is focused onto the first detector element and creates a charge packet. At T_2 the image has moved to the next detector. Simultaneously, the charge packet is clocked to the next pixel site. Here, the image creates additional charge that is added to the stored packet (created by the first detector). After four stages, the signal increases fourfold.

For this concept to work, the charge packet must always be in synch with the moving image. If there is a mismatch between the image scan velocity and pixel clock rate, the output is smeared and this adversely affects the in-scan MTF (discussed in Section 9.7. *TDI*). The accuracy to which the image velocity is known limits the number of useful TDI stages.

Although Figure 3-23 illustrates two readout registers, an array may have multiple readout registers and may be as large as 2048x32 TDI elements. Additional readouts are required to avoid charge well saturation. The multiple readout outputs may be summed in electronics external to the TDI device or may be placed in a horizontal serial register for summing (Figure 3-25).

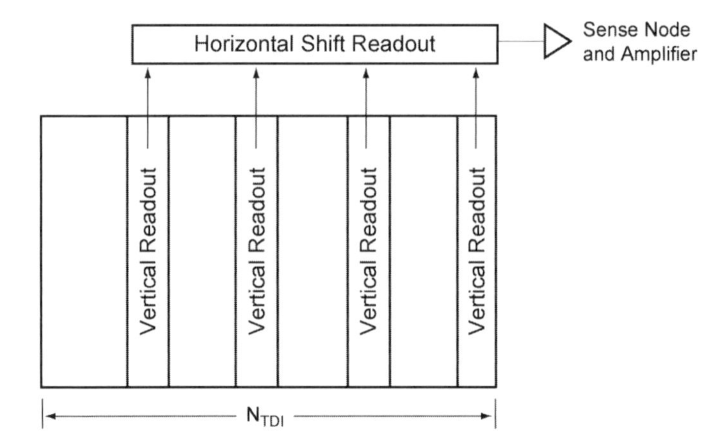

Figure 3-25. The array height must match the image height. To avoid saturation, multiple vertical register readouts may be used.

The vertical readout register must clock out the data before the next line is loaded into the register. The amplifier bandwidth or analog-to-digital conversion time limits the pixel rate. For example, if the amplifier can support 10^7 pixels/s and the register serves 500 vertical pixels, then the maximum (in-scan) line rate is 20,000 lines/s. If the amplifier can only handle 2×10^6 pixels/s and the line rate is maintained at 20,000 lines/s, then the vertical transfer register must be divided into five registers – each connected to 100 pixels. These registers and associated outputs operate in parallel (Figure 3-26).

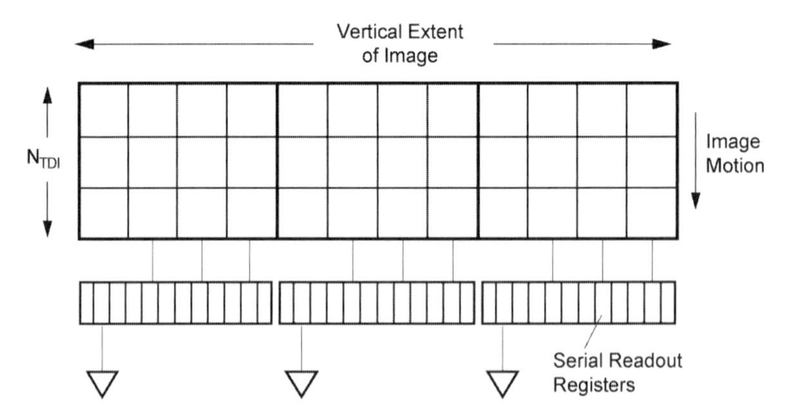

Figure 3-26. The vertical direction may be divided into subarrays to increase the line rate. The transfer register typically is a three-phase device.

TDI improves responsivity. Variations in individual detector responsivities in the TDI direction are averaged by $1/N_{TDI}$. Multiple outputs are used to increase the effective pixel rate. Each output is amplified by a separate amplifier that has unique gain characteristics. If the gains are poorly matched, the image serviced by one amplifier will have a higher contrast than the remaining image. In TDI devices, streaking occurs when the average responsivity changes from column-to-column.[15] Careful selection of arrays will minimize these nonuniformities. They can also be minimized though processing techniques employed external to the device.

Example 3-3
TDI LINE RATE

The pixels in a TDI are 15 μm square. The object is moving at 2 m/s and the object detail of interest is 150 μm. What is the lens magnification and what is the pixel clock rate?

Each pixel must view 150 μm on the target. Because the pixel size is 15 μm, the lens magnification is $M_{OPTICS} = 15/150 = 1/10$. The TDI pixel clock rate is (2 m/s)/150 μm = 133 Khz.

Example 3-4
TDI CAMERA

Bottles move on a conveyor belt at the rate of 1200 ft/min. An image processing system must detect defects without any human intervention. The bottle must be scrapped if a defect is larger than 0.1 inches. The camera sits 48 inches from the conveyor belt. If the detectors are 20 μm square, what focal length lens should be used? What is the camera line rate? If the bottle is 10 inches tall, what should the array size be?

Image processing software accuracy increases as the number of pixels on the target increases. While one pixel on the target may be sufficient for special cases, signal-to-noise ratio considerations and phasing effects typically require that the object cover at least three pixels. Thus, the image of the defect must be at least 60 μm.

The optical magnification is

$$M_{OPTICS} = \frac{R_2}{R_1} = \frac{\mathbf{60\,\mu m}}{\mathbf{(0.1\,inch)(25{,}400\,\mu m/inch)}} = \mathbf{0.0236} \tag{3-2}$$

where R_1 is the distance from the lens to the object and R_2 is the lens to array distance. The focal length is

$$fl = R_1 \frac{y_2}{y_1} = R_1 \frac{R_1 R_2}{R_1 + R_2} = \mathbf{1.1\ inches} \tag{3-3}$$

The image is moving at (1200 ft/min)(1 min/60 s)(304,800 μm/ft)(0.0236) = 143,865 μm/s. Because the detectors are 20 μm, the line rate is 7193 lines/s. This creates a detector dwell time of 1/7193 = 140 μs. That is, the edge of the defect has moved across the detector element in 140 μs. The pixel clock must operate at 7193 Hz and the integration time is 140 μs.

The image of the bottle is (10 inches)(25400 μm/inch)(0.0236) = 5994 μm. With 20 μm square detectors, the array must contain 300 detectors in the vertical direction. The number of TDI elements depends on the detector noise level, light level, and detector integration time. The signal-to-noise ratio increases by $\sqrt{N_{TDI}}$.

3.3.7. SUPER CCD

In 2000, Fujifilm introduced[16] a new CCD array layout. The octagonal detectors permit closer detector packing (Figure 3-27) due to shape and the zigzag wiring structure compared to a rectangular array. Compared to an "equivalent" interline the array has higher vertical and horizontal resolution (discussed in Section 10.3.3. *Super CCD*). Since scenes tend to contain less diagonal information, this layout seems matched to scene content which should lead to perceived higher image quality. To create standard imagery, the camera output data array must be rectangular. This is achieved by interpolation which allows Fujifilm to claim that a 6 megapixel array acts like a 12 megapixel array (Figure 3-28). Each pixel can be divided into two parts[17] to improve dynamic range by up to a factor of 4 (discussed in Section 6.2.2. *Super CCD*). Fujifilm labels the configuration illustrated in Figure 3-27 as Super CCD HR for high resolution and Figure 3-29 is labeled[18] as Super CCD SR for super dynamic range.

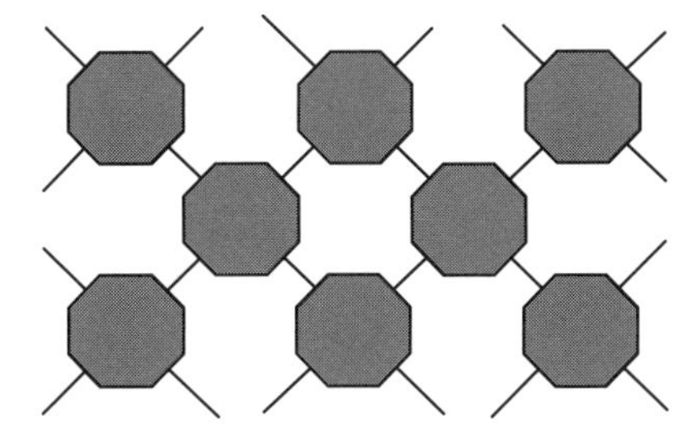

Figure 3-27. Fujifilm Super CCD HR detector layout. The array is at 45 degrees compared to the interline transfer configuration.

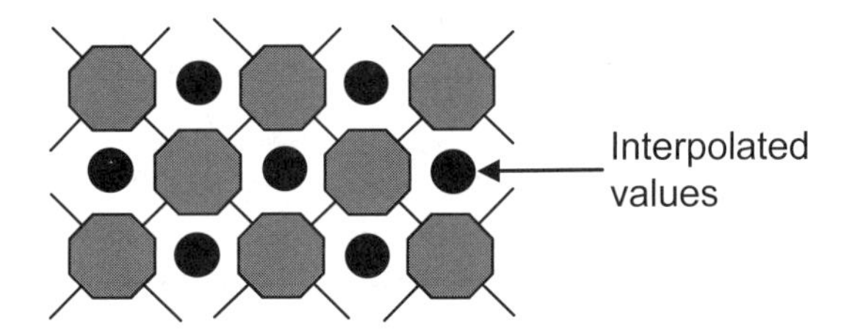

Figure 3-28. The pixel outputs must be interpolated at the circle locations to create a rectangular data array. One pixel is mapped into several datels.

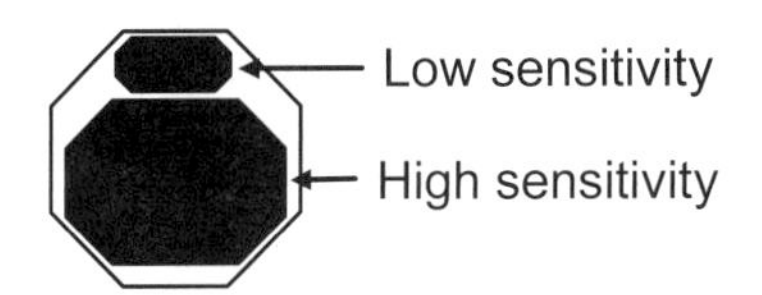

Figure 3-29. Since the area of the low sensitivity photodiode (R-pixel) is small, its output is less than the large area high sensitivity diode (S-pixel).

3.4. CHARGE CONVERSION (OUTPUT STRUCTURE)

Charge is converted to a voltage by a floating diode or floating diffusion. This circuit is sometimes called an electrometer. Figure 3-30 illustrates the last two gates of a four-phase system. The diode, acting as a capacitor, is precharged at a reference level. The capacitance, or sense node, is partially discharged by the amount of charge transferred. The difference in voltage between the final status of the diode and its precharged value is linearly proportional to the number of electrons, $\boldsymbol{n_e}$. The signal voltage after the source follower is

$$V_{SIGNAL} = V_{RESET} - V_{OUT} = n_e \frac{Gq}{C} \quad (3\text{-}4)$$

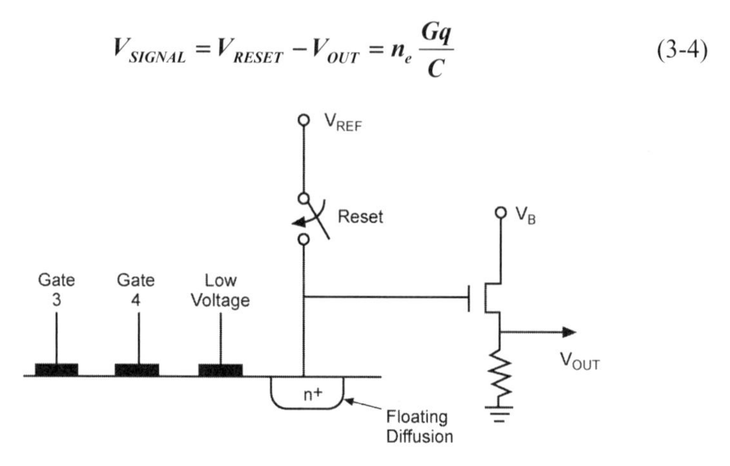

Figure 3-30. Output structure. The charge on the floating diffusion is converted to a voltage by the diffusion capacitance. The low voltage gate, which operates in a manner similar to a virtual well, allows charge transfer when the voltage on gate 4 is zero. Although the reset is shown as a switch, it is actually a FET.

The gain, G, of a source follower amplifier is approximately unity and q is the electronic charge (1.6×10^{-19} coul). The charge conversion is q/C. The output gain conversion or OGC is Gq/C. It typically ranges from 0.1 $\mu V/e^-$ to 10 $\mu V/e^-$. The signal is then amplified and processed by electronics external to the CCD sensor.

3.5. CORRELATED DOUBLE SAMPLING

After a charge packet is read, the floating diffusion capacitor is reset before the next charge packet arrives. The uncertainty in the amount of charge remaining on the capacitor following reset appears as a voltage fluctuation.

Correlated double sampling (CDS) assumes that the same voltage fluctuation is present on both the video and reset levels. That is, the reset and video signals are correlated in that both have the same fluctuation. The CDS circuitry may be integrated into the array package and this makes processing easier for the end user. For this reason it is included in this chapter.

CDS may be performed[19,20] in the analog domain by a clamping circuit with a delay-line processor or by the circuit illustrated in Figure 3-31. When done in the digital domain, a high-speed ADC is required that operates much faster than the pixel clock. Limitations of the ADC may restrict the use of CDS in some applications. The digitized video and reset pulses are subtracted and then clocked out at the pixel clock rate.

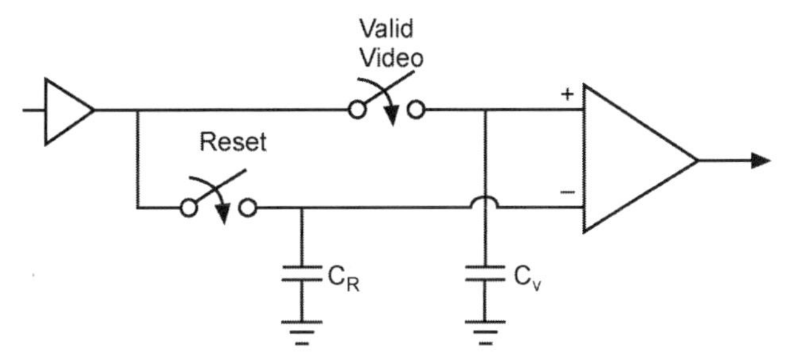

Figure 3-31. Correlated double sampling circuitry. The switches close according to the timing shown in Figure 3-32. The switches are actually FETs (After Reference 19).

The operation of the circuit is understood by examining the timing signals (Figure 3-32). When the valid reset pulse is high, the reset switch closes and the signal (reset voltage) is stored on capacitor C_R. When the valid video pulse is high, the valid video switch closes and the signal (video voltage) is stored on capacitor C_V. The reset voltage is subtracted from the video voltage in a summing amplifier to leave the CDS corrected signal. CDS also reduces amplifier 1/f noise.

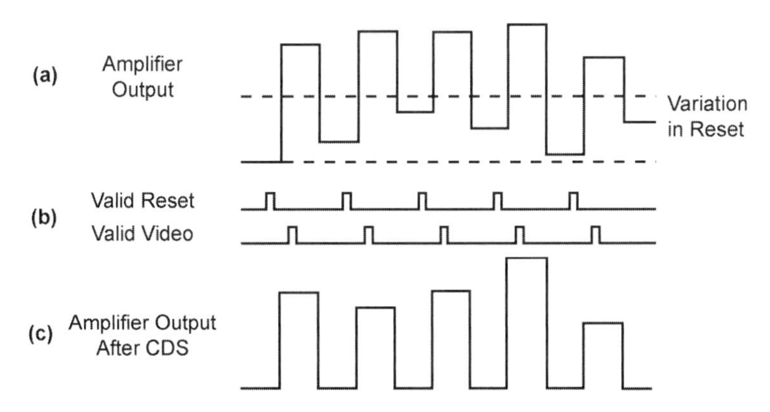

Figure 3-32. Correlated double sampling. (a) Amplifier output signal before CDS, (b) timing to measure video and reset levels, and (c) the difference between the video and reset levels after CDS. The dashed lines indicate the variation in the reset level.

3.6. ANTIBLOOM DRAIN

When a well fills, charge spills over into adjacent pixels in the same column, resulting in an undesirable overload effect called blooming. Channel stops prevent spillover onto adjacent columns. Overflow drains or antibloom drains prevent spillover. The drain can be attached to every pixel or may only operate on a column of pixels. Any photoelectron that is swept into the drain is instantly removed. Thus, the location of the drain is important.

If the drain is buried, the long-wavelength quantum efficiency is reduced because long-wavelength charge is generated below the wells and near the buried drain. A buried drain is also called a vertical overflow drain (VOD). Here, vertical refers to the *depth* of the device whereas the horizontal and vertical *directions* define the pixel size. If the drain is next to the detector, it is a lateral antibloom drain (Figure 3-33). This takes up real estate and lowers the fill-factor.

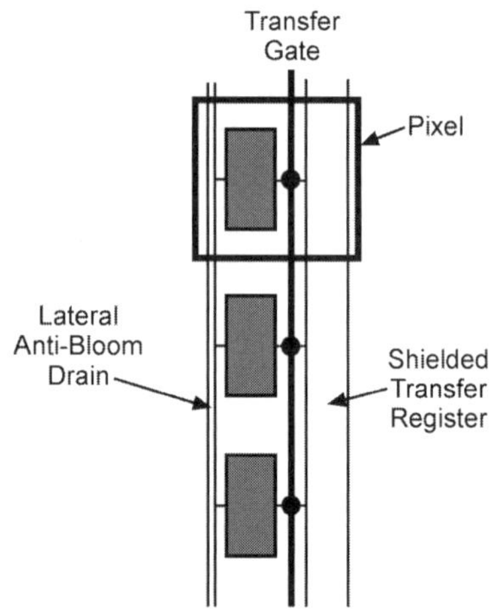

Figure 3-33. Lateral antibloom drain. The transfer gate controls the desired integration time.

Figure 3-34 illustrates the potential well created by a lateral drain. When the charge well fills over the drain barrier, charge flows into the drain. The barrier height can be controlled by an external voltage so that variable integration times are possible. Here, the antibloom drain is used for exposure control. The exposure control can work at any speed. The drain, acting as an electronic shutter, may be used synchronously or asynchronously. Asynchronous shuttering is important for use in processes where the motion of objects does not match the array frame timing or in high light level applications. The charge well is drained, reset, and then photoelectrons are integrated for the desired time.

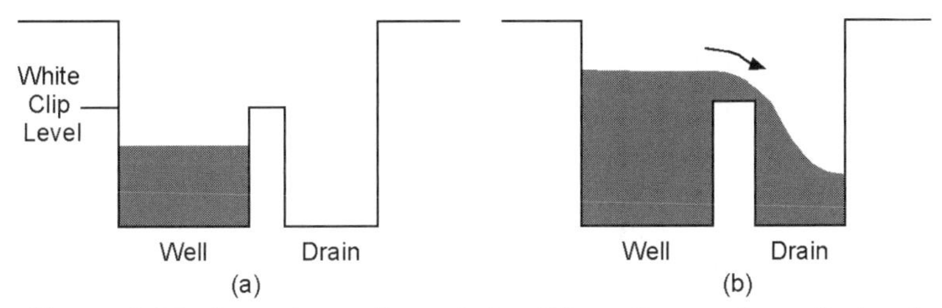

Figure 3-34. Lateral overflow drain. When the number of stored electrons exceeds the saddle point, the excess electrons flow into the drain.

In a perfect system, the output is linearly related to the input up to the antibloom drain limit. In real arrays, because of imperfect drain operation, a knee is created[21] (Figure 3-35). The response above the knee is called knee slope or knee gain. While scientific cameras operate in a linear fashion, there is no requirement to do so with consumer cameras. The knee is a selling feature in the consumer marketplace. The advantage of the knee is that it provides some response above the white clipping level. Each CCD manufacturer claims various competitive methods of handling wide scene contrasts. These methods usually involve manipulating the drain potential.

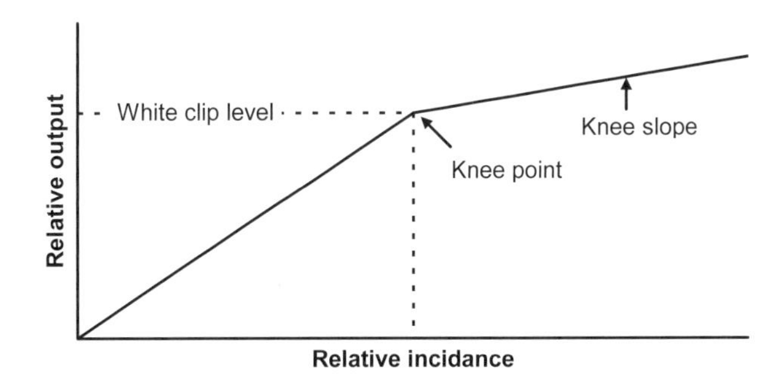

Figure 3-35. Characteristic knee created by an overflow drain.

Interline transfer devices may have the drain at the top of a vertical shift register. The excess charge cannot be drained while image information is in the vertical shift registers (Figure 3-36). This limits the range of integration times available. Figure 3-37 illustrates the variable integration afforded by the drain.

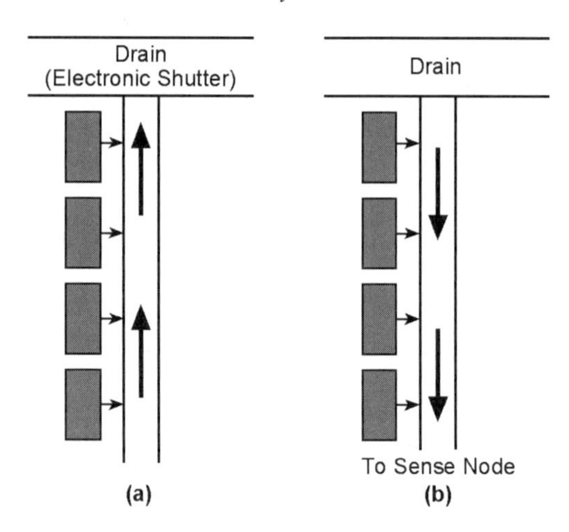

Figure 3-36. Interline transfer device where the drain is at the top of the vertical shift register. (a) Unwanted charge is sent to the drain and (b) desired charge is read out.

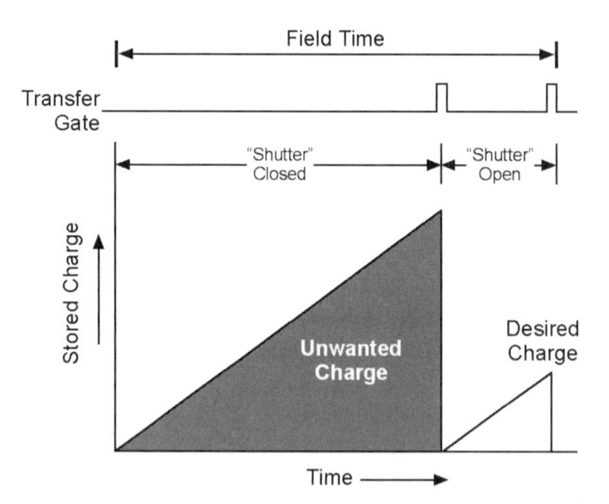

Figure 3-37. By judiciously activating the transfer gate, any integration time can be achieved. A variable integration time is equivalent to a variable electronic shutter.

3.7. LOW LIGHT LEVEL DEVICES

Signal-to-noise ratio issues limit the lowest practical detectable signal level for "standard" CCDs. Low-light-level detection is possible with an intensified CCD, electron-bombarded CCD, or electron-multiplying CCD. Each has its advantages and disadvantages. Although low-light-level televisions can be used for many applications, they tend to be used for military and scientific applications.

3.7.1. INTENSIFIED CCDS (ICCD)

The term *image intensifier* refers to a series of special imaging tubes[22] that have been designed to amplify the number of photons internally. These tubes increase the number of incident photons by several orders of magnitude. When photons hit the photocathode, electrons are liberated and thereby create an electron image. The electrons are focused onto the microchannel plate (MCP), and about 80% enter the MCP. The MCP consists of thousands of parallel channels (glass hollow tubes) that are about 10 μm in diameter. Electrons entering the channel collide with the coated walls and produce secondary electrons. These electrons are accelerated through the channel by a high potential gradient. Repeated collisions provide an electron gain up to several thousand. Emerging electrons strike the phosphor screen and a visible intensified image is created. The intensifier is an analog device and an observer can view the resultant image on the phosphor screen.

The intensifier output can be optically linked[23] to a solid state array with either a relay lens or a fiberoptic bundle (Figure 3-38). When coupled to a CCD, it is called an ICCD. The intensified camera spectral response is determined solely by the intensifier photocathode responsivity. Generally, the window material controls the short wavelength cutoff whereas the photocathode material determines the long wavelength cutoff.

Although many phosphors are available for the intensifier output, a phosphor should be selected whose peak emission is at the peak of the solid state array spectral responsivity. A color camera is not a good choice because it has lower quantum efficiency due to the color-separating spectral filters. Furthermore, single-chip color cameras have unequal sampling densities and this can lead to a loss in resolution.

The intensified camera can provide high-quality, real-time video when the illumination is greater than 10^{-3} lux (corresponding to starlight or brighter). For comparison, standard cameras (non-intensified) are sensitive down to about

1 lux. Thus, intensified cameras offer high quality imagery over an additional three orders of magnitude in illumination. A three-order magnitude increase in sensitivity implies that the MCP gain should be about a thousand. However, it must be an order of magnitude larger to overcome the transmission loss in the fiberoptic bundle or relay lens. For very low illumination levels it may be necessary to increase the integration time, and then real-time video is no longer possible.

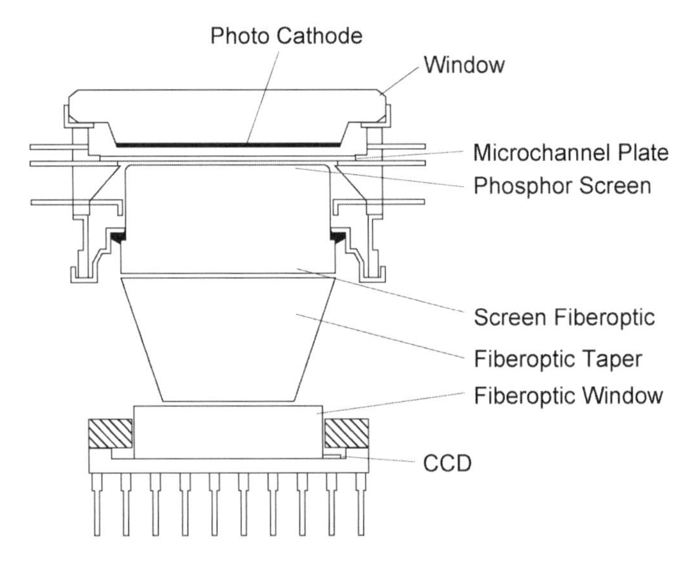

Figure 3-38. CCDs can be coupled to an image intensifier using either fiberoptic bundle or relay lens. The tapered fiberoptic bundle (shown here) tends to be the largest component of the ICCD.

By simply changing the voltage on the image intensifier, the gain can be varied. Thus, operation over a wide range of luminance levels ranging from starlight to daylight is possible. Gating is probably the biggest advantage of the intensified camera. Electronic shutter action is produced by pulsing the MCP voltage. Gating allows the detection of low-level signals in the presence of interfering light sources of much greater energy by temporal discrimination. Gate times as short as 5 ns FWHM (full width at half maximum) are possible. This allows the detection of laser diagnostic pulses against very intense continuous sources. Because full-frame transfer or frame transfer CCDs are used for ICCD applications, the inherent gating capability of the intensifier is used as a shutter during transfer time. This eliminates image smear.

Because the detector size is typically smaller than the intensifier's resolution, the intensifier's image must be minified. Fiberoptic bundles reduce camera size and weight (lenses are relatively heavy); but, fiberoptic coupling is complex because it requires critical alignment of the fibers to the individual detectors. The fiber area and pitch ideally should match the detector size and pitch for maximum sensitivity and resolution. However, to alleviate critical alignment issues, the fiber diameter is typically one-half the pixel pitch. This reduces sensitivity somewhat.

The lens-coupled system avoids the alignment problems associated with fibers and does not introduce moiré patterns. It offers more flexibility because an image intensifier with an output lens can be connected to an existing solid state camera. The lens coupling represents a cost-effective way to add gating capability to an existing camera. However, a lens-coupled system provides a much lower throughput (2× to 10× lower) than a fiber-coupled system.

The intensifier tube typically limits the intensified camera resolution. Increasing the array resolution may have little effect on the system resolution. The array is typically rectangular whereas the intensifier is round and the array cannot sense the entire amplified scene. Most commercially available intensified cameras are based on the standard military 18 mm image intensifier. Larger tubes (25 mm and 40 mm) are available but are more expensive and increase coupling complexity.

Intensifiers are electron tubes. Tubes can bloom so that scenes with bright lights are difficult to image. For example, it is difficult to image shadows in a parking lot when bright street lights are in the field of view. Intensifiers are subject to the same precautions that should be observed when using most photosensitive coated pickup tubes. Prolonged exposure to point sources of bright light may result in permanent image retention.

The intensifier specifications alone do not accurately portray how the system will work. Only a detailed analysis will predict performance. Resolution of the array will not describe system resolution if the image intensifier provides the limiting resolution. While the array photoresponse may be quite uniform, the overall nonuniformity is limited by the image intensifier. Standard military image intensifier photocathodes have nonuniformities approaching ≈ 30%. Only carefully selected intensifiers with low nonuniformity are used as part of the intensified system.

3.7.2. ELECTRON BOMBARDED CCD (EBCCD)

With the electron-bombarded CCD[24,25] (EBCCD), the CCD is placed within a vacuum tube. Photoelectrons are liberated by the photocathode and a large voltage accelerates the proximity focused electrons, which induce gain in the silicon. Gains can reach 1200, with typical values of 100 to 200. Additional gain is possible by adding an avalanche diode (Figure 3-39). The direct bombardment tends to degrade the CCD over time. An alternate approach is to place the CCD within an intensifier tube. The MCP provides the gain and the less-energetic MCP output bombards the CCD. As with the ICCD, the spectral responsivity is depends on the photocathode and not the CCD spectral response. Without the microchannel plate, screen or fiberoptic taper associated with the ICCD, the EBCCD appears to provide a better quality image.

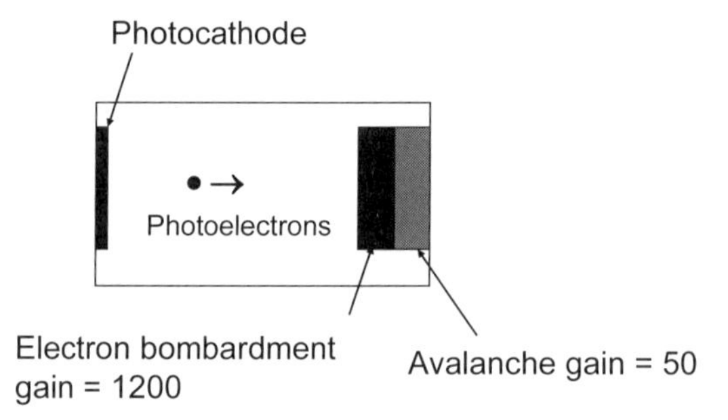

Figure 3-39. Combination of electron bombardment with the gain of an avalanche diode. Gains of 60,000 are possible. Actual system gain may be considerably less when considering the SNR.

3.7.3. ELECTRON MULTIPLYING CCD (EMCCD)

The electron-multiplying CCD (EMCCD) produces gain through impact ionization[26-28]. A multiplication register or gain register is added to the output register (Figure 3-40); this arrangement is often called on-chip multiplication. Designed for low light levels, it is important to maximize responsivity and therefore the gain register is typically placed on a full-frame transfer CCD. The multiplication register is similar to readout shift register except that one gate is held at a fixed voltage and another has a high clock voltage to induce gain (Figure 3-41). As illustrated in Figure 3-42, the gain is very dependent upon the applied voltage and sensor temperature. While the gain per cell may only be 1.015, with 500 multiplying cells, the total gain is $(1.015)^{500} \approx 1700$. The advantage of the EMCCD is that gain is applied before the readout thus minimizing the deleterious effect of readout noise in low light conditions.

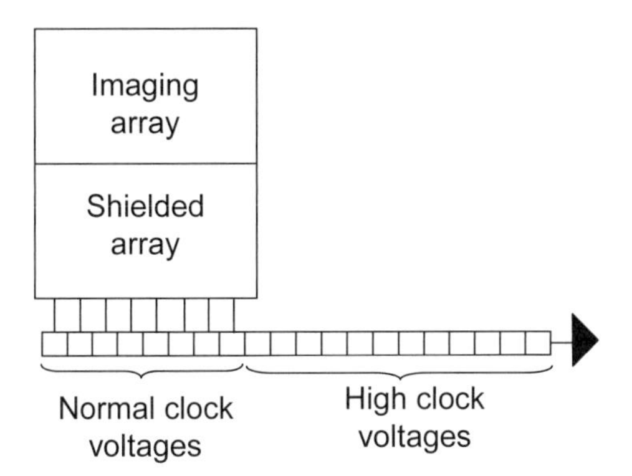

Figure 3-40. CCD with a multiplication register (high clock voltages).

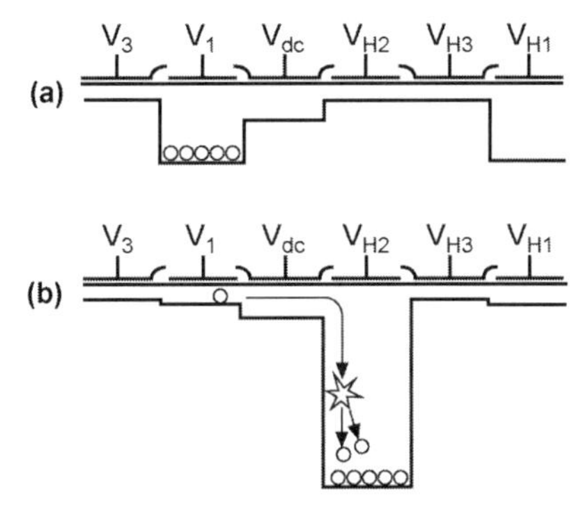

Figure 3-41. Impact ionization. (a) Low voltage and (b) a high clock voltage creates impact ionization within the well.

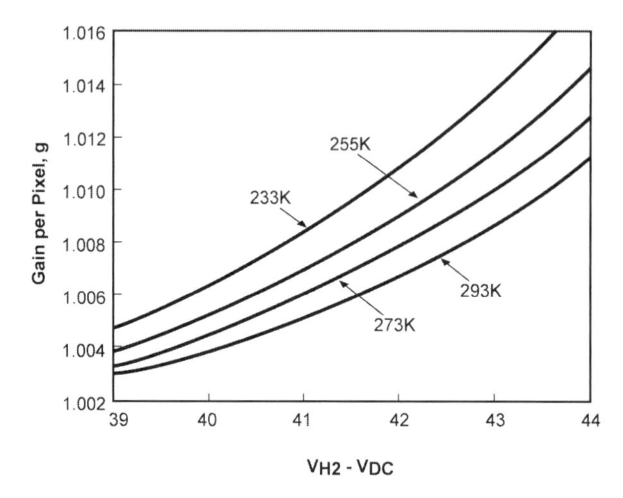

Figure 3-42. Gain versus voltage. The applied voltage may be as high as 50V. (From reference 29).

3.8. CHARGE INJECTION DEVICE (CID)

A charge injection device consists of two overlapping MOS capacitors sharing the same row and column electrode[30]. Figure 3-43 illustrates the pixel architecture.

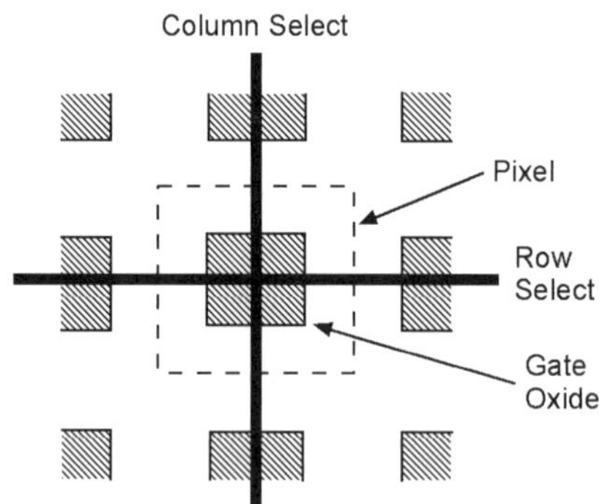

Figure 3-43. CID pixel architecture.

Although the capacitors are physically orthogonal, it is easier to understand their operation by placing them side-by-side. Charge is stored under the horizontally or vertically oriented capacitor. The nearly contiguous pixel layout provides a fill-factor of 80% or greater.

Figure 3-44 illustrates a functional diagram of an array and Figure 3-45 illustrates the pixel operation. In Figure 3-45a, a large voltage is applied to the columns and photogenerated carriers (usually holes) are stored under the column gate. If the column voltage is brought to zero, the charge transfers to the row gate (Figure 3-45b). The change in charge causes a change in the row gate potential which is then amplified and outputted. If V_1 is reapplied to the columns, the charge transfers back to the column gate. This is non-destructive readout and no charge is lost. Now additional charge integration is possible. Reset occurs by momentarily setting the row and column electrodes to ground. This injects the charge into the substrate (Figure 3-45c).

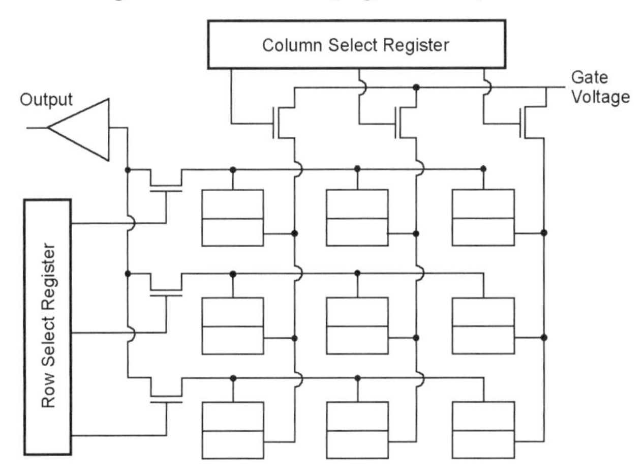

Figure 3-44. Functional layout of a 3×3 CID array. The row and column select registers are also called decoders. The select registers and readout electronics can be fabricated with CMOS technology.

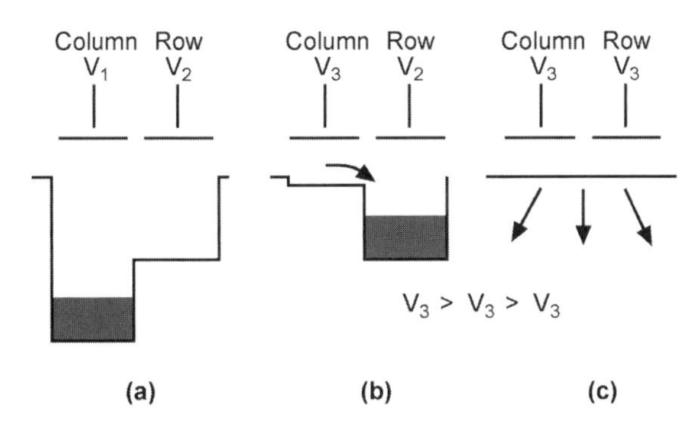

Figure 3-45. CID pixel operation. (a) Integration, (b) readout, and (c) injection.

The conversion of charge into voltage depends upon the pixel, readout line, and amplifier capacitance. The capacitance is high because all the pixels on a given row are tied in parallel. Therefore, when compared to a CCD, the charge conversion is small, yielding a small signal-to-noise ratio.

Because each pixel sees a different capacitance, CIDs tend to have higher pattern noise compared to CCDs. However, off-chip algorithms can reduce the amount of pattern noise. With no charge transfer, CIDs are not sensitive to charge transfer inefficiency. Without multiple gates, CIDs have larger well capacities than comparably sized CCDs. CIDs inherently have antibloom capability. Because charge is limited to a single pixel, it cannot overflow into neighboring pixels. Perhaps the greatest advantage of CIDs is random access to any pixel or pixel cluster. Subframes and binned pixels can be read out at high frame rates.

3.9. REFERENCES

1. J. Janesick, T. Elliott, R. Winzenread, J. Pinter, and R. Dyck, "Sandbox CCDs," in *Charge-Coupled Devices and Solid State Optical Sensors* V, M. M. Blouke, ed., SPIE Proceedings Vol. 2415, pp. 2-42 (1995).
2. J. Janesick and T. Elliott, "History and Advancement of Large Area Array Scientific CCD Imagers," in *Astronomical Society of the Pacific Conference Series*, Vol. 23, Astronomical CCD Observing and Reduction, pp. 1-39, BookCrafters (1992).
3. M. J. Howes and D. V. Morgan, eds., *Charge-Coupled Devices and Systems*, John Wiley and Sons, New York, NY (1979).
4. C. H. Sequin and M. F. Tompsett, *Charge Transfer Devices,* Academic Press, NY (1975).
5. E. S. Yang, *Microelectronic Devices*, McGraw-Hill, NY (1988).
6. E. L. Dereniak and D. G. Crowe, *Optical Radiation Detectors*, pp. 186-269, John Wiley and Sons, New York, NY (1984).
7. A. J. P. Theuwissen, *Solid-State Imaging with Charge-Coupled Devices*, Kluwer Academic Publishers, Dordrecht, The Netherlands (1995).
8. M. Kimata and N. Tubouchi, "Charge Transfer Devices," in *Infrared Photon Detectors,* A. Rogalski, ed., pp. 99-144, SPIE Press, Bellingham, WA (1995).
9. J. Janesick, *Scientific Charge-Coupled Devices,* SPIE Press, Bellingham, WA (2001).
10. Numerous articles can be found in the Proceedings of the SPIE conferences: *Sensors, Cameras, and Systems for Scientific/Industrial Applications* and *Semiconductor Photodetectors* SPIE Press, Bellingham WA
11. A. J. P. Theuwissen, *Solid-State Imaging with Charge-Coupled Devices*, pp. 317-348, Kluwer Academic Publishers, Dordrecht, The Netherlands (1995).
12. A. J. P. Theuwissen, *Solid-State Imaging with Charge-Coupled Devices*, pp. 54-66, Kluwer Academic Publishers, Dordrecht, The Netherlands (1995).
13. A. J. P. Theuwissen, *Solid-State Imaging with Charge-Coupled Devices*, pp. 161-165, Kluwer Academic Publishers, Dordrecht, The Netherlands (1995).
14. H.-S. Wong, Y. L. Yao, and E. S. Schlig, "TDI Charge-Coupled Devices: Design and Applications," *IBM Journal Research Development*, Vol. 36(1), pp. 83-106 (1992).

15. T. S. Lomheim and L. S. Kalman, "Analytical Modeling and Digital Simulation of Scanning Charge-Coupled Device Imaging Systems," in *Electro-Optical Displays*, M. A. Karim, ed., pp. 551-560, Marcel Dekker, New York (1992).
16. H Tamayama, O. Saito, and M. Inuiya, "High-definition Still Image Processing System Using a New Structure CCD Sensor," in *Sensors and Camera Systems for Scientific, Industrial, and Digital Photography Applications*; M. M. Blouke, N. Sampat, G. M. Williams, Jr., and T. Yeh, eds., SPIE Proceedings Vol. 3965, pp. 431-436 (2000).
17. K. Takemura, K. Oda, T. Nishimura, H. Tamayama, Y. Takeuchi, and T. Yamada "Challenge for Improving Image Quality of a Digital Still Camera," in *Sensors and Camera Systems for Scientific, Industrial, and Digital Photography Applications,* IV, M. M. Blouke, N. Sampat, and R, J. Motta, eds., Proceedings of SPIE Vol. 5017, pp. 385-392 (2003)
18. T. Ashida, H. Yamashita, M. Yoshida, O. Saito, T. Nishimura, and K. Iwabe, "Signal Processing and Automatic Camera Control for Digital Still Cameras Equipped with a New Type CCD': in *Sensors and Camera Systems for Scientific, Industrial, and Digital Photography Applications*, IV, M. M. Blouke, N. Sampat, and R, J. Motta, eds., Proceedings of SPIE Vol. 5310, pp. 42-50 (2004).
19. A. J. P. Theuwissen, *Solid-State Imaging with Charge-Coupled Devices*, pp. 228-231, Kluwer Academic Publishers, Dordrecht, The Netherlands (1995).
20. T. W. McCurnin, L. C. Schooley, and G. R. Sims, "Signal Processing for Low-light-level, High-precision CCD Imaging," in *Camera and Input Scanner Systems*, W. Chang and J. R. Milch, eds., SPIE Proceedings Vol. 1448, pp. 225-236 (1991).
21. S. Kawai, M. Morimoto, N. Mutoh, and N. Teranishi, "Photo Response Analysis in CCD Image Sensors with a VOD Structure," *IEEE Trans on Electron Devices*, Vol. 40(4), pp. 652-655 (1995).
22. See, for example, I. Csorba, *Image Tubes*, Howard Sams, Indianapolis, Indiana (1985).
23. Y. Talmi, "Intensified Array Detectors," in *Charge-Transfer Devices in Spectropscopy*, J. V. Sweedler, K. L. Ratzlaff, and M. B. Denton, eds., Chapter 5, VCH Publishers, New York (1994).
24. C. B. Johnson, "Review of Electron-bombarded CCD Cameras," in *Image Intensifiers and Applications; and Characteristics and Consequences of Space Debris and Near-Earth Objects*, C. B. Johnson, T. D. Maclay, and F. A. Allahdadi, eds., SPIE Proceedings Vol. 3434, pp 45-53 (1998).
25. X. Xu and J. Guo, "New EBCCD with Transferred Electron Photocathode for Range-gated Active Imaging System," in *Advanced Materials and Devices for Sensing and Imaging*, Yao and Y. Ishii, eds., SPIE Proceedings Vol. 4919, pp 536-544 (2002).
26. J. Hynecek, "Impactron – a New Solid State Image Intensifier," *IEEE Transactions on Electron Devices*, Vol. 48, pp. 2238-2241 (2001).
27. J. Hynecek and T. Nishiwaki, "Excess Noise and other Important Characteristics of low Light level imaging Using Charge Multiplying CCDs," *IEEE Transactions on Electron Devices*, Vol. 50, pp. 239-245 (2003).
28. D. J. Denvir and E. Conroy, "Electron multiplying CCDs," in *Opto-Ireland 2002: Optical Metrology, Imaging, and Machine Vision,* A. Shearer, F. D. Murtagh, J. Mahon, and P. F. Whelan, eds., SPIE Proceedings Vol. 4877 pp. 55-68 (2003).
29. M. S. Robbins and B. J. Hadwen, "The Noise Performance of Electron Multiplying Charge-coupled Devices," *IEEE Trans Electron Devices*, Vol. 50, pp 1227-1232 (2003).
30. S. Bhaskaran, D. Baiko, G. Lungu, M. Pilon, and S. VanGorden, "SpectraCAM SPM: A Camera System with High Dynamic Range for Scientific and Medical Applications" in *Focal Plane Arrays for Space Telescopes* II, T. J. Grycewicz and C. J. Marshall, eds., paper 59020D (2005).

4
CMOS FUNDAMENTALS

In the 1990s, monolithic complementary metal oxide silicon (CMOS) arrays emerged as a serious alternative to charge-coupled device (CCD) arrays. This is a natural outcome of improvements in silicon CMOS manufacturing technology, which have been technically and economically driven by digital applications (microprocessors, memory devices, and floating point gate arrays, etc.).

The fact that CMOS arrays can be manufactured using fabrication processes that are compatible with the generic CMOS industry (with its huge production volumes), suggests economies of scale that should translate to lowered CMOS array production costs. However, in the non-imaging silicon microelectronics world, some of the technological advances are not friendly to silicon CMOS array technology. There are silicon processing steps used in CCD manufacturing that are not simply not part of the processes used in standard or even array-tailored CMOS processing.

The key differences between CMOS and CCD array technologies are evident in terms of pixel-level characteristics, the method of multiplexing pixel data into a video format, sensor chip manufacturing requirements, ease of integration of on-chip digital features that enhance functionality and flexibility of array use, power consumption, and performance. CMOS pixel electronics are typically described by a shorthand notation that simply counts the number of transistors found inside a pixel (i.e., 3T, 4T, 5T, etc.). As the number of transistors increases, functionality and operating flexibility increase. When is comes to performance, 4T or higher pixels offer significant noise reduction.

Lo[1] has reviewed the historical developments and accomplishments of both CCD and CMOS imaging technologies through the end of the 1990s. Volume 50 of the IEEE Transactions on Electron Devices[2] is devoted to CMOS array technology and provides a comprehensive update in a series of papers authored by leading researchers. The Proceedings of the 2005 IEEE Workshop on CCDs and Advanced Image Sensors, provides the latest update to the developments in the visible imaging sensor field[3].

4.1. CCD AND CMOS ARRAYS: KEY DIFFERENCES

There are numerous functional and performance differences between CCD and CMOS arrays. Understanding the key differences allows intelligent selection of one over the other. This section also describes the advantages and limitations of each type of device.

4.1.1. CCD ARRAY CHARACTERISTICS

In standard front-side illuminated CCD technology, imaging pixels consist of simple MOS capacitor structures that are adjacent and overlapping (Figures 3-1 and 3-2). A bias on these structures creates an electric field. Each absorbed photon creates an electron-hole pair. The majority carriers (holes for a silicon substrate that is p-doped; electrons for a silicon substrate that is n-doped) are pushed by the electric field and/or pixel channel implant (buried-channel CCD devices) into the substrate. This creates a depletion region (actually a volume) below each pixel which is devoid of majority carriers. The opposite or minority carriers get attracted toward the gate of the CCD and are electrostatically stored in this depletion region. The minority carriers survive (i.e., are stored) since there are no majority carriers with which to recombine.

Signal charge packet size is dependent on the photon flux incident on a particular pixel, the pixel collection efficiency, and the integration time. At the end of an integration time, the signal charge packets are collectively transported through the very gates that were used to generate and confine them, by manipulation of the gate voltages. The packets are transported to the edge of the device to a one-dimensional parallel-to-serial shift register, one parallel row at a time (Figure 3-4). This serial register is usually located under a light shield since its only function is charge transport. The serial shift register then transports the pixel charge packets to an output circuit where it is converted to a voltage.

The charge-to-voltage conversion process takes place in circuit called an electrometer or simply the output circuit. The typical electrometer circuit consists of a MOS field-effect transistor arranged in a source-follower mode with a reset transistor connected to its gate (Figure 3-30). The charge packet is placed on the source-follower gate by the serial shift register. Upon generation of a voltage step from a charge packet, the electrometer is reset and the charge packet is removed, thereby preparing the electrometer to accept the next pixel charge packet from the serial shift register. The source follower electrometer configuration has the ability to drive a substantial capacitive load, such as a length of co-axial cable (e.g., up to several feet, depending on electrometer design and video data rate).

CCD architecture and operation has several notable characteristics (Figure 4-1). Charge packet creation is assumed noiseless. Electronic noise occurs in the electrometer circuit. Noise sources include white noise from the source follower transistor channel, 1/f noise from the source follower gate, and kTC noise created by resetting the source follower gate (noise is discussed in Section 6.5. *Noise*).

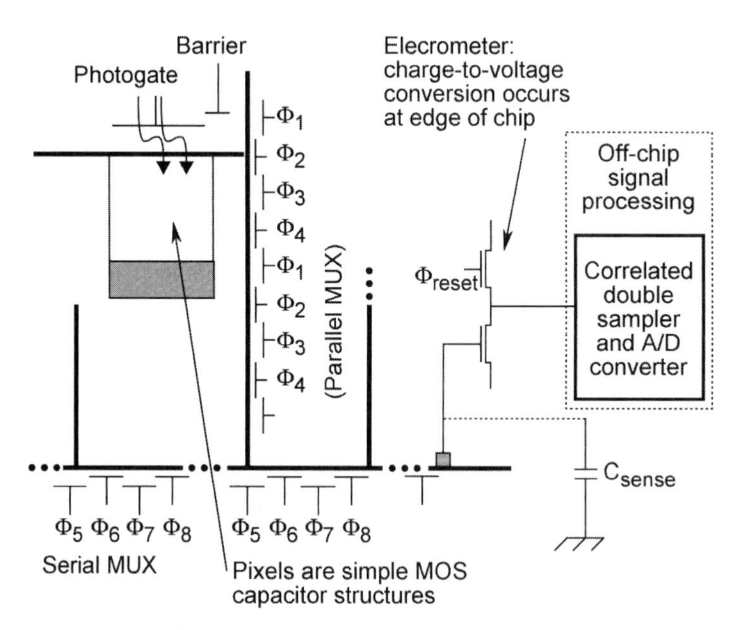

Figure 4-1. Typical CCD architecture including the pixel structure, parallel/serial multiplexing and output electrometer at the edge of the chip. The gate voltage phases are labeled as Φ1, …., Φ8. This is a composite of Figures 3-1, 3-2, 3-3, 3-30 and 3-31.

Charge transport must be nearly perfect from gate to gate, especially if the transport lengths are large as is the case for very large area CCD arrays. This efficiency is quantified by the charge transfer efficiency parameter and typically has a value of 0.99999 to 0.999999 per pixel (discussed in Section 9.6. *Charge transfer efficiency*). The CCD device must be defect-free. If a defect occurs at a transport site, all of the pixel signals before the defective pixel will be blocked.

There is no way to individually address and read out a small group of pixels (also called sub-window, sub-array, or region of interest) without clocking the entire parallel and serial gate structure. Relatively high gate voltages (8 to 15 V) are required to create the depletion wells under each pixel; these voltages must then be rapidly turned from high to low and vice versa to affect the pixel charge transport during the multiplexing process. The high gate voltages and the clocking of the gates across the entire array result in relatively high power consumption.

4.1.2. CMOS ARRAY CHARACTERISTICS

In CMOS arrays the charge-to-voltage conversion function occurs inside every pixel (Figure 4-2). The pixel area is shared by the photodiode, the pixel electrometer, and addressing/output connection circuitry. The pixel signals are converted from electrons to volts, buffered by a source follower transistor and then transferred to thin metal output buses through MOS transistor switches located inside each pixel.

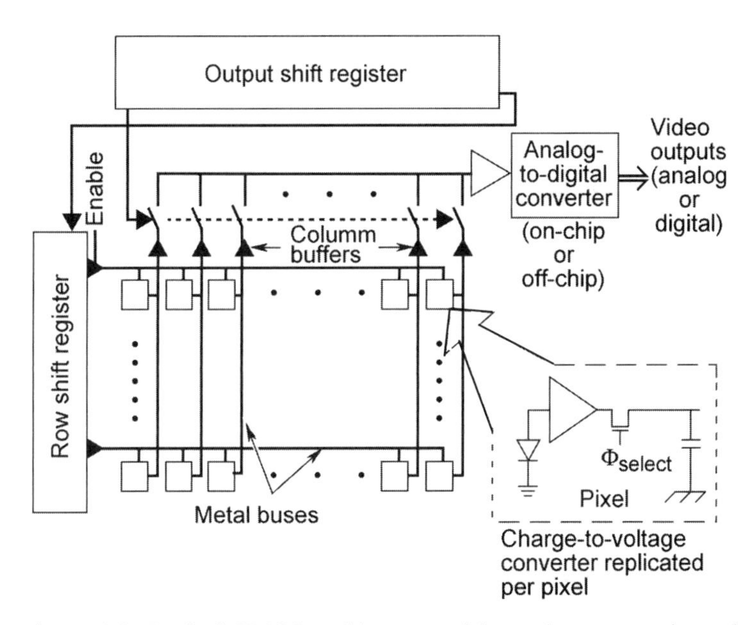

Figure 4-2. Typical CMOS architecture with an electrometer in each pixel. The parallel and serial busses address each pixel.

The parallel column busses are connected via column buffer circuits to corresponding serial buses with the aid of serial select switches for the final parallel-to-serial multiplexing function that generates the output video stream. The buffering provided by the pixel level electrometer circuits easily handle the bus capacitance.

There are several notable characteristics regarding the architecture and operating characteristics of a CMOS array. The voltages needed for the pixel reset and select transistor operations are generally lower those used for CCDs (by an order of magnitude in some cases). Furthermore, only one row (or column) is addressed at a time during readout. This results in much lower power consumption for a CMOS array compared to a CCD when these are compared at video frame rates. It is straightforward to include various digital timing and control signals directly on-chip along with other digital image processing functions. With appropriate control of the pixel select switches, random access of pixels, data windowing, variable frame rate operation within chosen windows, and interrupt-driven asynchronous array operation can all be supported. The sophistication and flexibility of these control operations increases the complexity of the required digital signals and the corresponding on-chip circuitry. However, this increased complexity will most likely lower performance.

Digitization can take place on-chip. There may be further multiplexing (or de-multiplexing) of the analog signals, as well as their amplification, before injection into one or more on-chip ADCs. The ADC video output(s) organization and multiplexing is generally specific to a given array application and very strongly influenced by the device pixel count and frame rate. This additional capability is very difficult to implement in CCDs. These functions are usually performed off-chip in CCD cameras. With on-chip functions, the CMOS camera can be much smaller that a comparable CCD camera.

4.2. CMOS ARRAYS: PREDICTIONS AND REALITY

In 1993, the visible sensor community was reminded of the advancements and associated potential of CMOS array technology in a paper by Fossum[4] entitled, "Active Pixel Sensors: Are CCDs Dinosaurs?" While this title seems to imply the imminent demise of CCD technology, the real focus of the paper was to highlight the logical development path for CMOS arrays and set expectations for progress in performance, functionality, and manufacturability.

Along with pixel sizes decreasing, feature sizes were expected to decrease (Figure 4-3). Smaller feature size means that the electronics (transistors, contacts, interconnections, and thinner metal buses) will fit within a pixel. In a scenario wherein a minimum desired pixel size is reached or where larger pixels can be used, the finer CMOS feature size allows for improved performance and/or functionality. In 2007, feature sizes as small as 0.065 microns (i.e., 65 nm) are possible, and CMOS arrays with pixel dimensions approaching

2 microns are routinely manufactured. Moore's law[5] still seems to be valid (size reduction by a factor of two every year).

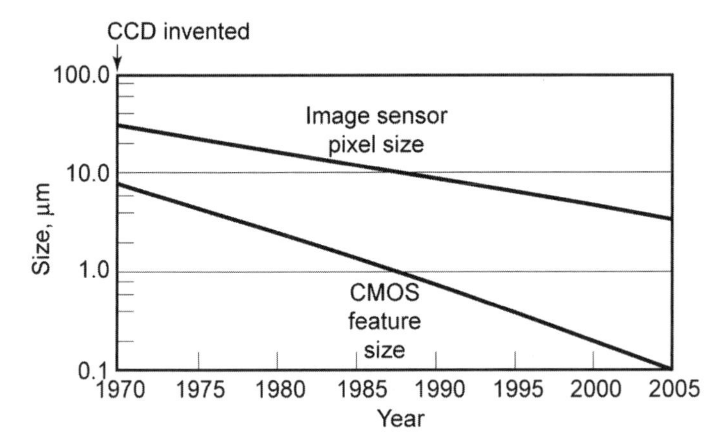

Figure 4-3. CMOS minimum feature size and typical pixel dimension. The chart was created in 1993 and therefore the sizes after 1993 were projections (after Reference 4).

As minimum feature size decreases, pixel circuitry can be made smaller thereby allowing smaller pixel dimensions for a fixed photodetector fill-factor or enhanced pixel functionality and performance of a fixed pixel size. However, there are semiconductor-scaling effects that must be considered with decreasing feature sizes. In 1996, Wong[6] discussed the performance impact of scaling minimum feature sizes below the level of 0.5 µm. His analysis showed that only relatively minor modifications to standard CMOS processing would be needed to ensure reasonable array performance down to feature sizes of 0.18 µm. But, CMOS foundries must be willing to depart from "standard" CMOS process technologies to ensure adequate array performance.

4.2.1. SMALLER PIXEL POTENTIAL LIMITATIONS

As the minimum feature size decreases, substrate doping must increase, diffusion lengths get smaller (due to the increased doping), gate oxides get thinner, transistor source/drain junction depths get shallower, transistor isolation methods change to shallow trench isolation (STI) and silicon-on-insulator (SOI), threshold voltage and threshold voltage variations are reduced, and supply voltages (and hence the maximum allowed signal voltage swing) decrease.

Reduced depletion depths due to higher doping result in more photocharge being collected from field-free regions, thereby increasing diffusion (discussed in Section 9.4. *Diffusion MTF*). To combat this effect, junction and the depletion depths under the photodiode or photogate region must be made sufficiently wide. For SOI technology, the substrate thickness (200 nm down to 25 nm) will compromise detector quantum efficiency particularly at the longer wavelengths. Dielectric isolation allows for better pixel isolation, thereby reducing diffusion MTF effects. With SOI technology, the smaller parasitic capacitances compared to bulk substrates offer advantages in terms of fixed pattern noise, power consumption, and clock feed-through.

Silicided metals are used to maintain high conductivity. However, silicided materials [e.g., cobalt-silicide ($CoSi_2$) or titanium-silicide ($TiSi_2$)] have reduced light transmission compared to polysilicon as illustrated in Figure 4-4. This increased opacity results in essentially no light transmission. It is essential to interrupt silicided metals in pixel regions that require light transmission. Fortunately it is straightforward to employ "silicide blocking" in selected areas, such as window-regions just above the optically active area.

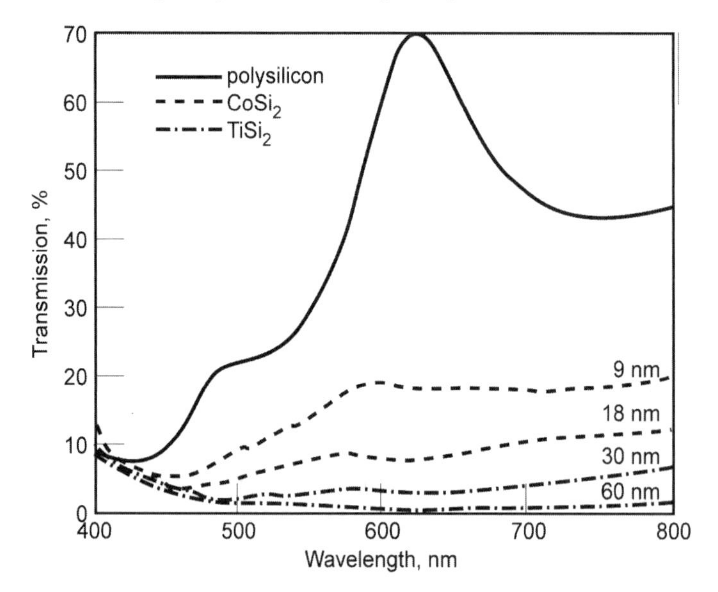

Figure 4-4. Silicided metals have high conductivity but low optical transmission. Measured spectral transmission of thin films deposited on quartz wafers. 150 nm polysilicon, 150 nm polysilicon reacted with 9 nm and 18 nm of cobalt silicide, and 150 nm polysilicon reacted with 30 nm and 60 nm of titanium silicide. (From reference 7).

Parasitic currents tend to increase as feature size decreases. For example, inter-band tunneling is increased due to the increased substrate doping and the "off-current" of MOS transistors also increases. However, pixel design techniques and the use of n-channel devices can help mitigate this effect.

A perhaps more serious issue is the reduced signal swing that results from the systematically lower CMOS supply voltages. The reduced dynamic range forces unpleasant camera system solutions (i.e., frame averaging) for sensor users that require dynamic ranges comparable to what can be achieved with CCDs (discussed in Section 6.10. *Lux transfer*). Smaller pixels imply better spatial resolution. However, reducing pixel size also impacts the optical design (discussed in Chapter 11, *Image quality*). Reducing the pixel size by a factor of two does not necessarily increase the spatial resolution by a factor of two.

4.2.2. SMALLER PIXEL DESIGN EVOLUTION

Wong[7] was the first to experimentally explore smaller CMOS pixel architecture and technology issues. He used a 0.25 μm feature size to create 7 μm square pixels with a 33% fill factor and powered by a 1.8 supply voltage to understand technology-scaling impacts on device performance. Conventional four-transistor photogate and photodiode pixels with an in-pixel reset switch, transfer gate, select switch, and source-follower amplifier were fabricated and tested. It is interesting to note that current CMOS imager technologies that use 0.18 μm feature size are based on 3.3 supply voltage.

His design departed substantially from the earlier 5-volt CMOS technologies. The specific CMOS design features were: 4 nm thick gate oxides, 0.12 μm nFET effective channel length, retrograde channel doping by high-energy implants, ultra-shallow source/drain implants to control short channel effects, self-aligned silicide gate and source/drains to reduce resistances, shallow trench isolation, tungsten-filled trench local interconnects, and tight-pitched borderless contacts.

The experimental results showed an extremely small floating diffusion node (~3.5 fF) with an output gain conversion of 46 μV/electron, which was impressive in 1988. The measured dark current was several times greater than that obtained with the best CCD technology. For the photogate pixel (a surface channel device) the dominant dark current generating mechanism was due to interface states at the Si/SiO_2 interface. This is the same effect exhibited by surface channel CCDs. When the photogate was biased into accumulation (thereby "pinning" the potential at the interface) the dark current was significantly reduced. This suggested that "pinning implants" used in CMOS arrays with photogate or photodiode pixels could reduce dark current.

These experiments did not look at important performance metrics such as modulation transfer function, pixel linearity, and pixel-to-pixel gain non-uniformity. However there were circuit-level work-arounds to the off-state leakage problem. The small well capacity is simply related to the limited voltage swing and minimum source follower gate area size, and points to a fundamental problem as supply voltages decrease and/or pixel size decreases.

In 2000, Lule et.al.[8], discussed CMOS design rule scaling effects. The trend in Figure 4-3 suggests pixel size of about 4 μm by the year 2005. However, smaller-pixel arrays[9-12] were available in 2005 with emphasis on digital still camera and cell phone camera applications. Current cell phones camera design philosophy suggests even smaller pixels requirements.

The observation by Wong[6] that CMOS-tailored manufacturing processes are essential to the continued development of array technology has proven absolutely true. CMOS arrays are routinely manufactured using CMOS feature sizes as small as 0.13 μm. Tailored CMOS processes[9] with 0.18, 0.15, and 0.13 μm feature sizes require supply (maximum) voltages of 3.3, 2.8, and 1.8 V, respectively. However the 3.3-volt supply is being maintained for pixel electronics even though feature sizes continue to shrink.

4.3. PIXEL ELECTRONICS

Every CMOS pixel has identical "pixel electronics" that determine the readout noise, linearity, fill factor, image lag, and well capacity. The electronics are described by a shorthand notation[13,14] that simply enumerates the number of transistors (Table 4-1) found inside a pixel. The confusion caused by such transistor counting is apparent. Not all components are true transistors. Alternatively, pixel electronics can be dictated by photodetector design (photodiode, pinned photodiode, or photogate). In reality, the wide variety of pixel designs that are available reflect combinations of these features and as such, the terminology can be confusing. It is prudent to examine a detailed circuit diagram of a CMOS pixel to understand its performance and operating characteristics.

Table 4-1
3T, 4T, 5T, and 6T "TRANSISTORS"
The transfer gate, global reset, and photogates are not true transistors

Pixel design	Detector types	"Transistors"
3T	Photodiode (PD) Deep-implant photodiode Pinned photodiode (PPD)	Source-follower transistor Reset transistor Row select transistor
4T	Pinned photodiode (PPD)	Source-follower transistor Reset transistor Row select transistor Transfer gate
5T	Pinned photodiode (PPD)	Source-follower transistor Reset transistor Row select transistor Transfer gate Global reset
6T	Photogate (PG)	Source-follower transistor Reset transistor Row select transistor Transfer gate Global reset Photogate

4.3.1. 3T CMOS PIXEL FAMILY

The 3T CMOS pixel[13,14] represents the simplest circuit. The three transistors are the field-effect transistor arranged in a source-follower configuration, a reset transistor, and a row select transistor. The gate of the reset transistor is pulsed on and then off, thereby removing signal information from the pixel at the end of an integration period, t_{INT}. The "signal" is the sum of the photocurrent and dark current (generated mainly by the n+ photodiode contact). Starting immediately after the reset gate is turned off (the reset transistor behaves like an open switch), this current, i_S, is accumulated on the photodiode intrinsic capacitance, the source follower gate capacitance, and the sense node diffusion capacitance (labeled sense node). The parallel combination of these capacitances is called the total "integration node" capacitance, C_{NODE}. The voltage is

$$V_{SIGNAL} = \frac{i_S t_{INT} q}{C_{NODE}} \tag{4-1}$$

At the end of an integration period this voltage is "latched" to a column bus using a row select transistor. These select switches connect the pixels to the column buses (Figure 4-2). That these are called "row select switches" indicates that all of the pixels along a given horizontal row are placed onto their corresponding column buses at the same time by the row select command. When this transfer of data onto the column buses is completed, the pixel is then reset by pulsing the reset gate. This entire cycle is then repeated.

Figure 4-5 illustrates the three-transistor structure for three different photodiodes[15]. Figure 4-5a contains a standard photodiode (PD) formed by a pn diode with the p-side located in a thin epitaxial layer chosen to be sufficiently thick to allows adequate photo-absorption and hence acceptable quantum efficiency. This pixel is popular due to its simplicity, small transistor count, and general compatibility with standard CMOS processing. It is used in high-volume applications that mandate low cost (e.g., cell phone cameras).

This pixel suffers from high dark current due to the n+ contact. Standard CMOS doping of the photodiode n-implant can result in a thin depletion region (typically less than 1 μm). This causes significant carrier diffusion for long-wavelength photons. Depending on the pixel node capacitance, kTC noise can significant. In contrast, scientific CCDs kTC noise is removed by CDS, thereby allowing extremely low noise operation (read noise is a few electrons rms). Removal of kTC noise in a 3T CMOS pixel requires digital CDS external to the imaging chip.

Figure 4-5b illustrates a PD structure with a deep n-implant. The deep implant PD improves long-wavelength MTF performance by increasing the pixel depletion depth and thereby decreasing carrier diffusion effects. Figure 4-5c illustrates a pinned photodiode (PPD). The special p+ pinning implant is very near the photodiode surface, which greatly reduces dark current generation from this interface. However, since the n+ contact is the dominant source of dark current for any 3T pixel, the true low dark current advantage of the PPD requires implementation in a 4T pixel design wherein the n+ contact is eliminated (to be discussed in Section 4.3.3).

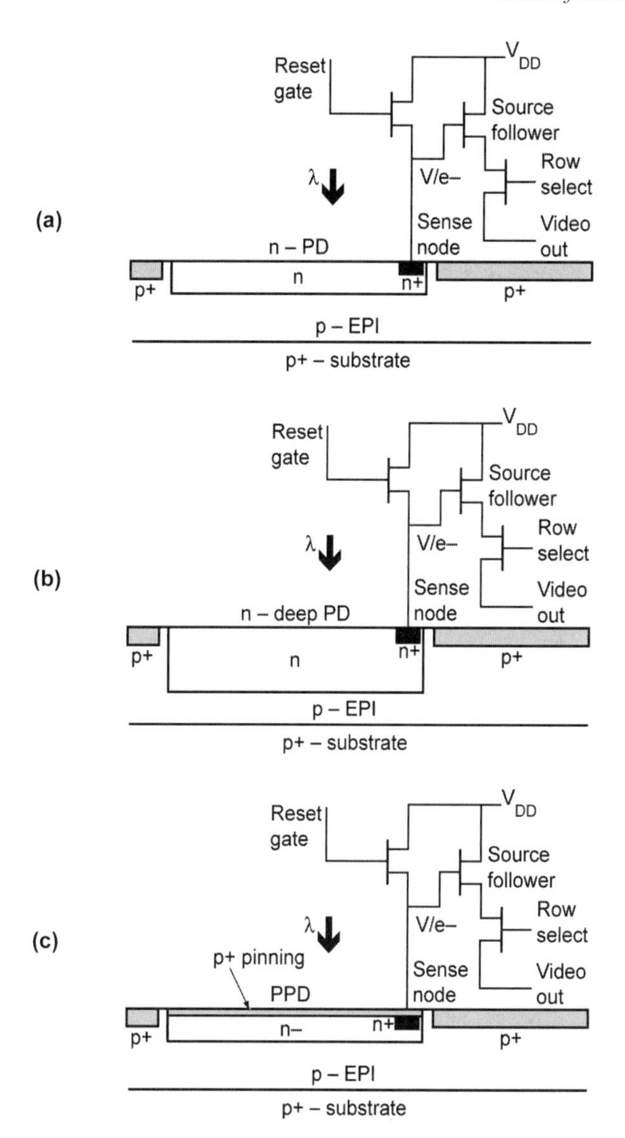

Figure 4-5. Three transistor (3T) pixels. (a) Photodiode, (b) deep-implant PD, and (c) pinned photodiode. Each design has advantages and limitations.

Proprietary pixel designs combine the p+ pinning implant and the deep depletion for optimal low noise and minimal diffusion. The PPD pixel allows for reduced sense node capacitance and hence high output gain conversion ($\mu V/e^-$). However, the depletion depth is a function of the signal charge stored on this node (the depletion region collapses as stored signal charge is increased); this can produce a substantial nonlinearity in the pixel response. The deep implant PD pixel exhibits higher read noise (no p+ pinning implant), can be designed to handle higher well capacity, and has less nonlinearity compared to a PPD pixel. The standard PD pixel has read noise, well capacity, and nonlinearity performance that are intermediate between deep n-well and PPD pixels.

A common problem with the 3T pixel family is image lag. A bright image in one frame produces a residual or ghost image in the next frame due to incomplete reset operation[15]. Lag is minimized by a "hard reset." Here, the reset gate voltage is greater than V_{DD} + V_{TH}, where V_{TH} is the transistor threshold voltage (typically a few tenths of a volt) and V_{DD} source follower transistor bias level (typically 1 to 2 volts).

The 3T pixel family has a substantial deficiency in terms of pixel readout. There is no isolation between the photodiode, sense node, and the gate of the source follower. The pixel signal formed during an integration period must to be latched to a column bus (by pulsing the row select transistor switch), followed by immediate closure of the reset switch to allow use of the pixel in the subsequent integration period. This problem is overcome with 4T and higher configurations.

4.3.2. 3T ROLLING SHUTTER

The integration period of a 3T array is defined as the time between successive pixel resets. If all of the pixels are reset simultaneously, then all of the pixels will be ready for latching to the row readout buses at the same time. However, a given output bus can only hold the signal voltage of one pixel at a time. The solution to this problem is to stagger the start of the integration time of the successive pixels while maintaining an equal integration period for each pixel. This stagger in pixel reset times allows for the time-sequential latching of pixel outputs onto row readout buses. The stagger time is the integration time divided by the number of pixels that are connected to a given column bus. The column buses feed sample-and-hold circuits that then connect to serial readout busses for the final multiplexing and readout. This readout mode does not have a common nomenclature and is called rolling shutter, ripple, or staggered readout.

Rolling shutter readout produces image distortion and causes image smear when there is sufficient scene motion during an integration time. Geyer et. al. provide[16] a detailed treatment of rolling shutter geometric image distortion effects with motion smeared imagery provided in Reference 17.

4.3.3. 4T, 5T, AND 6T CMOS PIXELS

In traditional pixel architectures, adding more transistors to a CMOS pixel results in greater array functionality and in most cases improved performance, but at the expense of lower fill factor and more pixel complexity. The 4T, 5T and 6T configurations are provided Figure 4-6 with the "transistors" listed in Table 4-1.

The 4T PPD CMOS pixel is identical to the 3T PPD (Figure 4-5) but has a transfer gate between the photodiode and the sense node. During an integration period the transfer gate is clocked low so that the signal charge accumulates on the PPD. The p+ surface implant ensures that the PPD region is completely depleted. At end of integration, the transfer gate is clocked high and the signal charge is transferred to the sense node. The coupled depletion regions (under the PPD and transfer gate) ensure complete charge transfer and no image lag. However, image lag can still result from incomplete reset of the sense node. The p+ surface implant reduces dark current. Since the PPD is decoupled from the sense node by the transfer gate, the 4T PPD pixel has superior linearity.

The disadvantage of the 4T PPD pixel is that it has a lower well capacity owing to required voltages applied to the PPD, transfer gate, and sense node. The p+ PPD and buried-channel under the transfer gate are formed by bombardment of this silicon region by heavy ions. This process is common in CMOS manufacturing and is called ion implantation. The result is called an "implant." The depth of the implant is controlled by the implant dose (essentially the ion velocity). However a special non-standard implant (different ion species, different implant energy) is needed to form the p+ PPD region. Also, the implants' precise doses and alignment are very critical. This had made the 4T design less desirable from a manufacturing viewpoint. However, CMOS foundries are making these custom implants more available and this limitation is rapidly disappearing.

Figure 4-6b illustrates a 5T PPD pixel. It is simply a 4T PPD pixel with the addition of a global reset gate that connects the PPD region to an n+ region (left side) and then a p+ region. When the global reset gate is pulsed, the signal charges stored under the PPD region is transferred to the n+ node and then removed to the p+ implant. This pixel has all of the benefits and detriments described for the 4T PPD pixel. This additional "transistor" allows for additional useful operating modes and flexibilities (discussed in the next section). Unfortunately, the added "transistor" further reduces the well capacity. For small pixel CMOS designs, this global gate is usually implemented with an implant usually for anti-blooming control. The area associated with an extra gate is hard to fit into really small pixel topologies.

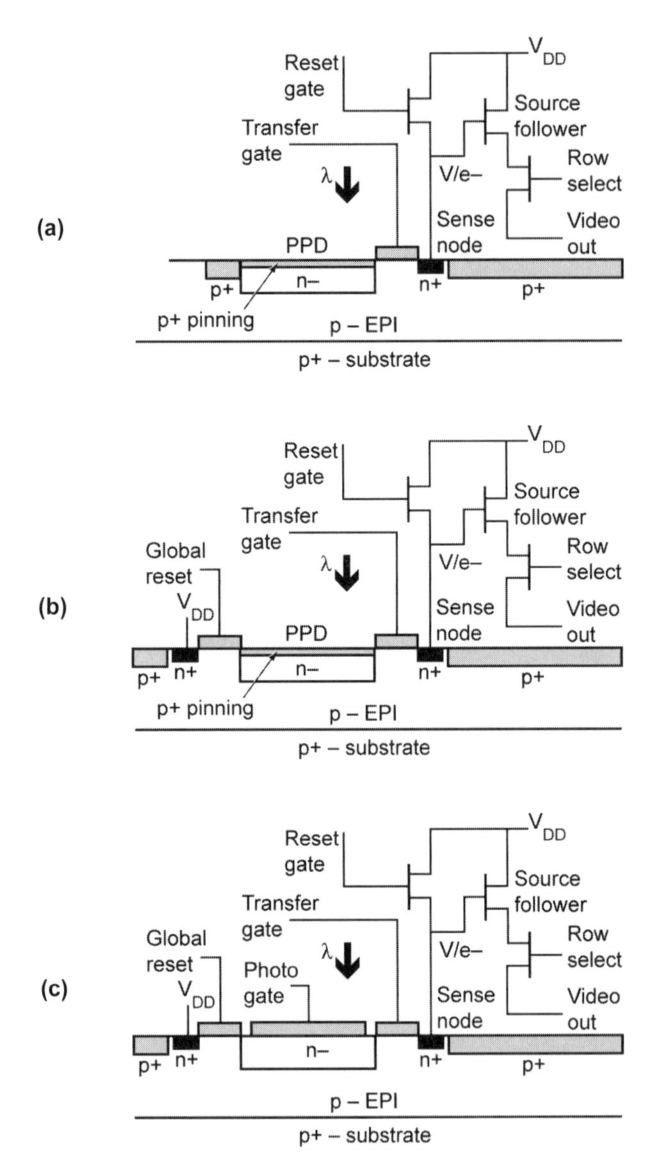

Figure 4-6. CMOS pixel electronics. (a) 4T pinned photodiode, (b) 5T PPD, and (c) 6T photogate.

Replacing the PPD with a photogate creates the 6T PG pixel (Figure 4-6c). The photogate produces a depletion region directly underneath it, just as in a CCD photogate. The transfer gate must to be clocked to produce a deeper well than under the PG and the well under the n+ diffusion must be deeper to ensure complete signal charge transfer from the photogate to the sense node and thereby avoid image lag. In addition, the relative opacity of the PG gate structure reduces the quantum efficiency of the 6T PG pixel. In the past, the PG frontside illuminated pixel was viewed with favor since it avoided the custom implant associated with the 5T PPD pixel. However, with the expanding availability of such custom implants in CMOS imager processing this concern is diminishing. The PG pixel approach is now a better fit with thinned, backside-illuminated approaches where light does not have to propagate through this gate.

4.3.4. SNAPSHOT OR SIMULTANEOUS AND GLOBAL SHUTTER READOUT MODE

The ability to reset all of the pixels in 4T, 5T and 6T CMOS at the same time allows the start and stop of the integration period to be exactly the same for every pixel. This operating mode is called the simultaneous integration or snapshot imaging mode. This synchronization avoids the geometrical distortion effects associated with the 3T rolling shutter readout. The global shutter feature shown for the 5T and 6T pixels also allows for global exposure control. In the operation of the 5T PPD pixel, signal charge integration of a given frame commences immediately after the charge from the previous frame is transferred to the integration node with the turning off of the transfer gate. When the global shutter gate is on, any signal charge generated in the photodetector region is removed and no signal charge accumulates. The start of signal charge integration occurs when the global shutter gate is turned off. The integration period therefore corresponds to the difference in time from the global shutter gate turn-off and transfer gate turn-on. This time difference can be programmed by gating the global shutter gate, thereby allowing, at least in principle, exposure control on a frame-by-frame basis. With proper feedback this rapid exposure control can allow for dynamic range management and automatic gain control (AGC) directly within a CMOS array.

This type of flexibility makes the 5T pixel appealing. For applications where larger pixels are allowed, the extra transistor count inside a CMOS pixel can turn into very useful enhanced functionality. A downside to the snapshot operation is that the signal is stored on the sensor node (n+ contact) with its high dark current. The dark current associated with this n+ contact can be 10 to 20 times higher than that is generated by the photodiode.

Another important feature of the 4T, 5T and 6T pixels is that is allows for time correlated pixel readout that enables noise reduction through correlated double sampling (discussed in Section 6.5.2. *Reset noise*). For CDS to be possible, the output of a given pixel just before and just after the reset operation, must be measured and preserved for subsequent correlated subtraction. This operation is called "true snapshot" operation. This is not possible with the rolling shutter approach required for 3T operation.

4.4. CMOS ARCHITECTURES

Arrays require complex electronics to create a viewable image. The bias and timing signals must be applied to the pixels on a row-by-row basis to cause the multiplexing of pixels signals onto column busses, then to serial multiplexing busses, and finally to output video ports. Figure 4-7 shows a block diagram of the row and column control signals that get applied at the pixel level.

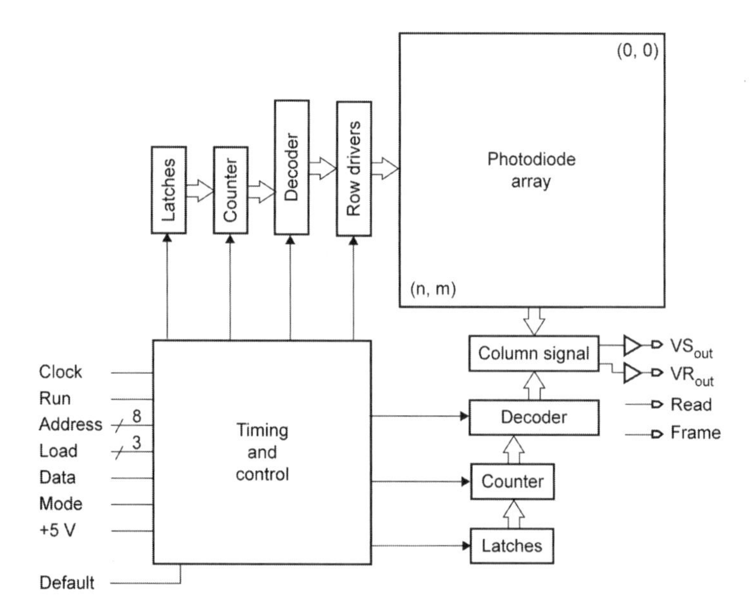

Figure 4-7. "Analog" CMOS imaging chip architecture with row and column decoding.

The compatibility of CMOS pixel processing and lower-power CMOS digital circuitry makes integration of analog-to-digital conversion directly onto the CMOS imaging chip very attractive. This level of integration opens the door to many applications that require low power, digital functionality, and the avoidance of low noise analog signals and interfaces. Figure 4-8 illustrates a digital CMOS imaging chip with "photons-in to bits-out" architecture.

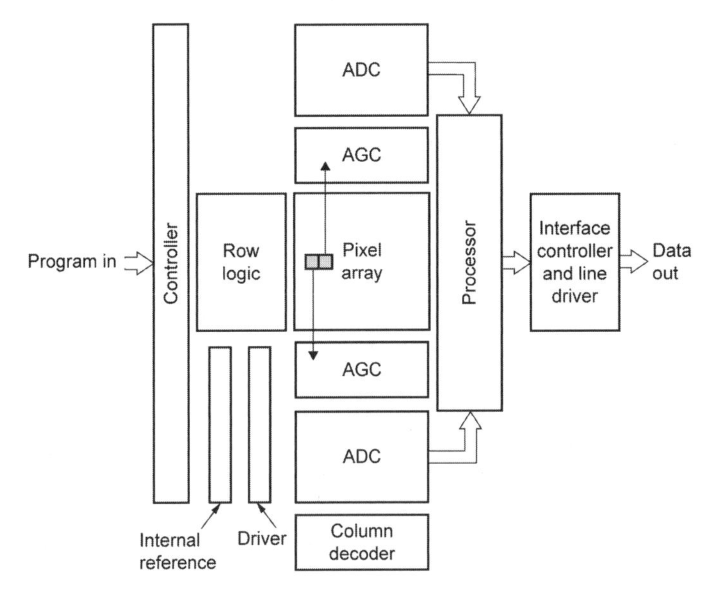

Figure 4-8 "Digital" CMOS Imaging chip with on-chip gain control, ADC and other digital features.

For special applications such as arrays operating at ultra-high frame rates, an on-chip ADC is almost mandated. The number of analog video leads needed to maintain a reasonable video rate can grow very large. The simplifications to the external camera electronics by having on-chip ADC and digital control are notable. In Figure 4-8 the entire chip runs off a single external power line. It also has on-chip programmable biases, frame rate, exposure control, electronic pan and zoom, and power management[18]. There are typically small on-chip control registers that can be programmed, often on a frame-by-frame basis, through a two-wire serial digital interface. The digitals outputs usually employ low-voltage digital signal (LVDS) technology for reduced power consumption and interface compatibility.

The typical CMOS array architecture with on-chip ADC involves an ADC per column approach (massively parallel). A low-power CMOS array with column-parallel successive approximation approach ADC was reported by Zhou et. al.[19] in 1997. In 2003 Krymski et. al. reported on the development of a 5000 frames/sec, 512×512 pixel CMOS array with a 16 μm pixel pitch[20] and a 240 frames/sec, 2352×1728 pixel array, with a 7 μm pixel pitch[21]. The latter used a total of 2352, 10-bit, column parallel on-chip ADCs with a total on-chip data rate of 9.75 Gbits/sec.

Equally interesting is a CMOS imaging sensor developed explicitly for high-definition television (HDTV) camera applications with a 2/3 diagonal format and with 2.1 Mpixels. The key benefit of this image sensor is its exceptional functionality and ability to reduce HDTV camera complexity, as well as camera mass, power, and volume[22].

4.5. CMOS FUTURE

With a fixed physical size, as the array format grows (more pixels per array), smaller pixels are mandated. There is also a strong economic incentive to make CMOS arrays as small as possible. As arrays are made smaller, more can be produced on a single wafer (8-inch diameter is typical). Generally, there are a fixed number of defects distributed randomly over a silicon wafer. As the number of arrays per wafer increases, the probability of having a defect in a particular array decreases and yield increases. The combination of these two factors results in manufacturing costs that are an extremely steep function of array size. Simply stated, smaller array sizes cost less.

There appears to be an impetus for continuing to reduce pixel dimensions (i.e., 2 μm and smaller), particularly in commercial imaging applications. CMOS processing enhancements, and the use of high-resistivity silicon and deep n-well isolation, offer the opportunity to maintain some of the key CMOS array performance characteristics at the smaller pixel dimensions. The implications of this from an image quality perspective are discussed in Chapter 11. *Image Quality*. However, there are some things that will clearly suffer in the drive to the use of smaller CMOS pixels. The fill factor will be difficult to maintain as pixel dimensions continue to shrink due to perimeter effects. Storage capacity and hence pixel dynamic range will be compromised due to less available pixel area. However, the need to retain supply voltages at 3.3 volts to ensure adequate signal swing is widely recognized in the CMOS array industry.

Fossum[12] discusses the ability to construct CMOS pixels with dimensions smaller than 1 μm, using 65 nm feature size, and points out the severe dynamic range and hence signal-to-noise ratio limitations that will result with the use of such pixels. He suggests that in spite of the seeming impracticality of such ultra-small pixels, if it is possible to build them, they will be built. Smaller pixels place a burden on the optical designer. Figures 11-3 through 11-5 illustrate noiseless imagery for small pixels with practical focal ratios. As the pixel size decreases, the optical blur diameter makes the image somewhat blurry. The acceptability of this imagery is scene dependent.

Although much of the discussion has focused on smaller pixels, there are many imaging applications wherein the use of larger pixels represents an acceptable tradeoff. For these cases adding more transistors to the CMOS pixel unit cell have a lesser relative impact on things like fill-factor reduction and loss of well capacity. The combinations of 4T or higher designs, coupled with the varieties of possible photodetector schemes, results in a plethora of possible advanced CMOS pixel designs. These large numbers of possible designs align with a corresponding number of performance requirements, "desirements," unique applications, and practical constraints.

The principal advantages of CMOS arrays have been stated many times: low-voltage operation, low power consumption, ease of integration of on-chip control, the possibility of random access to the image data, and potentially lower camera system costs. Integrating control and digitizing functions directly onto the imaging chip reduces (off-chip) camera electronic complexity. It appears that CMOS array fabrication using 0.13 μm features design rules can be fabricated with relatively minor modifications without too many performance regrets. To continue to exploit even finer-feature CMOS manufacturing technologies (e.g., 0.1 and 0.064 μm feature sizes), custom mixed-mode (i.e. analog and digital) "boutique" CMOS fabrication with flexible design and array-friendly features, will need to be readily available.

As discussed in Section 6.10. *Lux transfer*, the SNR drops significantly when the pixel size is less than 1 μm. The SNR will ultimately limit practical pixel sizes. As of this writing, CMOS imagers for cell phone cameras have mostly migrated to the 2X Super VGA format (1600×1200 pixels) with 2.2 μm square pixels. The diagonal dimension of these arrays allows them to fit into a 1/4-inch optical format with some margin.

Janesick summarizes advanced CMOS pixel concepts applied to scientific CCD technology[14,15]. His description focuses on larger pixel dimensions (i.e., around 8 μm) as well as backside-illuminated and hybrid array constructs. Pain has also reported on backside illuminated CMOS array development[23]. For these situations, substantial pixel area is available to implement pixel electronic, optical, and semiconductor features that are key to substantially improving pixel performance in the direction of the performance advantages enjoyed by scientific CCDs. Figure 4-9 illustrates the pixel architecture of a thinned, backside-illuminated CMOS array. The pixel photodetector and electronics are "vertically stacked." The photodiode is formed in a deep n-well and above it in n- and p-wells are the three transistors for a 3T implementation. The application of a negative substrate bias makes this thinned device fully depleted, a crucial factor in ensuring good MTF performance.

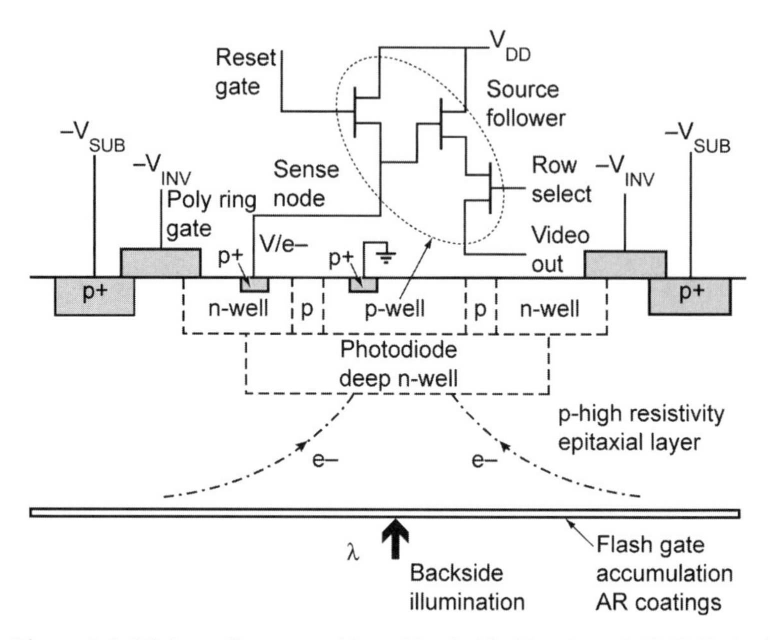

Figure 4-9. High-performance thinned backside illuminated CMOS pixel.

As time progresses, CMOS imagers will evolve to match the capability of scientific CCD imagers. Janesick claims that this is essentially true today. Backside thinning of CMOS arrays is an active research area. This will result in high (>90%) quantum efficiencies over the silicon spectral bandpass without the need for microlenses. The use of silicon-on-insulator (SOI) techniques will allow the backside thinning to be much less expensive, because the etching will stop at the oxide interface – thereby avoiding critical precision thinning steps. In order to avoid the effects of carrier diffusion MTF degradation, nearly intrinsic high-resistivity silicon must be used. If the quality of the silicon is good enough, silicon thicknesses of 50 microns can be fully depleted with voltages of 3.3 volts.

For ultra-small pixels, "random telegraph signal" (RTS) noise is comparable to the readout noise. This exotic and annoying noise mechanism is likely due to interface states at the silicon-oxide interfaces. Processing methods to reduce interface state density, and hence RTS, are being researched. Digital correlated-double sampling can be used in any digital still camera application and in many scientific imaging applications to allow extremely low pixel amplifier integrated noise levels. Work is underway to optimize pixel source follower geometries to reduce the white noise component of the pixel noise. Dark current levels of pinned photodiodes are limited by the properties of bulk silicon (a few picoamps/cm^2 at room temperature). The much higher dark currents in 3T and snapshot mode CMOS pixels are still an issue. The transition to higher frame rates is natural to CMOS design. High frames rates, low noise operation will be enabled by parallel output port architectures. Extremely low image lag is possible by the use of low-energy, self-aligned diode implants to the transfer gate regions. Finally, in the area of radiation effects (important for space applications), the so-called flat-band characteristics (i.e. threshold shifts due to oxide charging), are several orders of magnitude better in CMOS technology compared to CCD technology. In the area of radiation-induced dark current increase, CMOS and CCD technology are comparable.

Shared-pixel architectures are being pursued where high sensitivity and wide dynamic range inside of small pixels are desired. The concepts involve the sharing of transistor and readout circuitry between several pixels, to increase the effective pixel fill-factor. Sharing[24] can be carried out with various pixel patterns: 4×1, 2×1 and 2×2. Clearly this approach requires that these several photodiodes time-share pixel electronics; this will have implications in high frame rate applications. At extremely small pixel sizes, even in commercial applications, this approach is essentially mandated.

CMOS arrays seem to be developing along three distinct paths, with the help of semi-custom CMOS fabrication with processes tailored for imaging applications:

Custom specialty, non-color applications

CMOS arrays are being developed for ultra-high frame rate imaging, target tracking/dynamic windowing, and adaptive framing with variable resolution. The pixel dimensions do not necessarily have to be ultra-small. CCD technology is not compatible with the application requirements.

Digital still camera, HDTV, camcorder, and cell-phone camera applications

Very small pixels are desired along with rather large formats, hence the pixel amplifier designs need to be simple (e.g., 3T design). The pixel performance requirements are moderate and the limited dynamic range due to pixel well capacity limitations is handled by integration time gating as a method of exposure/gain control. The high degree of chip integration (i.e., camera on-a-chip with on-chip control and analog-to-digital conversion) and the low power consumption facilitates compact, lightweight, easy-to-use camera systems with long battery life.

Scientific and DoD applications

CMOS arrays are competing with scientific CCDs primarily in the high-end space and ground-based astronomy applications, where the space radiation environment must be contended with. These devices come with somewhat larger pixel dimensions. Important performance parameters include quantum efficiency, low noise, adaptive frame rates, low nonlinearity, high well capacity, and in some cases, high frame rate.

The scientific applications category implies a scenario where the custom CCD fabrication facilities, used for developing scientific backside thinned CCDs, go out of business due to economic anemia and lack of commercial/government support. The scientific CMOS arrays described by Janesick[14,15], or the hybrid CMOS array[25], becomes the backup approach to scientific CCD technology by default.

4.6. REFERENCES

1. Y. Lo, "Solid-State Image Sensor: Technologies and Applications," in *Input/Output and Imaging Technologies*, Y. T. Tsai, T.-M. Kung, and J. Larsen, eds., SPIE Proceedings Vol. 3422, pp. 70-80, (1998).

2. *IEEE Transactions on Electron Devices*, Volume 50, No.1, January, 2003.

3. *Proceedings of the 2005 IEEE Workshop on Charge-Coupled Devices and Advanced Image Sensors*, IEEE and IEEE Electron Devices Society, Nagano, Japan (2005).

4. E. R. Fossum, "Active Pixel Sensors: Are CCDs Dinosaurs?" in *Charge-Coupled Devices and Solid State Optical Sensors* III, M. M. Blouke, ed., SPIE Proceedings Vol. 1900, pp. 1-12 (1993).

5. See http://en.wikipedia.org/wiki/Moores_law. Wikipedia is a "living" document. The material may change from the date of this text (March 2007).

6. H. S. Wong, "Technology and Device Scaling Considerations for CMOS Imagers," *IEEE Transactions on Electron Devices*, Vol. 43, pp. 2131-2142, (1996).

7. H. S. Wong, R.T. Chang, E. Crabbe, and P.D. Agnello, "CMOS Active Pixel Image Sensors Fabricated Using a 1.8-V, 0.25-μm CMOS Technology," *IEEE Transactions on Electron Devices*, Vol. 45, pp. 889-894 (1998).

8. T. Lule, S. Benthein, H. Keller, F. Mutze, P. Rieve, K. Seibel, M. Sommer, and M. Bohm, "Sensitivity of CMOS Based Imagers and Scaling Perspectives," *IEEE Transactions on Electron Devices*, Vol. 47, pp. 2110-2122 (2000).

9. G. Agranov, T. Gilton, R. Mauritzson, U. Boettiger, P. Altice, J. Shah, J. Ladd, X. Fan, F. Brady, J. McKee, C. Hong, X. Li, and I. Patrick, "Optical-Electrical Characteristics of Small, Sub-4 μm and Sub-3 μm Pixels for Modern CMOS Imagers," in *Proceedings of the 2005 IEEE Workshop on Charge-Coupled Devices and Advanced Image Sensors,* IEEE and IEEE Electron Devices Society, Nagano, Japan, pp. 206-209, (2005).

10. H. Noh, K. Lee, D. Lee and K. Kim, "The Design and Characterization of CMOS Imager Sensor with Active Pixel Array of 2.0 3 m Pitch and Beyond," in *Proceedings of the 2005 IEEE Workshop on Charge-Coupled Devices and Advanced Image Sensors,* IEEE and IEEE Electron Devices Society, Nagano, Japan, pp. 197-200 (2005).

11. M. Murakami, M Masuyama, S. Tanaka, M. Uchida, K. Fujiwara, M. Kojima, Y. Matsunaga, and S. Mayumi, "2.8 μm-Pixel Imager Sensor Maicovicon," in *Proceedings of the 2005 IEEE Workshop*

on Charge-Coupled Devices and Advanced Image Sensors, IEEE and IEEE Electron Devices Society, Nagano, Japan, pp. 13-14 (2005).
12. E. Fossum, "What to do with Sub-Diffraction-Limit (SDL) Pixels? – A Proposal for a Gigapixel Digital Film Sensor (DFS)," in *Proceedings of the 2005 IEEE Workshop on Charge-Coupled Devices and Advanced Image Sensors*, IEEE and IEEE Electron Devices Society, Nagano, Japan, pp. 214-217 (2005).
13. J. Janesick, "Lux Transfer: Complementary Metal Oxide Semiconductors versus Charge-coupled Devices," *Optical Engineering*, Vol. 41, pp. 1203-1215 (2002).
14. J. Janesick, "Charge Coupled CMOS and Hybrid Detector Arrays," in *Focal Plane Arrays for Space Telescopes,* T. J. Grycewicz and C. R. McCreight, eds., SPIE Proceedings Vol. 5167, pp. 1-18 (2004).
15. J. Janesick, Sarnoff Corporation, private communication.
16. C. Geyer, M. Meingast, and S. Sastry, "Geometric Models of Rolling- Shutter Cameras," EECS Department, University of California, Berkeley (2006).
17. J. Janesick and G. Putnam, "Developments and Applications of High Performance CCD and CMOS Imaging Arrays," *Annual Review of Nuclear and Particle Science*, Vol. 53, pp. 263-300 (2003).
18. B. Pain, "Charge-Coupled Devices/CMOS Imaging Sensors and Applications", Professional short course, University of California, Los Angeles Extension, (February, 2006).
19. Z. Zhou, B. Pain and E.R. Fossum, "CMOS Active Pixel Sensor with On-Chip Successive Approximation Analog-to-Digital Converter," *IEEE Transaction on Electron Devices*, Vol. 44, pp. 1759-1763 (1997).
20. A. I. Krymski and N. Tu, "A 9-V/Lux-s 5000 Frames/s 512×512 CMOS Sensor," *IEEE Transaction on Electron Devices*, Vol. 50, pp. 136-143 (2003).
21. A. I. Krymski, N.E. Bock, N. Tu, D. Van Blerkom, and E. R. Fossum, "A High Speed, 240-Frames/s, 4.1 Mpixel CMOS Imager," *IEEE Transactions on Electron Devices*, Vol. 50, pp. 130-135 (2003).
22. L. Koslowski, G. Rossi, L. Blanquart, R. Marchesini, Y. Huang, and G. Chow, "Development Status of CMOS 1920×1080 Imaging System on-a-chip for 60p and 72p HDTV," *SMPTE Motion Imaging Journal*, Oct/Nov, 2004.
23. B. Pain, T. Cunningham, S. Nikzad, M. Hoenk, T. Jones, B. Hancock, and C. Wrigley, "A Back-Illuminated Megapixel CMOS Image Sensor," in *Proceedings of the 2005 IEEE Workshop on Charge-Coupled Devices and Advanced Image Sensors*, IEEE and IEEE Electron Devices Society, Nagano, Japan, pp. 35-38 (2005).
24. R. D. McGrath, H. Fujita, R. M. Guidash, T.J. Kenney, and W. Xu, "Shared Pixels for CMOS Image Sensors," in *Proceedings of the 2005 IEEE Workshop on Charge-Coupled Devices and Advanced Image Sensors*, IEEE and IEEE Electron Devices Society, Nagano, Japan, pp. 9-12 (2005).
25. Y. Bai, J. T. Montroy, J. D. Blackwell, M. Farris, L. J. Koslowski, and K. Vural, "Development of Hybrid CMOS Visible Focal Plane Arrays at Rockwell", in *Infrared Detectors and Focal Plane Arrays* VI, E. L. Dereniak, and R. E. Sampson, eds., SPIE Proceedings Vol. 4028, pp. 17-182 (2000).

5
ARRAY PARAMETERS

All detector arrays are fabricated for specific applications. Microlenses are used to increase the optical fill factor. Color filter arrays are filters that are placed over the detectors to create specific spectral responses that lend themselves to color imagery. For general video applications, the number of detectors is matched to the video standard to maximize bandwidth and image sharpness.

The symbols used in this book are summarized in the *Symbol List*, which appears after the *Table of Contents*.

5.1. WELL CAPACITY

The number of electrons stored in a well depends upon the pixels area, applied voltage, and array architecture. For back-of-the-envelope calculations, it is reasonable to assume that the storage is 1500 electrons/μm^2. A 10μm×10μm pixel could store about 150,000 electrons.

5.2. NUMBER OF DETECTORS

The EIA 170 video standard supports 485 lines (discussed in Section 7.5. *Video formats)*. However, one line is split between the even and odd fields so that there are only 484 continuous lines. Thus, detector arrays designed for EIA 170 compatibility tend to have 484 detectors in the vertical direction. For convenience, this has been reduced to 480 detectors. Image processing algorithms are more efficient with square pixels. With a 4:3 aspect ratio, the desired number of detectors is 640×480. Less expensive arrays will have a submultiple (for easy interpolation) such as 320×240. Several rows or columns are devoted to dark current and for light leakage. Therefore, an array may be 650×492 but the light sensitive part may be 640×480. It is manufacturer-dependent whether to specify the array size by the number of active pixels or the total number of pixels (which includes the dark pixels). Tables 5-1 and 5-2 provide photosensitive array sizes (without dark pixels). An extensive list is provided in Reference 1.

Beat frequencies are very obvious in color images. To avoid beat frequencies, the color pixel rate should be a multiple of the chrominance subcarrier frequency (3.579545 MHz). This results in arrays that contain 384,

576, or 768 horizontal detectors. While CFAs are designed for NTSC operation, the same chip can be used for monochrome video without the CFA. Therefore, many monochrome arrays also contain 384, 576, or 768 horizontal detectors. These arrays do not have square pixels. But, from a manufacturing point-of-view, this reduces the inventory of arrays offered.

Table 5-1
HDTV FORMATS
Aspect ratio 16:9 (H/V)

System	Pixels (H×V)	Number of detectors
S1	1280×720	921,600
S2, S3, S4	1920×1080	2,073,600

Table 5-2
COMPUTER VIDEO and TELECONFERENCING FORMAT

Acronym	Pixels (H×V)	Aspect (H/V)	Number of detectors
QCIF: Quarter common intermediate format	176×144	11/9	25,344
QVGA: Quarter video graphics array	320×240	4/3	76,000
CIF: Common image format	352×288	11/9	101,376
VGA: Video graphics array	640×480	4/3	307,200
SVGA: Super video graphics array	800×600	4/3	480,000
XGA extended graphics array	1024×768	4/3	786,400

Scientific array size tends to a power of 2 (e.g., 512×512, 1024×1024) for easy image processing. There is a perception that "bigger is better" both in terms of array size and dynamic range. Arrays may reach 8192×8192 with a dynamic range of 16 bits. This array requires (8192)(8192)(16) or 1.07 Gbits of storage for each image. Image compression schemes may be required if storage space is limited. The user of these arrays must decide which images are significant and through data reduction algorithms, store only those that have value. Otherwise, he will be overwhelmed with mountains of data.

While large-format arrays offer the highest resolution, their use is hampered by readout rate limitations. For example, consumer camcorder systems operating at 30 frames/s have a data rate of about 10 Mpixels/s. An array with 5120×5120 elements operating at 30 frames/s has a data rate of about 768 Mpixels/s. Large arrays can reduce readout rates by having multiple parallel ports servicing subarrays (Figure 3-14). The tradeoff is frame rate (speed) versus number of parallel ports (complexity of CCD design) and interfacing with downstream electronics. Because each subarray is serviced by different on-chip and off-chip amplifiers, the displayed image of the subarrays may vary in contrast and level. This is due to differences in amplifier gains and level adjustments.

There is no specific requirement that the digital still camera (DSC) aspect ratio match the 35-mm film format of 1.5 (Table 5-3). Popular DSC software programs allow the printed image to be any size. The physical limitation is the paper size. Acceptable image quality limits the print size (discussed in Section 11.8. *Viewing distance*).

Table 5-3
DIGITAL STILL CAMERA FORMATS

Acronym	Pixels (HxV)	Aspect (H/V)	Number of detectors
DSC1 "1.3 Megapixel"	1280×1024	1.25	1,310,720
DSC2 "2.0 Megapixel"	1600×1200	1.33	1,920,000
DSC3 "3.2 Megapixel"	1944×1672	1.16	3,250,368
DSC4 "6.3 Megapixel"	3072×2048	1.50	6,291,456

5.3. DARK PIXELS

Many arrays have "extra" photosites at the end of the array. These shielded photosites are used to establish a reference dark current level (or dark signal). The number of dark elements varies with device and manufacturer and ranges from a few to 25. The average value of the dark current pixels is subtracted from the active pixels leaving only photogenerated signal. For example, if the output of the dark pixels is 5 mV, then 5 mV is subtracted from the signal level of each active pixel.

Figure 5-1a illustrates the output from a single charge well. Light leakage may occur at the edge of the shield (see Example 3-1) and partially fill a few wells. Because of this light leakage, a few pixels (called dummy or isolation pixels) are added to the array. Figure 5-1b illustrates the output of three dark reference detectors, three isolation detectors, and three active pixels. The average value of the three dark pixels is subtracted from every active pixel value.

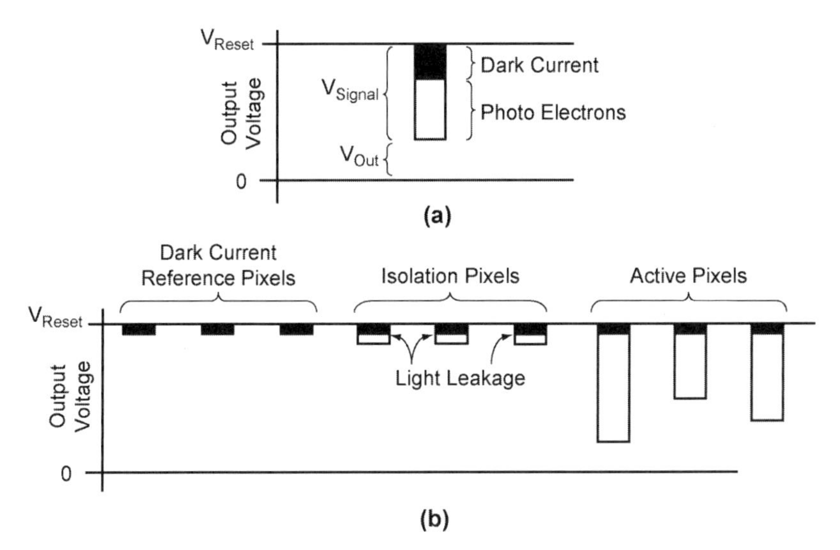

Figure 5-1. Typical output. (a) Single pixel and (b) active line with isolation and dark pixels. The varying voltage on the active pixels is proportional to the scene intensity.

While this process is satisfactory for general imagery and machine vision use, it may be unacceptable for scientific applications. Dark pixels have slightly different dark current than the active pixels. Furthermore, the dark current value varies from pixel to pixel. This variation appears as fixed pattern noise (discussed in Section 6.5.6. *Pattern noise*). Removing the average value does not remove the variability (dark current noise).

In critical scientific applications, only the dark value from a pixel may be removed from that pixel. That is, the entire array is covered (mechanical shutter in place) and the individual pixel values are stored in memory. Then the shutter is removed and the stored values are subtracted, pixel by pixel. This method ensures that the precise dark voltage is subtracted from each pixel. This, of course, increases computational complexity. Removal of dark current values allows maximum use of the analog-to-digital converter dynamic range. This method cannot remove the dark current shot noise. Dark current shot noise is reduced simply by reducing the dark current. Usually the array is cooled to minimize these thermally generated electrons.

The most common cooler is a thermoelectric cooler (TEC or TE cooler). Thermoelectric coolers are Peltier devices driven by an electric current that pumps heat from the CCD to a heat sink. The heat sink is cooled by passive air, forced air, or forced liquid (usually water). The final array temperature depends on the amount of heat generated in the array, the cooling capacity of the thermoelectric device, and the temperature of the TEC heat sink. Peltier devices can cool to about -60°C. Thermoelectric coolers can be integrated into the same package with the array (Figure 5-2). Cooling probably is only worthwhile for scientific applications. However, many devices contain TE coolers to stabilize the temperature so that the dark current does not vary with ambient temperature changes.

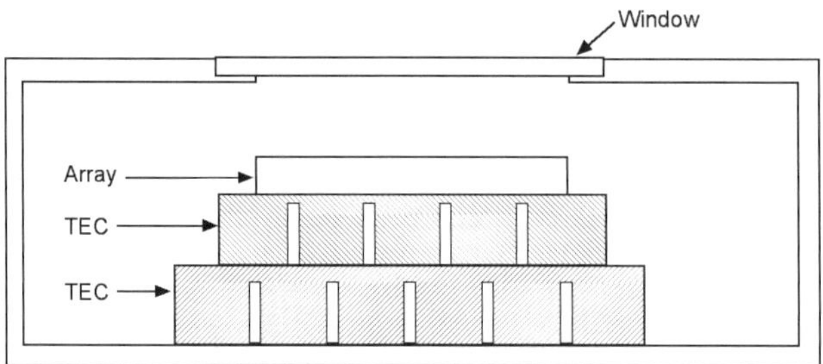

Figure 5-2. Cross section of a two stage TEC integrated into a package.

Although liquid nitrogen (LN, LN2, or LN_2) is at -200°C, the typical operating temperature is -120°C to -60°C because the charge transfer efficiency and quantum efficiency drop at very low temperatures. Condensation is also a problem, and the arrays must be placed in an evacuated chamber or a chamber filled with a dry atmosphere. Various cooling techniques and fabrication concepts are provided in Reference 2.

Additional isolation pixels may exist between the readout register and the sense node amplifier. These pixels reduce any possible interaction between the amplifier temperature and first column pixel dark current. They also reduce amplifier noise interactions.

5.4. VIDEO CHIP SIZE – OPTICAL FORMAT

Vidicon vacuum tubes were originally used for professional television applications. These were specified by the tube diameter. To minimize distortion and nonuniformities within the tube, the recommended[3] image size was considerably less than the overall tube diameter. When CCDs replaced the tubes, the CCD industry maintained the image sizes but continued to use the tube format nomenclature. The array diameter is related to the optical format by

$$\textit{Array diameter} \approx 0.63(\textit{optical format}) \qquad (5\text{-}1)$$

Everyone rounds the number of pixels, pixel size, and array size. This results in slightly different published numbers for "identical" chips. Table 5-4 provides the nominal relationship between optical format and array size. Lens systems are also specified by the same nomenclature to facilitate matching a lens to an array. As pixel sizes decreases, arrays may not fit into the standard values. The array manufacturers use the same formula, but normalize the format. For example, a 1/3.2-inch array is slightly smaller than a 1/3-inch array. Figure 5-3 illustrates active array size as a function of pixel size for different formats.

Table 5-4
ARRAY SIZE for STANDARD VIDEO FORMATS (4/3 aspect)

CAMERA FORMAT	STANDARDIZED ARRAY SIZE (H×V)	ARRAY DIAGONAL
1-inch	12.8 mm × 9.6 mm	16 mm
2/3-inch	8.8 mm × 6.6 mm	11 mm
1/2-inch	6.4 mm × 4.8 mm	8 mm
1/3-inch	4.8 mm × 3.6 mm	6 mm
1/4-inch	3.2 mm × 2.4 mm	4 mm
1/6-inch	2.4 mm × 1.8 mm	3 mm

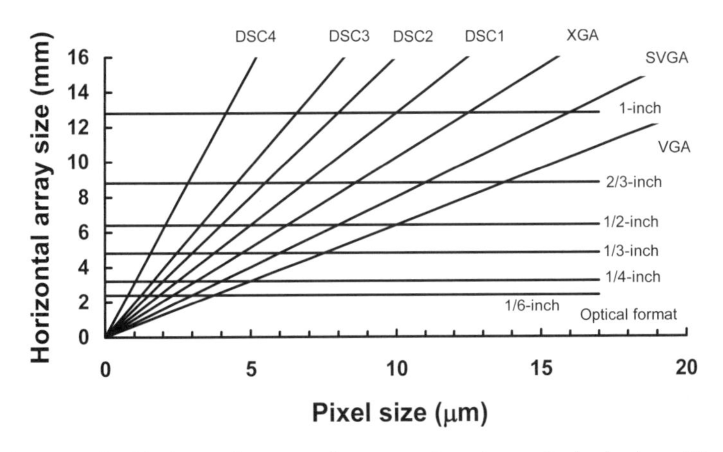

Figure 5-3. Horizontal array size as a function of pixel size. The optical format is for 4/3 aspect ratio (video formats). The optical format for the digital still cameras is slightly different due to the different aspect ratios (Table 5-3).

Using the small angle approximation, the camera horizontal field of view is

$$FOV \approx \frac{N_H d_H}{fl} \tag{5-2}$$

For fixed FOV, the focal length is directly proportional to the pixel size (Figure 5-4). As the pixel size decreases, the focal length must decrease. Sensitivity is proportional to A_D/F^2. The focal ratio must also decrease with decreasing pixel size to maintain the same sensitivity. Reducing the focal ratio places a burden on the optical designer and increases costs. The lowest practical focal ratio is zero (See Appendix for clarification).

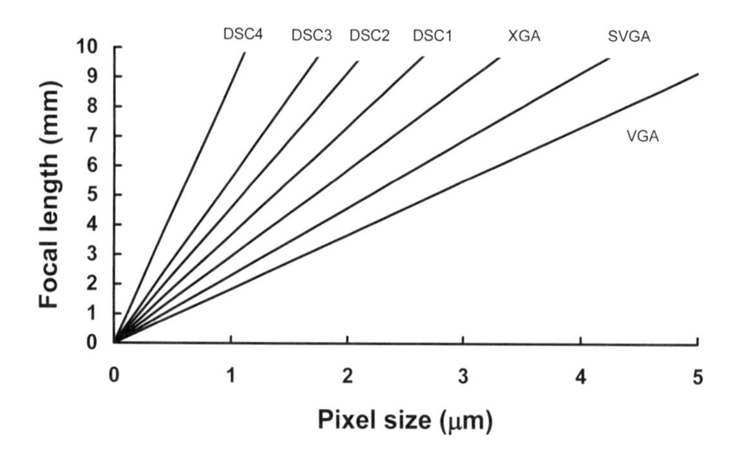

Figure 5-4. Focal length as a function of pixel size when HFOV = 20°.

5.5. MICROLENSES

Optical fill factor may be less than 100% due to manufacturing constraints in full transfer devices. For interline devices, the shielded vertical transfer register can reduce the fill factor to less than 20%. Microlens[4] assemblies (also called microlenticular arrays or lenslet arrays) increase the effective optical fill-factor (Figure 5-5), but may not reach 100% due to slight misalignment of the microlens assembly, imperfections in the microlens itself, nonsymmetrical shielded areas, and transmission losses. Most microlenses tend to form a circular image, whereas most detectors are square or rectangular. This reduces the effective optical fill factor. Fujifilm's Super CCD has octagonal detectors and this shape is better matched to the microlens image shape. As shown by the camera formula (Equation 2-18), the output is directly proportional to the detector area. Increasing the optical fill factor with a microlens assembly increases the effective detector size and, therefore, the output voltage.

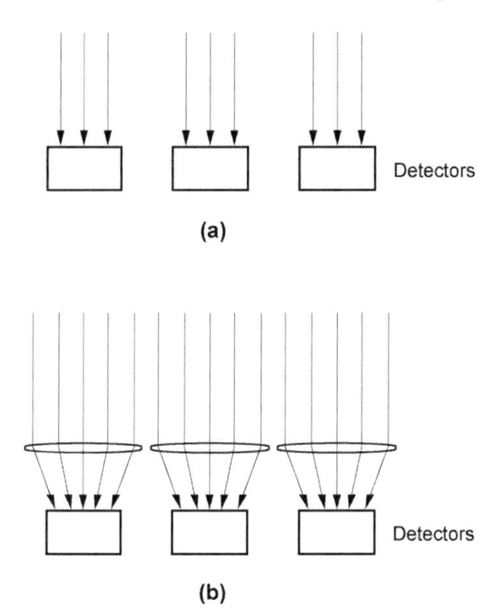

Figure 5-5. Optical effect of a microlens assembly. (a) With no microlens, a significant amount of photon flux is not detected. (b) The microlens assembly can image nearly all the flux onto the detector when a high focal ratio lens is used. These lenslets can either be grown on the array during the fabrication process or manufactured out of a material such as UV photoresist PMMA and placed on the array surface during packaging.

The photosensitive area is below the gate structure, and the ability to collect the light depends upon gate thickness. The cone of light reaching the microlens depends on the focal ratio of the camera lens. Figure 5-5 illustrates nearly parallel rays falling on the microlens. This case is encountered with high focal ratio lens systems. Low focal ratio primary camera lenses increase the cone angle[5], and the effective fill-factor decreases with decreasing focal ratio (Figure 5-6). Microlenses are optimized for most practical focal ratios. As the array size grows, off-axis detectors do not obtain the same benefit as on-axis detectors[6] thus compounding the cosine$^4\boldsymbol{\theta}$ roll-off effect.

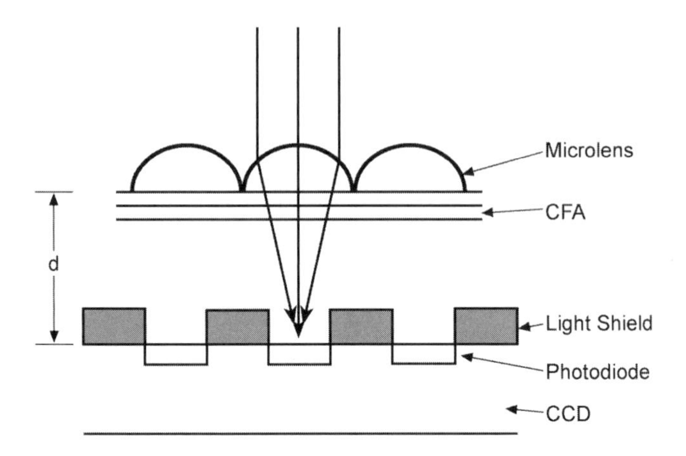

Figure 5-6. The polysilicon overcoats and light shielding structures act as a light tunnel that reduces the effectiveness of the microlens. As the number of layers increase, the distance ***d*** increases (typical of CMOS arrays) and exacerbates the tunneling problem. As the detector size decreases (less than 3 μm), microlenses become less effective. The birefringent crystal (required for a CFA) is not shown.

5.6. COLOR FILTER ARRAYS

The subjective sensation of color can be created from three primary colors. By adjusting the intensity of each primary (additive mixing), a full gamut (rainbow) of colors is experienced. This approach is used on all color displays. They have red, green, and blue phosphors that, when appropriately excited, produce a wide spectrum of perceived colors. The literature[7] is rich with visual data that form the basis for color camera design.

The "color" signals sent to the display must be generated by three detectors, each sensitive to a primary or its complement. The primary additive colors are red, green, and blue (R, G, B) and their complementary colors are yellow, cyan and magenta (Ye, Cy, Mg). For high-quality color imagery, three separate detectors (discussed in Section 7.4.3. *Color Correction*) are used whereas for consumer applications, a single array is used. The detectors are covered with different filters that, with the detector response, can approximate the primaries or their complements (Figure 5-7). A single array with filters is called a color filter array (CFA). The CFA finite transmittance is part of $\boldsymbol{T_{OPTICS}(\lambda)}$ in the camera formula. The vertically stacked color pixel is discussed in Section 6.1.1.

CCDs, CMOS, and EMCCDs and the 3CCD camera in discussed in Section 7.4.3. *Color correction*.

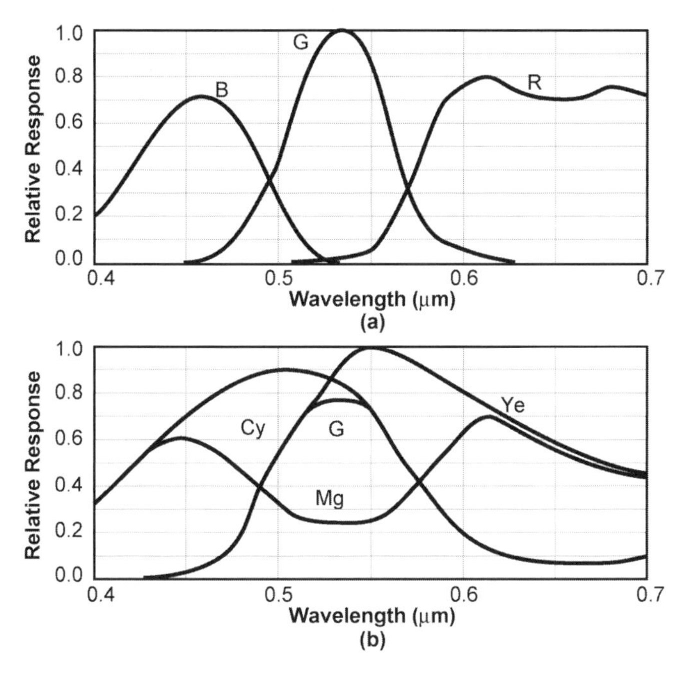

Figure 5-7. Desired spectral response for (a) the three primaries and (b) their complements.

While CFAs can have filters with the appropriate spectral transmittance to create the primaries, it is more efficient to create the complementary colors. Complementary filters have higher transmittances than the primary filters. The luminance channel is identical to the green channel and is often labeled Y. "White" (no color filter) is represented by W = R + G + B.

A linear set of equations relates the primaries to their complements. These equations are employed (called matrixing) in cameras to provide either or both output formats. Matrixing can convert RGB into other color coordinates:

$$\begin{aligned} \boldsymbol{Ye} &= \boldsymbol{R} + \boldsymbol{G} = \boldsymbol{W} - \boldsymbol{B} \\ \boldsymbol{Mg} &= \boldsymbol{R} + \boldsymbol{B} = \boldsymbol{W} - \boldsymbol{G} \\ \boldsymbol{Cy} &= \boldsymbol{G} + \boldsymbol{B} = \boldsymbol{W} - \boldsymbol{R} \end{aligned} \tag{5-3}$$

The basic color filter array patent[8] was granted to Bryce E. Bayer at Eastman Kodak in 1976. It consists of a repeating 2×2 pattern (Figure 5-8) and seems to be the most popular. There is nothing unique about the arrangement. Stripes and mosaic patterns are possible (Figure 5-9). The precise layout depends on the manufacturer's philosophy and his cleverness in reducing color aliasing.

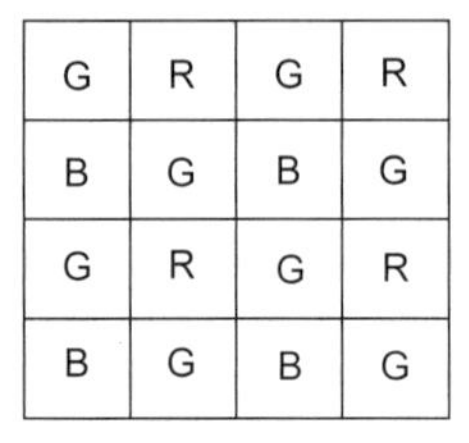

Figure 5-8. Bayer pattern.

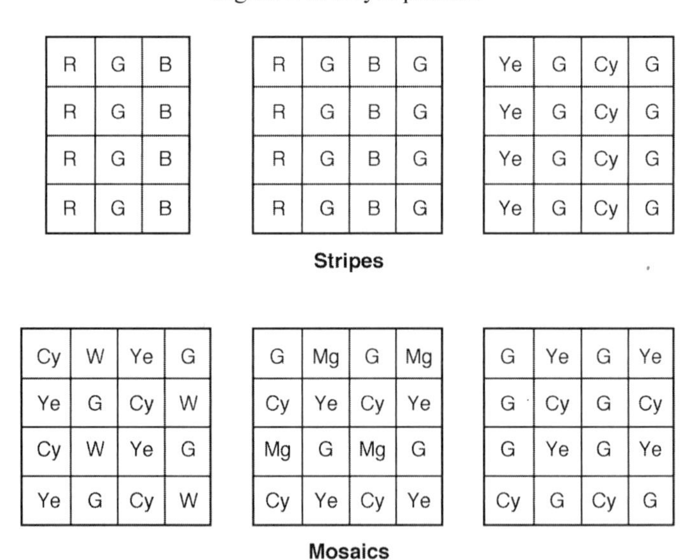

Figure 5-9. Representative stripe and mosaic arrays. Although shown covering a full frame device for clarity, they typically are used on interline transfer devices.

In many sensors, the number of detectors devoted to each color is different. The basic reason is that the human visual system (HVS) derives its detail information primarily from the green portion of the spectrum. That is, luminance differences are associated with green whereas color perception is associated with red and blue. The HVS requires only moderate amounts of red and blue to perceive color. Thus, many sensors have twice as many green as either red or blue detector elements. An array that has 768 horizontal elements may devote 384 to green, 192 to red, and 192 to blue. This results in an unequal sampling of the colors. Because the detectors are at discrete locations, target edges and the associated ambiguities are different for each color. A black-to-white edge will appear to have colors at the transition and monochrome periodic imagery may appear to have color. Aliasing is extremely annoying in color systems. A birefringent crystal (discussed in Section 10.4. *Optical low pass filter*), inserted between the lens and the array, accommodates the different sampling rates. The video signal from the CFA is embedded in the architecture of the CFA. The color video from a single chip CFA must be decoded (unscrambled) to produce usable R, G, and B signals. There are numerous algorithms available (discussed in Section 7.4.3. *Color Correction*), each supposedly provides the "best" imagery.

5.7. DEFECTS

Large arrays contain defects. Table 5-5 provides representative definitions. These definitions vary by manufacturer.

Table 5-5
ARRAY DEFECTS

DEFECTS	REPRESENTATIVE DEFINITION
Point defect	A pixel whose output deviates by more than 6% compared to adjacent pixels when illuminated to 70% saturation.
Hot point defects	Pixels with extremely high output voltages. Typically a pixel whose dark current is 10 times higher than the average dark current.
Dead pixels	Pixels with low output voltage and/or poor responsivity. Typically a pixel whose output is one-half of the others when the background nearly fills the wells.
Pixel traps	A trap interferes with the charge transfer process and results in either a partial or whole bad column (either all white or all dark).
Column defect	Many (typically 10 or more) point defects in a single column. May be caused by pixel traps.
Cluster defect	A cluster (grouping) of pixels with point defects.

Arrays with a few defects are more expensive than arrays with a large number. Depending on the application, the location of the defects may be important. For example, the center ("sweet spot") must be fully operational, with increasing defects allowed as the periphery is approached. Defects are regions of reduced sensitivity. They can also be regions of increased dark current. For scientific application such as spectroscopy, the variation in dark current can affect results – particularly if long integration times are used. These pixels may be labeled as "hot" or "warm." Hot and warm detectors do not follow the same temperature dependence as "normal" detectors.[9]

5.8. REFERENCES

1. See www.en.wikipedia.org/wiki/display_resolution. Wikipedia is a "living" document that is constantly updated by readers. The current list may be different from when this text was written (February 2007).

2. C. Buil, *CCD Astronomy*, pp. 125-152, Willmann-Bell, Richmond, VA (1991).

3. R. G. Neuhauser, "Photosensitive Camera Tubes and Devices," in *Television Engineering Handbook*, K. B. Benson, ed., page 11.34, McGraw-Hill, New York, NY (1986).

4. D. Daly, *Microlens Arrays*, CRC Press (2005).

5. J. Furukawa, I. Hiroto, Y. Takamura, T. Walda, Y. Keigo, A. Izumi, K. Nishibori, T. Tatebe, S. Kitayama, M. Shimura, and H. Matsui, "A 1/3-inch 380k Pixel (Effective) IT-CCD Image Sensor," *IEEE Transactions on Consumer Electronics*, Vol. CE-38(3), pp. 595-600 (1992).

6. M. Deguchi, T. Maruyama, F. Yamasaki, T. Hamamoto, and A. Izumi, "Microlens Design Using Simulation Program for CCD Image Sensor," *IEEE Transactions on Consumer Electronics*, Vol. CE-38(3), pp. 583-589 (1992).

7. See, for example, J. Peddie, *High-Resolution Graphics Display Systems*, Chapter 2, Windcrest/McGraw-Hill, New York, NY (1993) or A. R. Robertson and J. F. Fisher "Color Vision, Representation, and Reproduction," in *Television Engineering Handbook*, K. B. Benson, ed., Chapter 2, McGraw Hill, New York, NY (1985).

8. B. E. Bayer, U.S. patent #3,971,065 (1976).

9. As of February 2007, CCD specifications can be found at Princeton Instruments Reference Library at http://www.piacton.com/library/tutorials/detectors/. Other web-sites offer specifications.

6
ARRAY PERFORMANCE

The most common array performance measures are read noise, charge well capacity, and responsivity. From these the minimum signal, maximum signal, signal-to-noise ratio, and dynamic range can be calculated. Full characterization includes quantifying the various noise sources, charge transfer efficiency, spectral quantum efficiency, linearity, and pixel nonuniformity. These additional metrics are necessary for the most critical scientific applications. Because the array is the basic building block of the camera, camera terminology is often used for array specifications.

The magnitude of each noise component must be quantified and its effect on system performance must be understood. Noise sources may be a function of the detector temperature. Predicted system performance may deviate significantly from actual performance if significant 1/f noise or other noise is present. There are myriad factors involved in system optimization. It is essential to understand what limits the system performance so that intelligent improvements can be made.

Janesick[1] uses three methods to characterize CCD performance: 1) photon transfer, 2) x-ray transfer, and 3) photon standard. The photon transfer technique appears to be the most valuable in calibrating, characterizing, and optimizing arrays for scientific and industrial applications. The lux transfer technique[2] (photometric equivalent of the photon transfer) characterizes digital still cameras and video products. Additional information, necessary for characterizing scientific arrays, is obtained from the x-ray transfer and photon standard methods. Soft x-ray photons produce a known number of electron hole pairs in a localized area of the CCD. Because the number and location of electron hole pairs are known precisely, it is easy to determine the charge transfer efficiency. With the photon standard technique, the CCD array views a source with a known spectral distribution. Using calibrated spectral filters or a monochromator, the spectral quantum efficiency is determined.

Device specifications depend, in part, on the application. Arrays for general video applications may have responsivity expressed in units of V/lux. For scientific applications, the units may be in V/(J/cm^2) or, if a digital output is available, DN/(J/cm^2), where DN refers to a digital number. For example, in an 8-bit system the digital numbers range from zero to 255. These units are incomplete descriptors unless the device spectral response and source spectral characteristics are furnished.

The maximum output occurs when the charge well is filled. The exposure that produces this value is the saturation equivalent exposure (SEE). With this definition, it is assumed that the dark current produces an insignificant number of electrons so that only photoelectrons fill the well.

Sensitivity suggests something about the lowest signal that can be detected. It is usually defined as the input signal that produces a signal-to-noise ratio of one. This exposure is the noise equivalent exposure (NEE). The minimum signal is typically given as equivalent electrons rms. It is only one of many performance parameters used to describe system noise performance. These include the noise equivalent irradiance (NEI) and noise equivalent differential reflectance (NE$\Delta\rho$).

The symbols used in this book are summarized in the *Symbol List*, which appears after the *Table of Contents*.

6.1. QUANTUM EFFICIENCY

With CCD, CMOS, EMCCD, the Super CCD devices, photons are absorbed in silicon and converted to photoelectrons. The spectral quantum efficiency depends upon a variety of factors including coatings, array thickness, gate thickness, and well voltage. Low-light-level device (ICCD and EBCCD) quantum efficiency depends upon the photocathode selected.

There is some confusion on the definition of quantum efficiency. The pixel size is well-defined but the photosensitive area is not due to diffusion, vignetting by metal structures, and architecture. If $M_q(\lambda,T)$ photons/(s-m^2) fall on the photosensitive surface of A_D produce n_{PE} photoelectrons, then the spectral quantum efficiency is

$$QE(\lambda) = \frac{n_{PE}}{M_q(\lambda,T)\, t_{INT}\, A_D} \qquad (6\text{-}1)$$

Since the photosensitive area may not be known, it is convenient to specify an effective quantum efficiency which depends upon the pixel area

$$QE_{EFFECTIVE}(\lambda) = \frac{n_{PE}}{M_q(\lambda,T)\, t_{INT}\, A_P} \qquad (6\text{-}2)$$

For a well-defined photosensitive area, the ratio A_D/A_P is the area fill factor. Microlenses increase the effective fill factor and thereby increase the effective quantum efficiency. For CMOS pixels with an ill-defined photosensitive area, the fill factor must be obtained experimentally.

6.1.1. CCDs, CMOS, AND EMCCDs

For an ideal material, when the photon energy is greater than the semiconductor band gap energy, each photon produces one electron hole pair (quantum efficiency is one). However, the absorption coefficient in silicon is wavelength-dependent (Figure 6-1). Long-wavelength photons are absorbed deeper into the substrate than short wavelengths. Very long-wavelength photons may pass through the CCD and not be absorbed. Beyond 1.1 μm, the absorption is essentially zero because the photon energy is less than the band gap energy. For a doping concentration of 10^{17}cm^{-3}, the depth at which 90% of the (internal) photons are absorbed is given in Table 6-1.

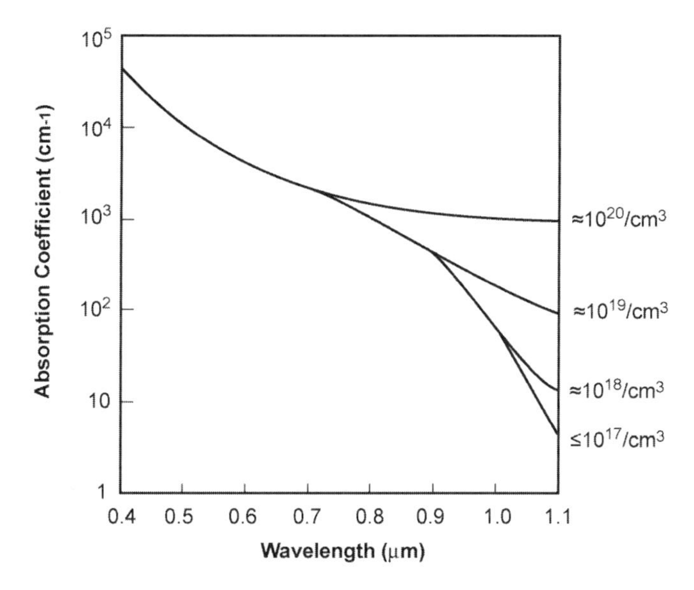

Figure 6-1. Silicon absorption coefficient as a function of wavelength and doping concentration at 25°C. Detectors typically have doping concentrations less than 10^{17}/cm^3. (From Reference 3).

Table 6-1
REPRESENTATIVE 90% ABSORPTION DEPTH

Wavelength (μm)	Depth (μm)
0.40	0.19
0.45	1.0
0.50	2.3
0.55	3.3
0.60	5.0
0.65	7.6
0.70	8.5
0.75	16
0.80	23
0.85	46
0.90	62
0.95	150
1.00	470
1.05	1500
1.10	7600

Foveon's direct image sensor[4,5] exploits the variable absorption depths by creating an array that has three stacked wells. As such, a single pixel can provide the three colors simultaneously (Figure 6-2). This avoids the color crosstalk and aliasing problem encountered with color filter arrays (discussed in Section 10.4. *Optical low pass filter*) and is less complicated than the 3CCD approach (Figure 7-5).

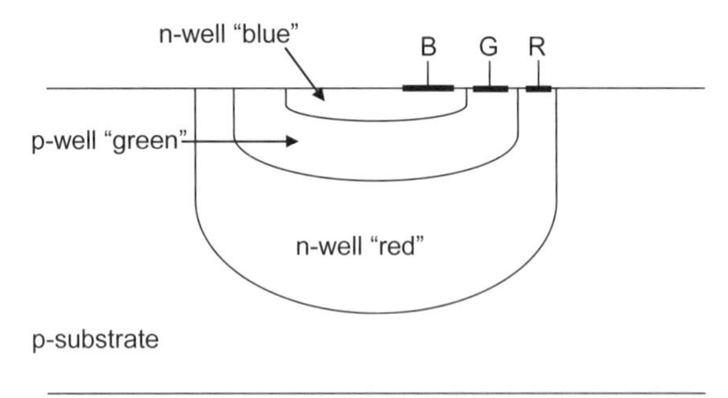

Figure 6-2. Vertically stacked three-color pixel. The specific well depths depend on the doping level. Foveon well depths are 0.2 μm, 0.6 μm, and 2 μm for the "blue", "green," and "red" wells, respectively.

A photon absorbed within the depletion region will create one photoelectron (internal quantum efficiency equals unity). However, the depletion region size is finite and long-wavelength photons will be absorbed within the bulk material. An electron generated in the substrate will experience a three-dimensional random walk until it recombines or reaches the edge of a depletion region where the electric field exists. If the diffusion length is zero, all electrons created within the bulk material will recombine immediately and the quantum efficiency approaches zero for these photons. As the diffusion length approaches infinity, the electrons eventually reach a charge well and are stored. Here, the internal quantum efficiency approaches one.

Doping controls the diffusion length and the quantum efficiency is somewhere between these two extremes (Figure 6-3). The quantum efficiency is dependent upon the gate voltage (low voltages produce small depletion regions) and the material thickness (long-wavelength photons will pass through thin substrates). Diffusion creates a responsivity that overlaps pixels (Figure 6-4). The maximum signal and other performance parameters defined in this chapter are for large area illumination. That is, a large number of detectors are illuminated and the average output is used. As the image size decreases, diffusion effects become more apparent.

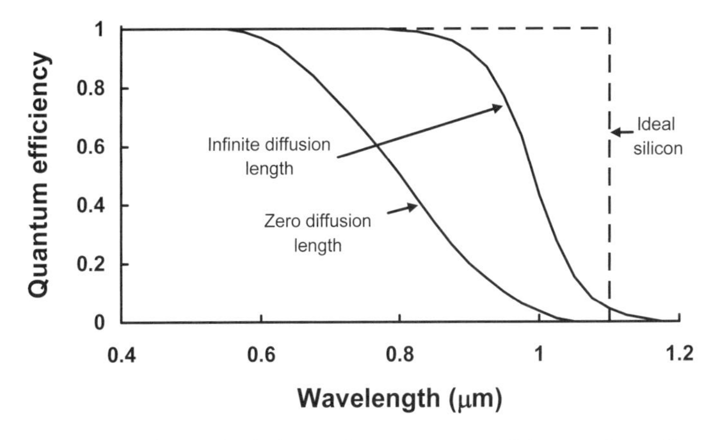

Figure 6-3. Theoretical internal quantum efficiency for silicon photosensors 250 μm thick. The long-wavelength quantum efficiency depends upon the thickness of the substrate and the diffusion length. Most arrays have a relatively large diffusion length so that the quantum efficiency tends to be near the infinite diffusion length curve. (From Reference 6.)

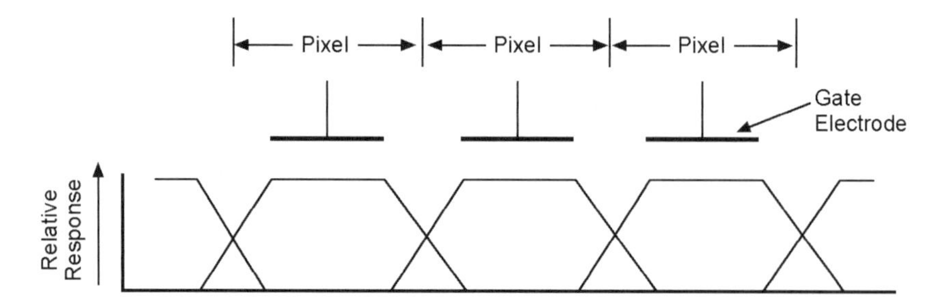

Figure 6-4. Idealized relative spatial response to long wavelength photons.

For front-illuminated devices with photogates, the spectral responsivity deviates from the ideal spectral response due to the polysilicon gate electrodes. The transmittance of polysilicon starts to decrease below 0.6 μm and becomes opaque at 0.4 μm (Figure 6-5). The transmittance depends upon the gate thickness and material. The gate structure is a thin film and therefore interference effects will cause variation in quantum efficiency that is wavelength-dependent. Interference effects can be minimized through choice of material and film thickness. Because different manufacturing techniques are used for each array type, the spectral response varies with manufacturer. If a manufacturer uses different processes, then the spectral response will vary within his product line. Replacing the polysilicon gates with more transmissive[7] indium tin oxide (ITO) increases the responsivity in the visible. ITO has a higher absorption in the UV leading to reduced UV quantum efficiency.

Even with no polysilicon overcoat, UV photons are absorbed at the surface. These electrons recombine before they reach a storage site. Hence, the quantum efficiency is very low in the UV. The UV response can be enhanced[8] with a UV fluorescent phosphor. These phosphors are deposited directly onto the array and emit light at approximately 0.54 to 058 μm when excited by 0.12 to 0.45 μm light. Phosphors radiate in all directions and only the light fluorescing toward the array is absorbed. Lumogen, one of many available organic phosphors, has an effective quantum efficiency of about 15%. It does not degrade the quantum efficiency in the visible region because it is transparent at those wavelengths. The coating does not change the spectral response of the detector elements. Rather, it converts out-of-band photons into in-band photons. Therefore, the detector with the coating appears to have a spectral response from 0.12 to 1.1 μm.

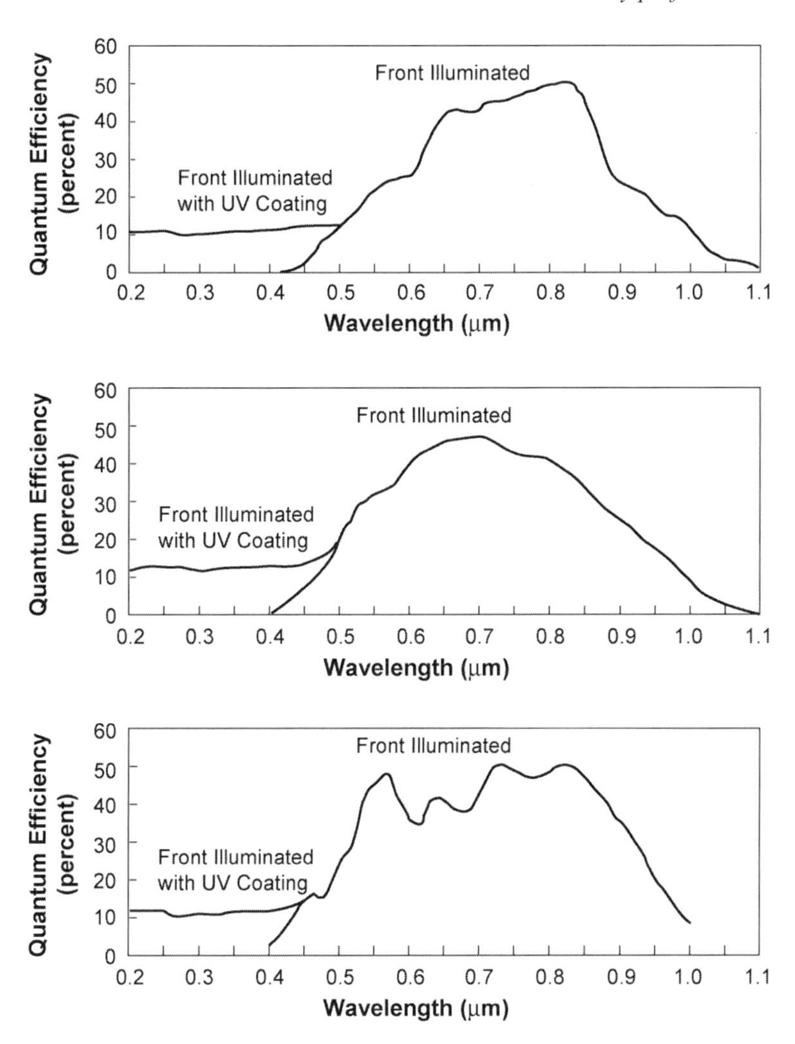

Figure 6-5. Typical quantum efficiencies provided by different manufacturers. Above 0.8 μm, the quantum efficiency drops because the electron hole pairs, generated deep in the array, may recombine before reaching a storage site. Below 0.6 μm, the polysilicon overcoat starts to become opaque (see Figure 4-4). The variation in spectral response may not be a concern for many applications. The ripples in the responsivity are caused by interference effects. Color filter arrays, with their finite transmission, reduce the responsivity. See Figures 6-7 and 6-8 for addition quantum efficiency curves.

Photodiodes, as used in most interline transfer devices, do not have polysilicon gates. Therefore, these devices tend to have a response that approaches the ideal silicon spectral response. That is, there are no interference effects and the blue region response is restored. With virtual phase devices, one polysilicon electrode is replaced by a shallow highly doped layer. Because there is no electrode over this layer, higher quantum efficiency occurs in the blue region. CIDs consist of two overlapping MOS capacitors sharing the same row and column electrode. Because the electrode only covers a part of the pixel, the area-averaged spectral response is higher than a comparable CCD.

Devices with vertical antibloom drains have reduced red response. The buried drain is in close proximity to where long photon absorption occurs. Any electron that enters the drain is instantly removed.

The signal is further reduced by the light shields in the interline transfer devices where the maximum fill factor is 50%. Illuminating the array from the back side (Figure 6-6) avoids the polysilicon and light shield problems, and the fill factor can reach 100%. Silicon is reflective so that an antireflection coating must be applied. Photons entering the back side are absorbed in the silicon and diffuse to the depletion region. However, short-wavelength photons are absorbed near the surface and these electron hole pairs recombine before reaching a storage site in a thick wafer. Therefore, the wafer is thinned to about 10 μm to maintain good quantum efficiency. In back-side thinned devices, the incident photon flux does not have to penetrate the polysilicon gate sandwich structure and interference effects are much easier to control.

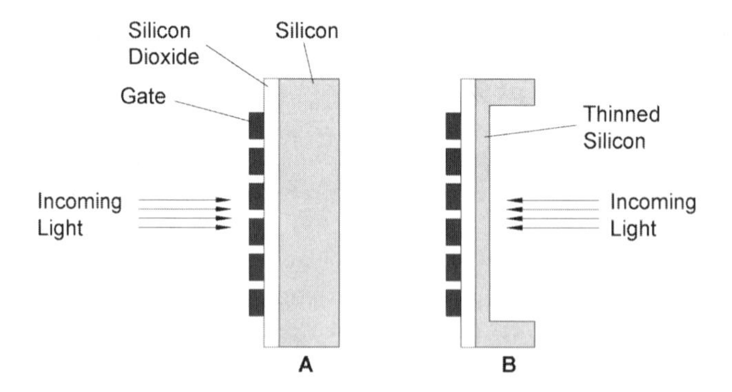

Figure 6-6. (a) Front illuminated and (b) back illuminated arrays. Owing to their manufacturing difficulty and fragile design, back-illuminated devices are usually limited to scientific applications that require high quantum efficiency.

With a proper anti-reflection coating, a quantum efficiency of over 90% is possible. However, silicon has an index of refraction that varies with wavelength, making it somewhat difficult to manufacture an anti-reflection coating that is effective across the entire spectrum. Coatings that optimize the response in the near IR have reduced effectiveness in the visible (Figure 6-7). Lumogen reduces the effectiveness of the anti-reflection coating. New anti-reflection coatings overcome this problem and some offer UV response without the use of lumogen.[9]

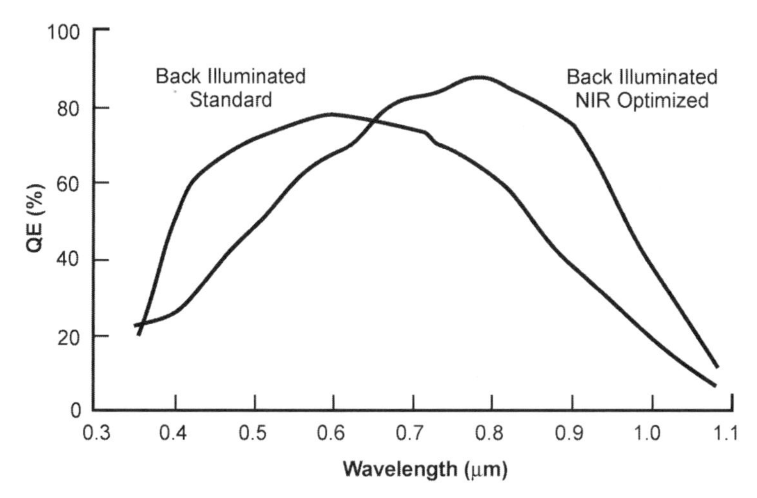

Figure 6-7. Effect of different anti-reflection coatings on back illuminated CCDs. (From Reference 10).

As the wavelength increases, the silicon becomes more transparent. With thin back-illuminated devices, the rear surface and silicon oxide layer create an optical etalon. This creates constructive and destructive interference that appear to modulate the spectral responsivity.[10] At 0.8 μm the modulation occurs approximately every 5 nm. This is bothersome in some spectroscopic applications. For other applications, the waveband is sufficiently wide to cover several etalon fringe cycles; this averages out the effect.

CCD arrays for general and industrial applications are within a sealed environment for protection. Light must pass though a window to reach the array. In addition to reflection losses at all wavelengths, the glass transmittance decreases for wavelengths below 0.4 μm. For scientific applications requiring high sensitivity, the glass can be coated with a broadband anti-reflection coating. For UV applications, a quartz UV transmitting window can be used.

Figure 6-8 illustrates the spectral quantum efficiency for several CCD arrays and a single CMOS array (Curve 7). This CMOS pixel has a fill factor of about 28% and no microlens. A microlens could improve the spectral quantum efficiency by a factor of about two to achieve an effective peak quantum efficiency of about 55%. A somewhat similar increase is seen for CCDs (compare Curve 2 with Curve 3).

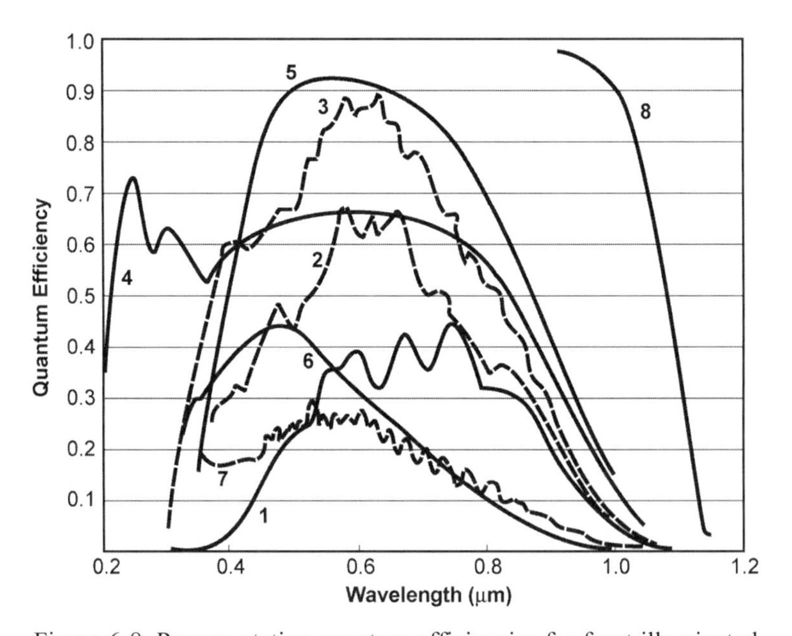

Figure 6-8. Representative quantum efficiencies for front illuminated and back illuminated arrays. (1) Front-illuminated CCD with polysilicon gates, (2) front-illuminated CCD with ITO gates, (3) front-illuminated CCD with ITO gates and microlens, (4) back-illuminated CCD with UV enhancement, (5) back-illuminated CCD with visible enhancement, (6) 7.4 μm interline transfer array with microelens, (7) 6.0 μm CMOS pixel, and (8) deep depletion 300 μm thick pixel. The thick deep depletion is similar to Figure 6-2. The spectral response depends on the process used to manufacture each device. Highest response is obtained by a backside-illuminated CCD array. The CMOS sensor exhibits lowest sensitivity because of a low fill factor and the thin epitaxial silicon. After Reference 9.

The photosensitive area of a CMOS detector is less well-defined. The photodiode region is typically a rectangular structure with a notch in the lower center. An opaque silicided contact covers the notch. Near the reset and select buses are various transistor structures and their associated contacts. The CMOS photosensitive area is not rectangular, but is a complex shape that depends on the location of the electronics, contacts, and bus lines. As a result, it is difficult to definitively specify a fill factor.

Experimentally, a CMOS photoresponse was mapped out using a focused helium-neon laser spot[11]. The spot was systematically stepped in increments of 0.3 μm over the 9 μm square surface (Figure 6-9).

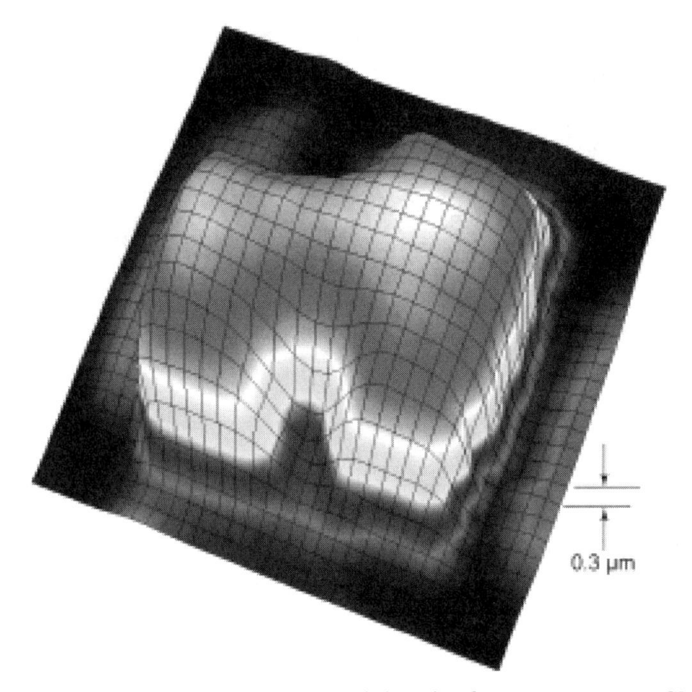

Figure 6-9. 3-D isometric responsivity plot for a 9-μm square CMOS pixel.

The architecture is laid over the response in Figure 6-10. It is clear that a CMOS pixel can be sensitive to light in regions other than inside the photodiode boundary (where light sensitivity is clearly expected) as long as these regions are not blocked by metal lines or opaque structures. The optical fill factor (ratio of photoresponse area to pixel area) is greater than the geometrical fill factor (ratio of notched rectangle to pixel area).

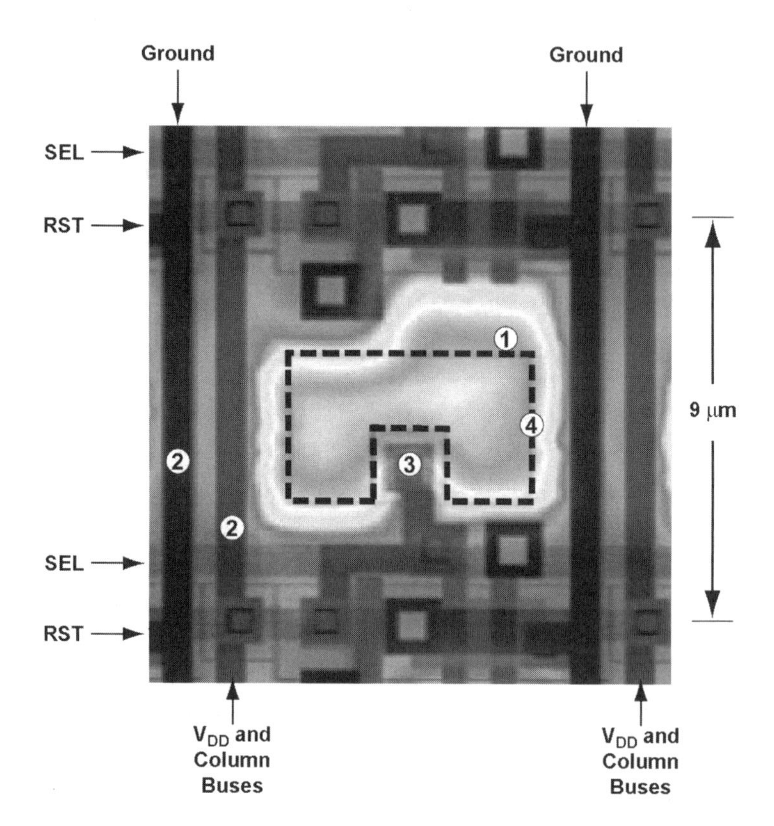

Figure 6-10. Overlay of pixel photoresponse (Figure 6-9) with CMOS architecture. **Region 1** shows peripheral response from a region of clear silicon above the photodiode boundary. **Region 2** identifies the power and ground lines. **Region 3** shows the highest opacity to light of any region in the photodiode due to a silicided contact (notch shaped). **Region 4** is the edge of the photodiode boundary.

Figure 6-11 illustrates the cross-section of a typical CMOS pixel, showing an end-on view of several metal lines (bias, ground, column buses, etc) that run across the array. These metal lines must be vertically isolated from each other by an oxide material to avoid dielectric breakdown. That is, the vertical distance between the metal lines cannot be reduced below a certain spacing for a given line voltage. Since this spacing must be maintained to ensure adequate pixel voltages, the vertical dimension $\boldsymbol{h}$ will shrink more slowly than the photodiode lateral dimension $\boldsymbol{w}$, as design rules get smaller. This means that the ratio, $\boldsymbol{h}/\boldsymbol{w}$, tends to increase as the pixel dimensions decrease. Looking down is like looking into a well with the photodiode at the bottom of the well. For sharply focused light, the incoming rays will be obstructed (vignetted) by the metal layers, resulting in a quantum efficiency loss[12] (also called pixel vignetting). Placing a microlens into the volume directly above photodiode will restrict its shape and hence it performance. As the ***h*/*w*** ratio increases, microlenses become less effective.

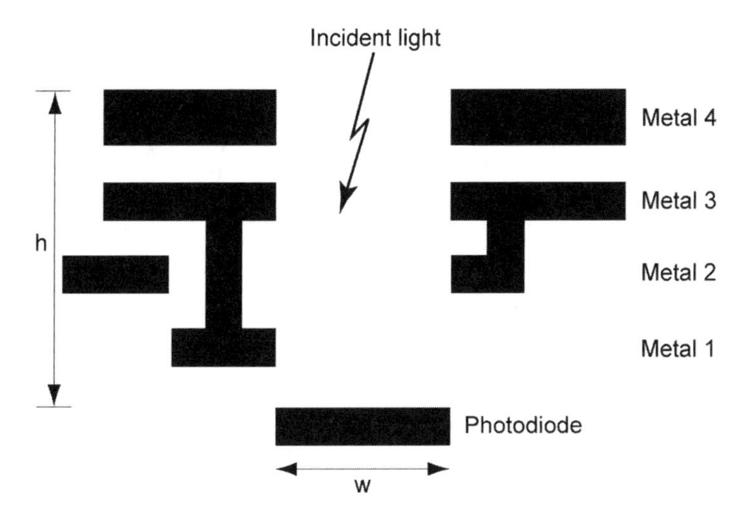

Figure 6-11. Typical cross-sectional diagram of a CMOS pixel. The layer labeled **Metal 4** acts as a light-shield. The other metal layers are used for ground, bias, and buses. Note that the vertical placement is to scale (layer thickness and layer separation).

6.1.2. ICCD AND EBCCD

The ICCD and EBCCD quantum efficiency depends on the photocathode. Image intensifiers have changed, with GenII tubes (multialkali photocathode) being replaced with GenIII tubes (GaAs photocathode). While Figure 6-12 illustrates three, a variety of photocathodes are available.[13] Reference 14 compares various photocathodes that could be used for ICCDs, EBCCDs, and EMCCDs.

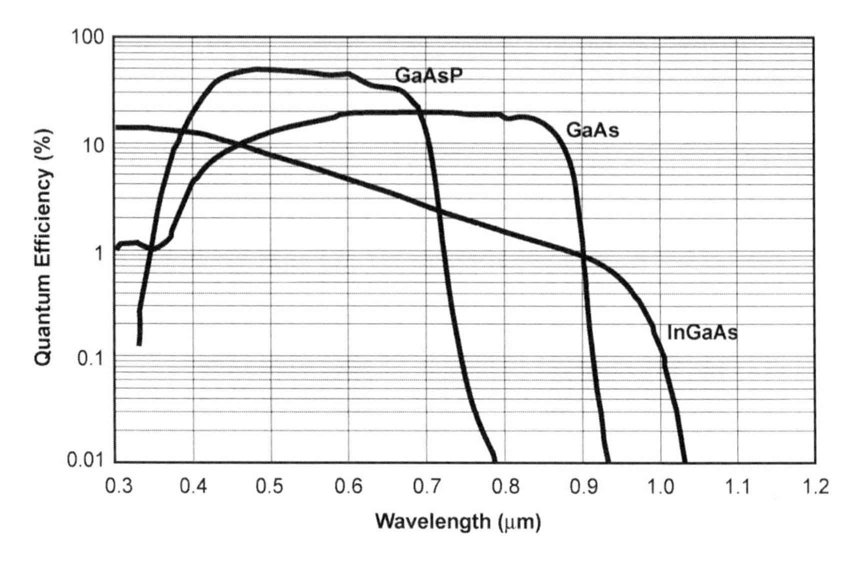

Figure 6-12. Quantum efficiencies for three photocathodes. The quantum efficiency depends upon the window material and photocathode material.

6.2. RESPONSIVITY

Figure 6-13 illustrates a signal transfer diagram. The shutter represents the integration time, t_{INT}. The detector converts the incident photons, $n_{DETECTOR}$, into photoelectrons at a rate determined by the quantum efficiency. A voltage is created when the electrons are transferred to the sense node capacitance. The signal after the source follower amplifier is

$$V_{SIGNAL} = n_e \frac{Gq}{C} \tag{6-3}$$

where n_e is the total number of electrons in the charge packet. It includes both photoelectrons and dark current electrons. The output gain conversion, Gq/C, is typically 0.1 μV/e⁻ to 10 μV/e⁻. The source follower amplifier gain, G, is near unity and, therefore, it is sometimes omitted from radiometric equations. Amplifiers, external to the CCD device, amplify the signal to a useful voltage. These amplifiers are said to be off-sensor or off-chip.

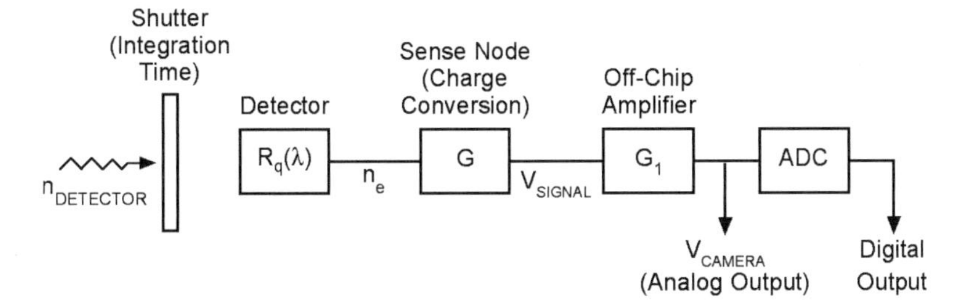

Figure 6-13. Representative signal transfer diagram. Many systems do not have a shutter, but it represents t_{INT}. The output may be specified as the number of electrons, output voltage of the source follower amplifier, or output of the off-chip amplifier. Some devices provide a digital output. V_{SIGNAL}, V_{CAMERA}, and DN can be measured whereas n_e is calculated.

The spectral quantum efficiency is important to the scientific and military communities. When the array is placed into a general video or industrial camera, it is convenient to specify the output as a function of incident flux density or energy density averaged over the spectral response of the array.

6.2.1. CCD, CMOS, and EMCCD

For an array with $E_e(\lambda)$ incidance, the total number of photoelectrons generated is

$$n_{PE} = \frac{A_D}{q} \int_{\lambda_1}^{\lambda_2} E_e(\lambda) R_e(\lambda) t_{INT} \, d\lambda \qquad (6\text{-}4)$$

where $R_e(\lambda)$ is the spectral response expressed in units of A/W. The array output voltage (after the source follow amplifier) is $(G\, q\, n_{PE})/C$

$$V_{SIGNAL} = \frac{G}{C} A_D \int_{\lambda_1}^{\lambda_2} E_e(\lambda) R_e(\lambda) t_{INT} \, d\lambda \qquad (6\text{-}5)$$

It is desirable to express the responsivity in the form

$$V_{SIGNAL} = R_{AVE} \int_{\lambda_1}^{\lambda_2} E_e(\lambda) t_{INT} \, d\lambda \qquad (6\text{-}6)$$

Multiplying the incidance by the integration time provides J/cm^2. The value R_{AVE} is an average response that has units of V/(J/cm^2), and the integral has units of J/cm^2. Note that detector responsivity is specified in the CGS system (cm^2). For a large extended source $E_e(\lambda) = M_e(\lambda)$. Combining the two equations provides

$$R_{AVE} = \frac{G}{C} A_D \frac{\int_{\lambda_1}^{\lambda_2} M_e(\lambda) R_e(\lambda)\, d\lambda}{\int_{\lambda_1}^{\lambda_2} M_e(\lambda)\, d\lambda} \quad \frac{\mathrm{V}}{\mathrm{J/cm^2}} \tag{6-7}$$

The variables R_e and R_{AVE} are the for array package. It is the detector responsivity multiplied by all the intervening optics (i.e., window, microlens, CFA, and optical low-pass filter). Usually the manufacturer includes the window transmittance in the quoted responsivity. Other components may be included with the array specification or specified separately as part of $T_{OPTICS}(\lambda)$.

Experimentally, the most popular approach to measure R_{AVE} is to vary the exposure (e.g., the integration time). The value R_{AVE} is the slope of the output/input transformation (Figure 6-14). By placing a calibrated radiometer next to the array (at distance R_1 from the source), the incidance, E_e, can be measured. These equations are valid over the region that the array output/input transformation is linear. It is imperative that the source color temperature remains constant. Assuming an incandescent bulb is used, reducing the bulb voltage reduces the output, but this changes the color temperature. Changing the source temperature changes the integrals and this leads to incompatible results (see Section 2.8. *Normalization issues*). The incidance can also be varied by changing the integration time, inserting neutral density filters or varying the viewing distance R_1.

The value R_{AVE} is an average type responsivity that depends on the source characteristics and the spectral quantum efficiency. While the source can be standardized (e.g., CIE illuminant A or illuminant D6500), the spectral quantum efficiency varies by device (Figures 6-5, 6-7, and 6-8). Therefore extreme care must be exercised when comparing devices solely by the average responsivity.

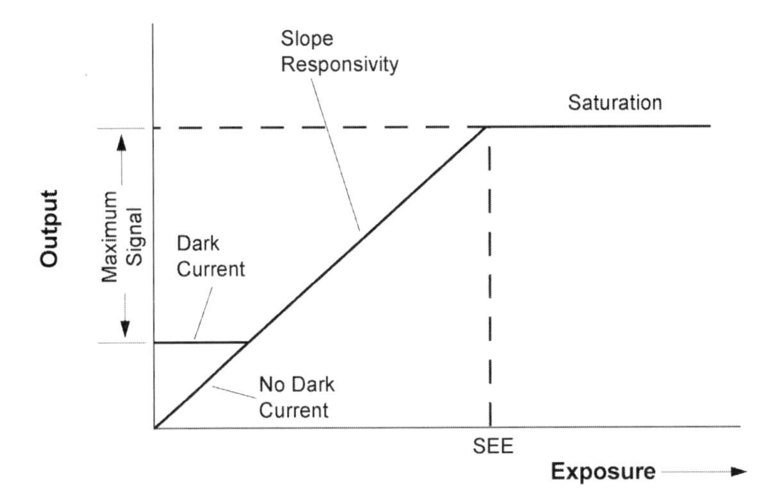

Figure 6-14. The average responsivity is the slope of the output-input transformation. The maximum input or the saturation equivalent exposure (SEE) is the input that fills the charge wells. Dark current limits the available signal strength.

Both $\boldsymbol{M_e(\lambda)}$ and $\boldsymbol{R_e(\lambda)}$ are functions of wavelength. If a very small wavelength increment is selected, $\boldsymbol{M_e(\lambda)}$ and $\boldsymbol{R_e(\lambda)}$ may be considered as constants and Equation 6-7 can be approximated as

$$\boldsymbol{R}_{AVE} \approx \frac{G}{C} A_D \frac{M_e(\lambda_o) R_e(\lambda_o) \Delta\lambda}{M_e(\lambda_o)\Delta\lambda} = \frac{G}{C} A_D R_e(\lambda_o) \tag{6-8}$$

The responsivity (with units of A/W) is related to the quantum efficiency $\boldsymbol{R_q}$ by $\boldsymbol{R_e} = (\boldsymbol{q\lambda/hc})\ \boldsymbol{R_q}$. If the wavelength is measured in micrometers, then $\boldsymbol{R_e} = (\lambda/1.24)\ \boldsymbol{R_q}$. At a specific wavelength

$$\begin{aligned} R_{SIGNAL}(\lambda_o) &= \frac{Gq}{C}\frac{\lambda_o}{hc} A_D R_q(\lambda_o) \\ &= \frac{G}{C} A_D R_e(\lambda_o) \end{aligned} \tag{6-9}$$

or

$$R_{SIGNAL}(\lambda_o) = \frac{G}{C}\frac{\lambda_o}{1.24} A_D R_q(\lambda_o) \quad \frac{\mathbf{V}}{\mathbf{J/cm^2}} \tag{6-10}$$

Here, the detector area is measured in square centimeters. Typically $R_q(\lambda_o)$ is evaluated at the wavelength that provides the peak quantum efficiency: $R_q(\lambda_P)$. The symbol η is often used for the quantum efficiency.

If the device has a digital output, the number of counts (integer value) is

$$DN = \mathbf{int}\left[V_{CAMERA}\frac{2^N}{V_{MAX}}\right] \tag{6-11}$$

where N is the number of bits provided by the analog-to-digital converter and ***int*** represents the integer value of the bracketed term. The value N is usually 8 for general video applications but may be as high as 16 for scientific cameras. The value V_{MAX} is the maximum output that corresponds to a full well, N_{WELL}:

$$V_{\mathrm{MAX}} = G_I\frac{G\,q}{C}\,N_{WELL} \tag{6-12}$$

The responsivity at the peak wavelength is approximated by

$$R_P \approx \left[\frac{2^N}{N_{WELL}}\frac{\lambda_P}{1.24}A_D R_q(\lambda_P)\right]\quad\frac{\mathrm{DN}}{\mathrm{J/cm^2}} \tag{6-13}$$

This assumes that the analog-to-digital converter input dynamic range exactly matches V_{MAX}. This may not be the case in real systems. If an antibloom drain is present, then the responsivity is defined up to the knee point (see Figure 3-35) and Equation 6-13 is modified accordingly.

When using arrays for general video, the output is normalized to photometric units. When viewing a source of "infinite" size ($r \gg R$ in Figure 2-12), the lux value is

$$N_{LUX} = 683\int_{0.38}^{0.75} V(\lambda)M_P(\lambda,T)\,d\lambda \tag{6-14}$$

It is convenient to express the array average response as

$$V_{SIGNAL} = R_{PHOTOMETRIC}\left[683\int_{0.38}^{0.75\,\mu m} M_P(\lambda)V(\lambda)\,d\lambda\right] \tag{6-15}$$

where $R_{PHOTOMETRIC}$ has units of V/lux and the bracketed term has units of lux. Combining Equations 6-5 and 6-15 yields

$$R_{PHOTOMETRIC} = \frac{\frac{G}{C} A_D \int_{\lambda_1}^{\lambda_2} M_e(\lambda) R_e(\lambda) t_{INT} d\lambda}{683 \int_{0.38}^{0.75\,\mu m} M_P(\lambda) V(\lambda) d\lambda} \quad \frac{\text{V}}{\text{lux}} \tag{6-16}$$

$R_{PHOTOMETRIC}$ depends on the integration time and is either 1/60 or 1/30 s depending on the architecture for EIA 170 or NTSC video standards. Experimentally, the photometric illuminance is measured with a calibrated sensor. The input intensity levels are changed either by moving the source (changing R_1) or by inserting neutral-density filters between the source and the array.

The responsivity depends on the spectral response and the spectral content of the illumination. The apparent variation in output with different light sources was discussed in Section 2.8. *Normalization issues*. Selecting an array based on the responsivity is appropriate if the anticipated scene illumination has the same color temperature as the calibration temperature. That is, if the photometric responsivity is measured with a CIE illuminant A and the scene color temperature is near 2856 K, then selecting an array with the highest photometric responsivity is appropriate. Otherwise, the average photometric responsivity is used for informational purposes only.

Example 6-1
MAXIMUM OUTPUT

A device provides an output gain conversion of 6 μV/e$^-$. The well size is 70,000 electrons. What is the maximum output?

The maximum output is the OGC multiplied by the well capacity or 420 mV. If the dark current creates 5,000 electrons, then the maximum signal voltage is reduced to (70,000-5,000)(0.006 mV) = 390 mV. This signal may be amplified by an off-chip amplifier.

Example 6-2
DIGITAL RESPONSIVITY

A device with a well capacity of 50,000 electrons has its output digitized with a 12-bit analog-to-digital converter (ADC). What is the digital responsivity?

The step size is the well capacity divided by the number of digital steps. $50{,}000/2^{12} = 12.2$ electrons/DN. Equation 6-2 provides the number of electrons as a function of input exposure. The relationship between the ADC maximum value and charge well capacity assumes an exact match between V_{MAX} and the ADC input range. If these do not match, an additional amplifier must be inserted just before the ADC.

Example 6-3
RESPONSIVITY

A 2/3-inch format interline transfer device provides an output gain conversion of 6 μV/e$^-$. If the quantum efficiency is 0.8 at 0.6 μm, what is the responsivity in units of V/(μJ/cm^2)?

From Table 5-4, a 2/3-inch format with 4:3 aspect ratio is 8.8 μm × 6.6 μm. If the array is 640×480, each pixel is 13.75μm × 13.75μm. An interline transfer device has a 50% fill factor, then $A_D = 9.45 \times 10^{-7}$ cm^2. The constant hc is 1.99×10^{-25} J-m/photon. Using Equation 6-9,

$$R_{SIGNAL}(\lambda_o) = (6\text{x}10^{-6})\left(\frac{0.6\text{x}10^{-6}}{1.99\text{x}10^{-25}}\right)(9.45\text{x}10^{-7})(0.8) = 13.7\text{x}10^{6}$$

$$\frac{\text{V}}{\text{J/cm}^2} = \left(\frac{\text{V}}{\text{e}^-}\right)\left(\frac{\text{m}}{(\text{J}-\text{m})/\text{photon}}\right)(\text{cm}^2)\left(\frac{\text{electrons}}{\text{photons}}\right) \tag{6-17}$$

or $R_{SIGNAL}(\lambda_o) = 13.6$ V/(μJ/cm^2).

Example 6-4
PHOTOMETRIC OUTPUT

An array designed for video applications has a responsivity of 250 μV/lux when measured with CIE illuminant A. What is the expected output if the input is 1000 lux?

If the color temperature does not change and only the intensity changes, then the output is (250 μV)(1000) = 250 mV. However, if the color temperature changes, the output changes in a nonlinear manner. Similarly, two sources may produce the same illuminance but V_{SIGNAL} may be quite different. The color temperature must always be specified when $R_{PHOTOMETRIC}$ is quoted.

Implicit in this calculation is that the integration time remains constant. Because photometric units are used for general video applications, it is reasonable to assume that t_{INT} is either 1/60 or 1/30 s for EIA 170 compatibility.

Example 6-5
PHOTOMETRIC CONVERSION

An array designed for video applications has a sensitivity of 20 V/(μJ/cm^2). What is the estimated photometric response? The relative spectral response of $R_{SIGNAL}(\lambda)$ is illustrated in Figure 6-15.

Substituting Equation 6-9 into Equation 6-16 provides

$$R_{PHOTOMETRIC} = t_{INT} R_{SIGNAL-MAX} \left[\frac{\int_{\lambda_1}^{\lambda_2} E_e(\lambda) R_{SIGNAL-RELATIVE}(\lambda) t_{INT} \, d\lambda}{683 \frac{1}{A_D} \int_{0.38}^{0.75\,\mu m} E_e(\lambda) V(\lambda) \, d\lambda} \right] \qquad (6\text{-}18)$$

Two different spectral responsivities are considered: one as shown in Figure 6-15, and the other has an additional infrared blocking filter that cuts off at 0.7 μm. Numerical integration provides the value of the bracketed term (Table 6-2). As the color temperature increases, the bracketed value may either increase or decrease depending upon the spectral response. With $R_{SIGNAL\text{-}MAX}$ = 0.2 V/(J-cm^2) and t_{INT}= 1/60 s $R_{PHOTOMETRIC}$ is given in Table 6-3.

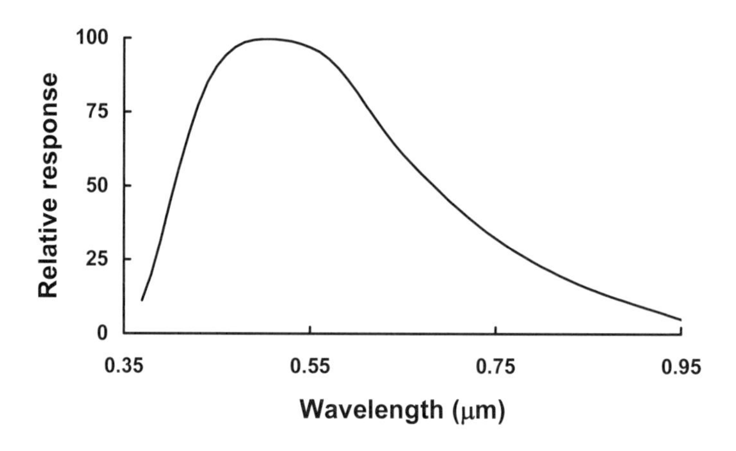

Figure 6-15. Relative spectral responsivity of $R_{SIGNAL}(\lambda)$.

Table 6-2
EQUATION 6-18 BRACKETED TERM

COLOR TEMPERATURE	NO INFRARED FILTER	WITH INFRARED FILTER
2856 K	0.00473	0.00319
6500 K	0.00391	0.00353

Table 6-3
PHOTOMETRIC RESPONSIVITY (V/lux)

COLOR TEMPERATURE	NO INFRARED FILTER	WITH INFRARED FILTER
2856 K	0.158	0.106
6500 K	0.130	0.117

6.2.2. SUPER CCD

The equations in Section 6.2.1 apply to the single-detector Super CCD. For the Super CD SR, the two separate photodiodes (Figure 3-29) have different sensitivities and when the signals are combined, offer a dynamic range that is 4 times greater then a conventional CCD. Automatic exposure control[15] combines the two signals

$$\textit{\textbf{Signal}} = k_{HIGH} n_{HIGH} + k_{LOW} n_{LOW} \qquad (6\text{-}19)$$

Generally, $k_{HIGH} + k_{LOW} = 1$. When $k_{HIGH} = k_{LOW} = 0.5$, the combined output is the average of the two values (Figure 6-16) and is less than one large detector. The output can, of course, be amplified.

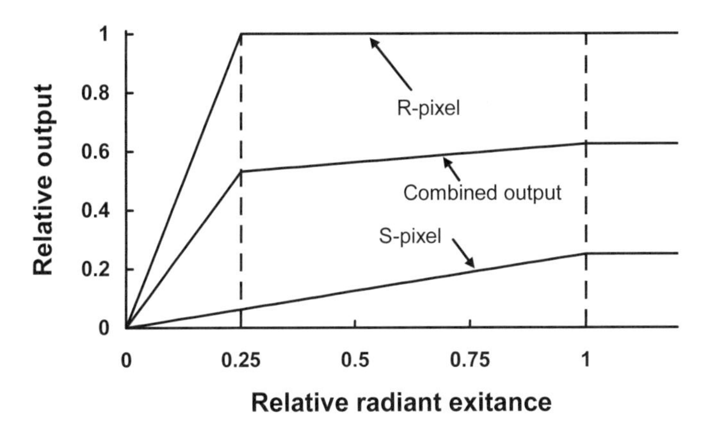

Figure 6-16. Super CCD SR output. The area of the low sensitivity photodiode (S-pixel) is 16 times smaller than the high sensitivity photodiode (R-pixel) area. The effective well capacity of the S-pixel is 4 times smaller.

6.2.3. ICCD and EBCCD

As illustrated in Figure 6-17, the photocathode creates photoelectrons that are amplified by the MCP. The amplified electrons are converted back into photons by the phosphor screen. These photons are relayed to the CCD by either a fiberoptic bundle or a relay lens. The CCD creates the photoelectrons that are measured. The quantum efficiency at each conversion is less than unity and each optical element yields a finite transmittance loss.

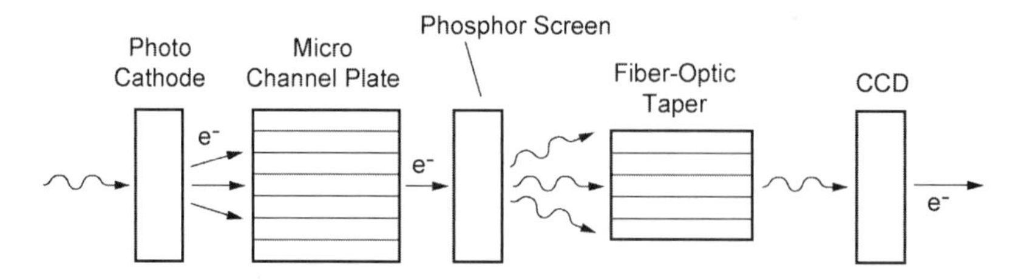

Figure 6-17. Multiple photon-electron conversions occur in an ICCD.

The ICCD uses a lens to image the target onto the intensifier faceplate. Omitting the wavelength notation for equation brevity, the number of signal electrons generated by the photocathode is

$$n_{CATHODE} = \frac{\eta_{CATHODE} M_q}{4F^2(1 + M_{OPTICS})^2} t_{INT} T_{OPTICS} A_{PIXEL} \tag{6-20}$$

The variable A_{PIXEL} is the projected area of a CCD pixel onto the photocathode. Geometrically, it is the CCD detector area multiplied by the magnification of the fiberoptic bundle. However, A_{PIXEL} depends on fiberoptic bundle characteristics and microchannel pore size. It may be somewhere between 50% of the CCD detector geometric projected area to several times larger.[16] As such, it represents the largest uncertainty in signal-to-noise ratio calculations. The Appendix provides a clarification when $F < 3$.

The microchannel plate amplifiers the number of electrons:

$$n_{MCP} = G_{MCP} n_{CATHODE} \tag{6-21}$$

These electrons are then converted to photons by the phosphor screen

$$n_{SCREEN} = \eta_{SCREEN} n_{MCP} \tag{6-22}$$

The photons travel to the CCD via a fiberoptic bundle whose effective transmittance is T_{fo}. This is a composite value that includes the fiberoptic transmittance and coupling losses.

$$n_{PHOTON-CCD} = T_{fo} n_{SCREEN} \tag{6-23}$$

The number of photoelectrons at each wavelength sensed in the CCD is proportional to its quantum efficiency

$$n_{PE} = \eta_{CCD} n_{PHOTON-CCD} \tag{6-24}$$

As M_{OPTICS} approaches zero, the total number of photoelectrons

$$n_{PE} = \int \eta_{CCD} \eta_{SCREEN} \eta_{CATHODE} T_{fo} G_{MCP} \left[\frac{M_q}{4F^2} t_{INT} A_{PIXEL} \right] d\lambda \qquad (6\text{-}25)$$

The limits of integration are defined by the photocathode spectral sensitivity. For a lens-coupled system, T_{fo} is replaced by an effective relay lens transmittance.[17] Each has its own merits. Transmission losses in a fiberoptic bundle increase proportionally to the square of the minification ratio. This leads to a practical minification of about 2:1. Light is lost because the screen is a Lambertian source, whereas the fiberoptic can only accept light over a narrow cone (acceptance angle). But lenses are large and bulky and may provide a throughput as low as 2%. Relay lenses also suffer from vignetting. Here, the "relayed" intensity decreases as the image moves from the center to the periphery of the image intensifier tube.

Because the lens coupling is less efficient than the fiberoptic coupling, the number of electrons produced is lower. The fiberoptic coupled system, with its higher effective transmittance and better sensitivity, can detect lower signals.

6.3. DARK CURRENT

The CCD output is proportional to the exposure, $L_q(\lambda)t_{INT}$. The output can be increased by increasing the integration time, and long integration times are generally used for low-light-level operation. However, this approach is ultimately limited by dark current leakage that is integrated along with the photocurrent. For a large 24 μm square pixel, a dark current density of 1 nA/cm^2 produces 36,000 electrons/pixel/s. If the device has a well capacity of 860,000 electrons, the well fills in 24 s. The dark current is only appreciable when t_{INT} is long. It is not usually a problem for consumer applications but is a concern for some scientific applications. As the dark current fills the well, there is less room for the photoelectrons. At saturation, $n_{PE} + n_{DARK} = N_{WELL}$.

There are three main sources of dark current: 1) thermal generation in the depletion region, 2) thermal generation and diffusion in the neutral bulk material, and 3) thermal generation due to surface states. Dark current densities vary significantly among manufacturers, with values ranging from 2 pA/cm^2 to 10 nA/cm^2. Dark current due to thermally generated electrons can be reduced by cooling the device. Surface-state dark current is minimized with multi-phase pinning[18,19] (MPP). Typically, MPP devices have charge wells that are two to three times smaller than conventional devices; these devices are more likely to saturate. Some devices operate in either mode. MPP is the most common architecture used in scientific and medical applications.

The number of thermally generated dark current electrons is proportional to the dark current density J_D

$$n_{DARK} = \frac{J_D A_{PIXEL} t_{INT}}{q} \tag{6-26}$$

The pixel area is $A_{PIXEL} = d_{CCH} d_{CCV}$. Dark current is generated across the entire pixel and not just the detector. For thermal generation in the depletion region, it is proportional to

$$J_D = aT^2 e^{-\frac{E_G - E_T}{kT}} \tag{6-27}$$

The variable E_G is the band gap and E_T is the impurity energy gap. With TDI, the pixels in a TDI column are summed and the dark current increases

$$n_{DARK} = \frac{J_D A_{PIXEL} t_{INT} N_{TDI}}{q} \tag{6-28}$$

In principle, dark current density can be made negligible with sufficient cooling. At room temperature, the dark current density decreases approximately twofold for every 8 to 9°C drop in array temperature. Conversely, the dark current density increases by a factor of two for every 8 to 9°C increase in array temperature. Because the temperature is in the exponent of Equation 6-27, the doubling temperature decreases as the array temperature decreases. At low temperatures, a 1°C change in temperature changes the dark current by a significant amount. This means that if cooling is desired, the temperature must be maintained with high precision. Table 6-4 provides the dark current for Texas Instruments chip TC271 using experimentally[20] obtained values of $E_G - E_T \approx 0.58$ eV and $J_D = 100$ pA/cm^2 at 27°C. Dark current can approach[21] 3.5 electrons/pixel/s at -60°C and 0.02 electrons/pixel/h at -120°C.

The output amplifier constantly dissipates power. This results in local heating of the silicon chip. Because dark current is dependent on temperature, a small variation in the device temperature profile produces a corresponding profile in the dark current. To minimize this effect, the output amplifier is often separated by several isolation pixels to move the amplifier away from the active sensors.

Table 6-4
DARK CURRENT VALUES for TI TC271
(Normalized to 100 pA/cm^2 at 27°C)

TEMPERATURE (°C)	DARK CURRENT (pA/cm^2)	PERCENT CHANGE for 1° C	DOUBLING TEMPERATURE
60	1136	6.9	10.8
40	276	7.7	9.5
20	55.8	8.9	8.4
0	9.02	10.2	7.2
-20	1.11	11.9	6.2
-40	9.58×10^{-2}	14.1	5.3
-60	5.33×10^{-3}	17.0	4.4
-80	1.66×10^{-4}	20.9	3.6

Example 6-6
DARK CURRENT

A CCD array has a well size of 96,000 electrons, dark current density of 10 pA/cm^2, and pixel size 8 μm square. The advertised dynamic range is 80 dB. Is the dark current noise significant? Assume that the integration time is 1/60 s (16.67 ms). Using Equation 6-26, the number of dark electrons is

$$n_{DARK} = \frac{\left(\mathbf{10x10^{-12}}\,\frac{\boldsymbol{A}}{\boldsymbol{cm^2}}\right)\mathbf{(0.0167\ s)(8x10^{-4}\ cm)^2}}{\mathbf{1.6x10^{-19}}\,\frac{\boldsymbol{coul}}{\boldsymbol{electron}}} = \mathbf{835}\ \boldsymbol{e^-} \qquad (6\text{-}29)$$

Assuming Poisson statistics, the dark current shot noise is √835 or 28.9 electrons rms. The dynamic range is the charge well capacity divided by the noise floor. With a dynamic range of 10,000:1, the noise floor is 9.6 electrons. Because the noise variances add, the total noise is

$$\langle n_{SYS} \rangle = \sqrt{\mathbf{28.9^2 + 9.6^2}} = \mathbf{30.4} \quad \boldsymbol{e^- rms} \qquad (6\text{-}30)$$

Therefore, the dark current will affect the measured SNR at low light levels. If the integration was extended, the dark current will saturate the well in 1.92 s.

Example 6-7
ARRAY COOLING

The dark current density decreases by a factor of 2 for every 8°C drop in temperature. What should the array temperature be so that the dark current noise is 5 electrons rms for the array described in Example 6-6?

The dark current noise must be reduced by a factor of 5/28.9 = 0.173 and the dark current must be reduced by $(0.173)^2$ or 0.03. The doubling factor approximately provides

$$\frac{n_{DARK-COOL}}{n_{DARK-AMBIENT}} = 2^{-\left(\frac{T_{AMBIENT}-T_{COOL}}{T_{DOUBLE}}\right)} \tag{6-31}$$

where T_{DOUBLE} = 8°C. Then, $T_{AMBIENT}$ - T_{COOL} = 40.5. That is, the array temperature must be 40.5°C below the temperature at which the dark current density was originally measured. Because the dark current is reduced by 0.03, the time to saturate increased by 1/0.03 or 33.3 times to 64 sec.

Although this varies by manufacturer, many use $T_{AMBIENT}$ = 20°C. Note that this is an approximation only and Equation 6-31 should be used to determine the required temperature. In reality, the user does not precisely select the temperature but, if possible, cools the array well below the desired temperature. The lowest possible temperature is obviously limited by the cooling device.

6.4. MAXIMUM SIGNAL

The maximum signal is that input signal that saturates the charge well and is called the saturation equivalent exposure (SEE). It is

$$SEE = \frac{V_{MAX}}{R_{AVE}} \quad \frac{\mathrm{J}}{\mathrm{cm}^2} \tag{6-32}$$

The well size varies with architecture, number of phases, and pixel size. If an antibloom drain is present, the maximum level is taken as the white clip level (Figure 3-35). Referring to Figure 6-14, the maximum value of V_{MAX} is

$$V_{MAX} = G_1 \frac{Gq}{C}(N_{WELL} - n_{DARK}) \tag{6-33}$$

For back-of-the-envelope calculations, the dark current is considered negligible ($n_{DARK} \approx 0$).

6.5. NOISE

Many books and articles[1,2,22-28] have been written on noise sources. The level of detail used in noise modeling depends on the application. Shot noise is due to the discrete nature of electrons. It occurs when photoelectrons are created and when dark current electrons are present. Additional noise is added when reading the charge (reset noise) and introduced by the amplifier (1/f noise and white noise). If the output is digitized, the inclusion of quantization noise may be necessary. Switching transients coupled through the clock signals also appear as noise. It can be minimized by good circuit design.

Figure 6-13 illustrated the signal transfer diagram. The noise transfer diagram is somewhat different (Figure 6-18). With different transfer functions, both the optical and electronic subsystems must be considered to maximize the system signal-to-noise ratio.

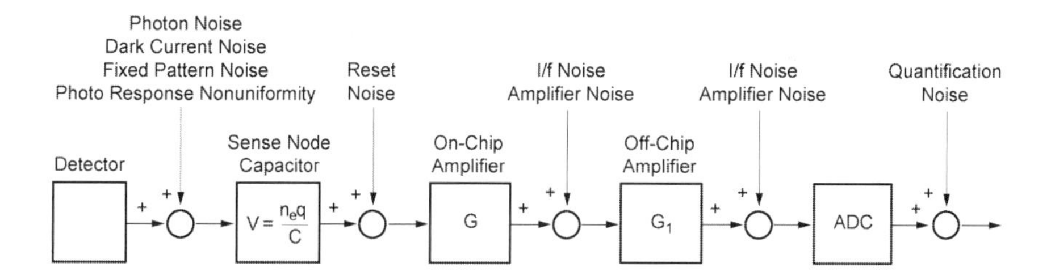

Figure 6-18. The various subsystems are considered as ideal elements with the noise introduced at appropriate locations.

Although the origins of the noise sources are different, they all appear as variations in the image intensity. Figure 6-19a illustrates the ideal output of the on-chip amplifier. The array is viewing a uniform source and the output of each pixel is identical.

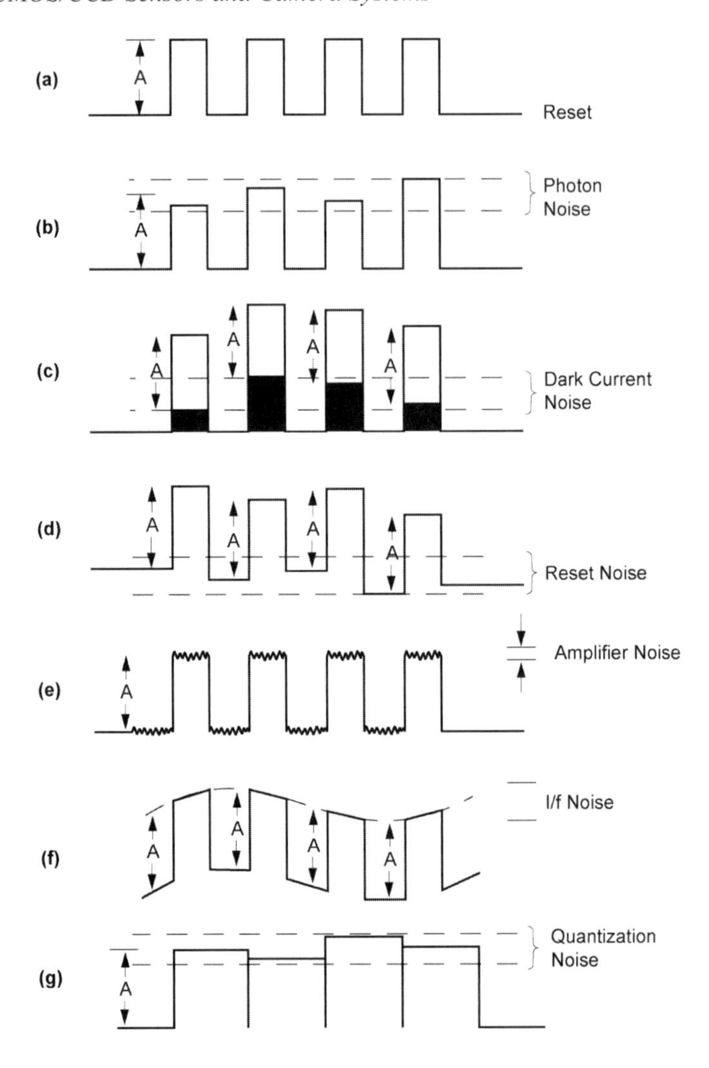

Figure 6-19. Noise sources affect the output differently. (a) though (f) represent the output after the on-chip amplifier. (g) is the output after the sample-and-hold circuitry. The array is viewing a uniform source that produces identical outputs (amplitude = A) for each pixel. (a) ideal output, (b) photoelectron shot noise, (c), dark current shot noise, (d) reset noise, (e) amplifier noise, (f) amplifier 1/f noise, and (g) quantization noise. All these processes occur simultaneously.

Under ideal conditions, each pixel would have the same dark current; this value is subtracted from all pixels leaving the desired signal. However, dark current also exhibits fluctuations. Even after subtracting the average value, these fluctuations remain and create fixed pattern noise (Figure 6-19c). The output capacitor is reset after each charge is read. Errors in the reset voltage appear as output signal fluctuations (Figure 6-19d). The amplifier adds white noise (Figure 6-19e) and 1/f noise (Figure 6-19f). Amplifier noise affects both the active video and reset levels. The analog-to-digital converter introduces quantization noise (Figure 6-19g). Quantization noise is apparent after image reconstruction. Only photoelectron shot noise and amplifier noise affect the amplitude of the signal. With the other noise sources listed, the signal amplitude, $\boldsymbol{A}$, remains constant. However, the value of the output fluctuates with all the noise sources. Because this value is presented on the display, the processes appear as displayed noise. These noise patterns change from pixel to pixel and on the same pixel from frame to frame.

Pattern noise refers to any spatial pattern that does not change significantly from frame to frame. Dark current varies from pixel to pixel; this variation is called fixed pattern noise (FPN). FPN is due to differences in detector size, doping density, and foreign matter getting trapped during fabrication. Photoresponse nonuniformity (PRNU) is the variation in pixel responsivity and is seen when the device is illuminated. This noise is due to differences in detector size, spectral response, and thickness in coatings.

Pattern "noises" are not noise in the usual sense. PRNU occurs when each pixel has a different average value. This variation appears as spatial noise to the observer. Frame averaging will reduce all the noise sources except FPN and PRNU. Although FPN and PRNU are different, they are sometimes collectively called scene noise, pixel noise, pixel nonuniformity, or simply pattern noise. They are minimized by off-chip look-up tables.

It is customary to specify all noise sources in units of equivalent electrons at the detector output. When quoting noise levels, it is understood that the noise magnitude is the rms of the random process producing the noise. Noise powers are considered additive. Equivalently, the noise sources are RSSed (root sum of the squares) or added in quadrature:

$$\langle \boldsymbol{n}_{SYS} \rangle = \sqrt{\langle \boldsymbol{n}_1^2 \rangle + \ldots + \langle \boldsymbol{n}_i^2 \rangle + \ldots + \langle \boldsymbol{n}_N^2 \rangle} \tag{6-34}$$

where $\langle \boldsymbol{n}_i^2 \rangle$ is the noise variance for source $\boldsymbol{i}$ and $\surd\langle \boldsymbol{n}_i^2 \rangle = \langle \boldsymbol{n}_{\mathrm{i}} \rangle$ is the standard deviation measured in rms units.

For system analysis, it may be sufficient to consider only the on-chip amplifier noise. The array manufacturer usually provides this value and may call it readout noise, mux noise, noise equivalent electrons, or the noise floor. The value varies by device and manufacturer. Reset noise can be reduced to a negligible level with CDS. CDS can also reduce the source follower 1/f noise. The off-chip amplifier is usually a low-noise amplifier such that its noise is small compared to the on-chip amplifier noise. Although always present, quantization noise is reduced by selecting the appropriated sized analog-to-digital converter (e.g., selecting an ADC with many quantization levels). Dark current is reduced through cooling. The noise model is

$$\langle n_{SYS} \rangle = \sqrt{\langle n_{SHOT}^2 \rangle + \langle n_{RESET}^2 \rangle + \langle n_{FLOOR}^2 \rangle + \langle n_{ADC}^2 \rangle + \langle n_{PATTERN}^2 \rangle} \quad (6\text{-}35)$$

The noise model does not include effects such as banding and streaking, which are forms of pattern noise found in linear and TDI arrays. Banding can occur with arrays that have multiple outputs serviced by different nonlinear amplifiers (see Figure 3-14). Streaking occurs when the average responsivity changes from column to column in TDI devices.[29] Only a complete signal-to-noise ratio analysis can determine which noise source dominates.

6.5.1. SHOT NOISE

Both photoelectrons and dark current contribute to shot noise. Using Poisson statistics, the variance is equal to the mean:

$$\langle n_{SHOT}^2 \rangle = \langle n_{PE}^2 \rangle + \langle n_{DARK}^2 \rangle = n_{PE} + n_{DARK} \quad (6\text{-}36)$$

These values should be modified by the charge transfer efficiency, ε:

$$\langle n_{SHOT}^2 \rangle = \langle n_{PE}^2 \rangle + \langle n_{DARK}^2 \rangle = \varepsilon^N n_{PE} + \varepsilon^N n_{DARK} \quad (6\text{-}37)$$

Because the CTE is high ($\varepsilon^N \approx 1$), it is usually omitted from most equations. However, CTE should be included for very large arrays. With TDI, the number of photoelectrons and dark current electrons increases with the number of TDI elements, N_{TDI}

$$\langle n_{SHOT}^2 \rangle = N_{TDI} n_{pe} + N_{TDI} n_{DARK} \quad (6\text{-}38)$$

6.5.2. RESET NOISE

The noise associated with resetting the sense node capacitor is often called kTC noise. This is due to thermal noise generated by the resistance, $\boldsymbol{R}$, within the resetting FET. The Johnson noise current variance is

$$\langle i_{RESET}^2 \rangle = \frac{4kT}{R} \Delta f_e \tag{6-39}$$

Because the resistance is in parallel with the sense node capacitor, the noise equivalent bandwidth is $\Delta f_e = RC/4$. Then

$$\langle i_{RESET}^2 \rangle = kTC \tag{6-40}$$

When expressed as equivalent electrons, the rms noise is

$$\langle n_{RESET} \rangle = \frac{\sqrt{kTC}}{q} \quad e^{-1}\ rms \tag{6-41}$$

The value $\langle n_{RESET} \rangle$ represents the uncertainty in the amount of charge remaining on the capacitor following reset.

A 0.2 pf capacitor produces about 126 e^- rms kTC noise at room temperature and a 0.01 pF capacitor produces about 40 electrons rms. Reducing the capacitance reduces the noise. This has an added benefit that the device output gain conversion, Gq/C, increases. While cooling reduces this noise source, cooling is used primarily to reduce dark current. That is, cooling reduces dark current noise exponentially whereas reset noise is reduced only by $\sqrt{T}$. kTC noise may be significantly reduced with correlated double sampling. It is therefore neglected in most analyses.

6.5.3. ON-CHIP AMPLIFIER NOISE

Amplifier noise consists of two components: 1/f noise and white noise. If f_{KNEE} is the frequency at which the 1/f noise equals the white noise, then the amplifier noise density is

$$V_{ON-CHIP} = V_{AMP-NOISE}\left(1 + \frac{f_{KNEE}}{f}\right) \quad \frac{\text{rms volts}}{\sqrt{\text{Hz}}} \tag{6-42}$$

The 1/f noise can be minimized[2] through correlated double sampling. When discussing individual noise sources, noise is usually normalized to unit bandwidth. When modified by electronic subsystems, the total system noise power is

$$\sigma_{SYS}^2 = \int_0^\infty S(f_e) \left| H_{SYS}(f_e) \right|^2 df_e \quad \text{volts} \tag{6-43}$$

where $S(f_e)$ is the total noise power spectral density from all sources. The subscript e is used to emphasize that this is electrical frequency (Hz). $H_{SYS}(f_e)$ is the frequency response of the system electronics. The noise equivalent bandwidth is that bandwidth with unity value that provides the same total noise power. Ignoring 1/f noise, the noise equivalent bandwidth is

$$\Delta f_{elec} = \frac{\int_0^\infty S(f_e) \left| H_{SYS}(f_e) \right|^2 df_e}{S_o} \tag{6-44}$$

The bandwidth Δf_e applies only to those noise sources that are white and cannot be applied to 1/f noise. Although common usage has resulted in calling Δf_e the noise *bandwidth*, it is understood that it is a power equivalency.

If t_{CLOCK} is the time between pixels, the bandwidth for an *ideal* sampled-data system is

$$\Delta f_{elec} = \frac{1}{2t_{CLOCK}} = \frac{f_{CLOCK}}{2} \tag{6-45}$$

When equated to electrons,

$$\langle n_{ON-CHIP} \rangle = \frac{C}{Gq} V_{ON-CHIP-AMP-NOISE} \sqrt{\Delta f_e} \quad e^- \, rms \tag{6-46}$$

As the clock frequency increases (for faster readout, t_{CLOCK} decreases), $\langle n_{ON-CHIP} \rangle$ increases. This is the noise floor (supplied by the manufacturer) and will be labeled as $\langle n_{FLOOR} \rangle$.

6.5.4. OFF-CHIP AMPLIFIER NOISE

The off-chip amplifier noise is identical in form to the on-chip amplifier noise (Equation 6-42). The 1/f knee value may be different for the two amplifiers. If f_{KNEE} is small, the equivalent number of noise electrons is

$$\langle n_{OFF-CHIP} \rangle = \frac{C}{GG_1 q} V_{OFF-CHIP-AMP-NOISE} \sqrt{\Delta f_e} \quad e^- \; rms \,. \qquad (6\text{-}47)$$

The amplifier gain G_1 is usually sufficiently high that off-chip amplifier noise can be neglected.

6.5.5. QUANTIZATION NOISE

The analog-to-digital converter produces discrete output levels. A range of analog inputs can produce the same output. This uncertainty, or error, produces an effective noise given by

$$V_{NOISE} = \frac{V_{LSB}}{\sqrt{12}} \quad \text{V rms} \qquad (6\text{-}48)$$

where V_{LSB} is the voltage associated with the least significant bit. For an ADC with N bits, $V_{LSB} = V_{MAX}/2^N$. When expressed in equivalent electrons,

$$\langle n_{ADC} \rangle = \frac{C}{GG_1 q} \frac{V_{LSB}}{\sqrt{12}} \quad e^- \; \text{rms} \qquad (6\text{-}49)$$

When the ADC is matched to the amplifier output, V_{MAX} corresponds to the full well. Then

$$\langle n_{ADC} \rangle = \frac{N_{WELL}}{2^N \sqrt{12}} \qquad (6\text{-}50)$$

Ideally, $\langle n_{ADC} \rangle$ is less than the noise floor. This is achieved by selecting a high resolution ADC (large N).

Example 6-8
TOTAL NOISE

An array has a noise floor of 20 electrons rms, well capacity of 50,000 electrons, and an 8-bit analog-to-digital converter. Neglecting pattern noise, what is the total noise?

The LSB represents $50{,}000/2^8 = 195$ electrons. The quantization noise is $195/\sqrt{12}$ or 56.3 electrons rms. The total noise is

$$\langle n_{SYS} \rangle = \sqrt{\langle n_{SHOT}^2 \rangle + \langle n_{FLOOR}^2 \rangle + \langle n_{ADC}^2 \rangle}$$
$$= \sqrt{\mathbf{50000 + 20^2 + 56.3^2} = \mathbf{231}} \quad e^- \ rms \tag{6-51}$$

With a full well, the quantization noise seems insignificant compared to the shot noise. But as the photoelectron intensity decreases, the shot noise decreases. The ADC noise is greater than the noise floor. This suggests that a 12-bit (4096 levels) should be used.

6.5.6. PATTERN NOISE

Fixed pattern noise (FPN) refers to the pixel-to-pixel variation[30,31] that occurs when the array is in the dark. It is primarily due to dark current differences. It is a signal-independent noise and is additive to the other noise powers. As a multiplicative noise, FPN is traditionally expressed as a fraction of the total number of dark current carriers. If U_{DARK} is the fixed pattern ratio, then

$$\langle n_{FPN} \rangle = U_{FPN} n_{DARK} \tag{6-52}$$

For well-designed arrays, $U_{FPN} \approx 1\%$.

PRNU is due to differences in responsivity (when light is applied). It is a signal-dependent noise and is a multiplicative factor of the photoelectron number. PRNU is specified as a peak-to-peak value or an rms value referenced to an average value. This average value may either be full well or one-half full well value. That is, the array is uniformly illuminated and a histogram of responses is created. The PRNU can either be the rms of the histogram divided by the average value or the peak-to-peak value divided by the average value.

Since the definition varies by manufacturer, the test conditions must be understood when comparing arrays. For this text, the rms value is used.

Because dark current becomes negligible by cooling the array, PRNU is the dominant pattern component for most arrays. Similarly, PRNU is expressed as a fraction of the total number of charge carriers. This approach assumes that the detectors are operating in a linear region with the only difference being responsivity differences. If U_{PRNU} is the fixed pattern ratio or nonuniformity, then

$$\langle n_{PRNU} \rangle = U_{PRNU} n_{PE} \tag{6-53}$$

For well-designed arrays, $U_{PRNU} \approx 1\%$. As with shot noise, ε^N should be added to the equation. Because $\varepsilon^N \approx 1$ for modest sized arrays, it is omitted from most equations. TDI devices, through their inherent averaging, reduce FPN and PRNU by $1/N_{TDI}$. PRNU is usually supplied by the manufacturer. Off-chip gain/level correction algorithms can minimize FPN and PRNU. For system analysis, the corrected pattern noise value is used. The pattern noise is

$$\langle n_{PATTERN} \rangle = \sqrt{\langle n_{FPN}^2 \rangle + \langle n_{PRNU}^2 \rangle} = \sqrt{(U_{FPN} n_{PE})^2 + (U_{DARK} n_{DARK})^2} \tag{6-54}$$

In Figure 6-20, only PRNU and photoelectron shot noise is plotted. The total noise value increases when PRNU is excessive. Because cost increases as the PRNU decreases, the optimum PRNU occurs at the knee of the curve. This occurs when the photoelectron shot noise is approximately equal to PRNU or $U_{PRNU} \approx 1/\sqrt{n_{PE}}$. For worst case analysis, the charge well capacity should be used for n_{PE}. Dark current and FPN will provide the same shaped curves as Figure 6-20. What is important is the relationship between $\langle n_{DARK} \rangle$, $\langle n_{PE} \rangle$, and $\langle n_{FLOOR} \rangle$.

For CCD arrays, the output circuit (electrometer) is at the edge of the CCD chip where it can occupy an area that is much larger than a pixel. This increased real estate allows special two-stage source followers circuits to be used, which in turn enable extremely linear photoresponse (small U_{PRNU}).

The situation for a CMOS pixel is very different. The electrometer must fit within the boundaries of a pixel. Since the photodiode junction capacitance (and hence the integration node capacitance) changes with signal level, the output response is nonlinear (see Equation 4-1). The second source of nonlinearity is due to the combination of the source follower and row or column select transistor. The amount of nonlinearity is very design specific and depends on parasitic capacitance effects, operating bias conditions and associated current levels, and other semiconductor circuit details. However, while the nonlinearity of CMOS will not be as good as a CCD, the nonlinearity will be very consistent from pixel to pixel because of the precision photolithography. As such, it can be easily corrected. Less expensive devices (e.g., mobile applications) may not employ nonuniformity correction and PRNU may be bothersome in these devices.

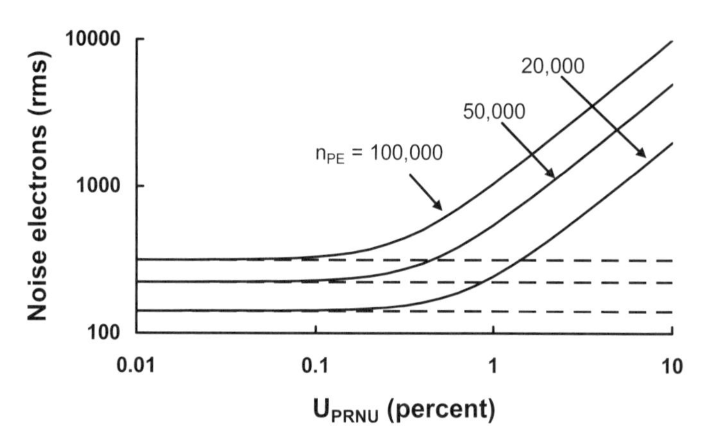

Figure 6-20. Noise as a function of U_{PRNU} for three signal levels. Shot noise is independent of PRNU (horizontal dashed lines). The value $1/\sqrt{n_{PE}}$ is 0.7%, 0.4% and 0.3% when n_{PE} is 20,000, 50,000 and 100,000 electrons, respectively. These values represent the "knee" in the curve.

6.5.7. SUPER CCD NOISE

The noise sources described in the preceding sections apply to the Super CCD. For the Super CCD SR, the two noise variances are added

$$\langle n_{SUPER\ CCD\ SR} \rangle = \sqrt{k_{HIGH}^2 \langle n_{HIGH}^2 \rangle + k_{LOW}^2 \langle n_{LOW}^2 \rangle} \qquad (6\text{-}55)$$

where $\langle n_{HIGH} \rangle$ is the shot noise and pattern noise associated with the high sensitivity detector and $\langle n_{LOW} \rangle$ is the noised in the low-sensitivity detector. When $k_{HIGH} = k_{LOW} = 0.5$, the system averages the two signals (Figure 6-21). While this appears to reduce noise, the signal is also reduced (Figure 6-16).

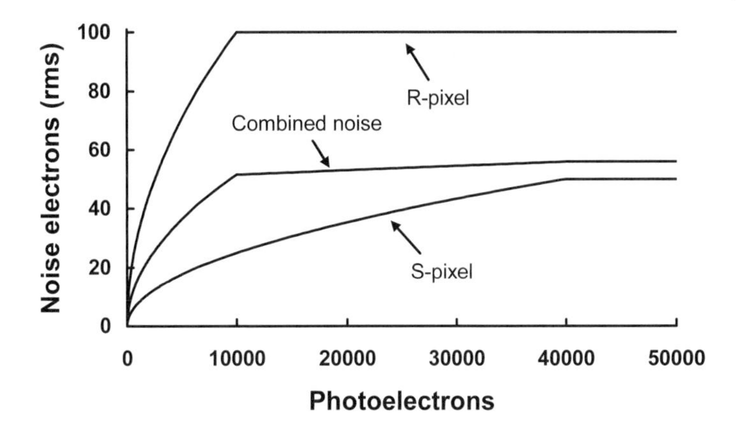

Figure 6-21. Super CCD SR combined photoelectron shot noise. The R-pixel charge well was arbitrarily selected to hold 10,000 electrons As $k_{HIGH} \rightarrow 1$, the combined noise approaches R-pixel noise. As $k_{HIGH} \rightarrow 0$, the combined noise approaches S-pixel noise.

6.5.8. EMCCD NOISE

The gain process in low-light imaging systems introduces excess noise and is defined by a noise factor

$$k_{EMCCD} = \frac{\langle n_{AFTER\ GAIN} \rangle}{M \langle n_{BEFORE\ GAIN} \rangle} \tag{6-56}$$

The overall gain depends upon the number of multiplying cells and the gain per cell: $M = (g)^N$. If the gain does not add noise, there is no excess noise ($k_{EMCCD} = 1$). The excess noise factor is[32]

$$k_{EMCCD}^2 \approx 2\frac{(M-1)}{M^{(N+1)/N}} + \frac{1}{M} \tag{6-57}$$

For large N (which creates a large M), $k_{EMCCD} \approx 1.4$ (Figure 6-22). Since the gain is before the sense node and amplifier, the output noise is

$$\langle n_{EMCCD} \rangle = \sqrt{k_{EMCCD}^2 M^2 (n_{PE} + n_{DARK}) + k_{EMCCD}^2 M^2 \langle n_{PATTERN}^2 \rangle + \langle n_{FLOOR}^2 \rangle} \tag{6-58}$$

Since the dark current noise is also amplified, it is desirable to cool EMCCDs to maximize the SNR. As an added benefit, the gain increases as the temperature decreases (Figure 3-42). The gain should be sufficiently large to minimize the noise floor contribution. If the amplified shot noise is 10 times the noise floor, then the required gain is

$$M > \sqrt{5\frac{\langle n_{FLOOR}^2 \rangle}{n_{PE}}} \tag{6-59}$$

As the scene intensity decreases (low light level conditions), $\boldsymbol{M}$ must increase. As shown in Figure 6-23, significant gain is required when the noise floor is large and the number of photoelectrons is small.

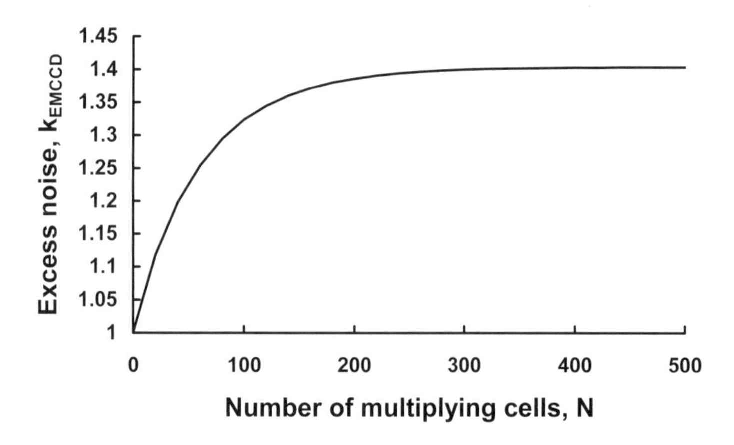

Figure 6-22. Excess noise as a function of $\boldsymbol{N}$ when $\boldsymbol{g} = 1.015$

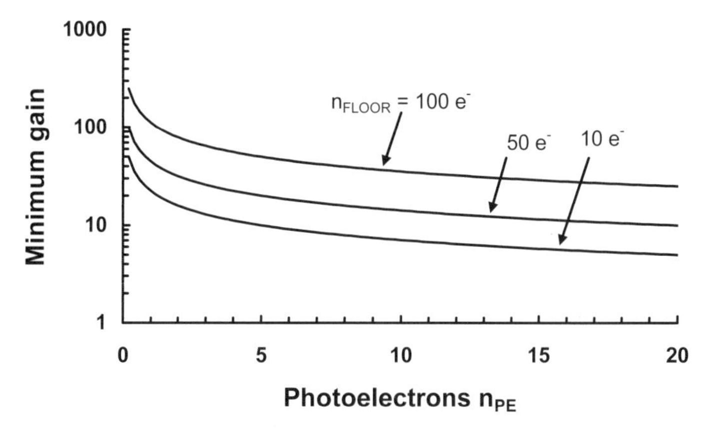

Figure 6-23. Required gain to minimize the readout noise contribution to the total noise.

6.5.9. ICCD NOISE

The photocathode produces both photoelectron noise and dark noise which is present both in ICCDs and EBCCDs

$$\left\langle n_{PC-SHOT}^2 \right\rangle = \left\langle n_{CATHODE}^2 \right\rangle + \left\langle n_{PC-DARK}^2 \right\rangle \tag{6-60}$$

The MCP amplifies these signals and adds excess noise so that

$$\left\langle n_{MCP}^2 \right\rangle = k_{MCP}^2 G_{MCP}^2 \left\langle n_{PC-SHOT}^2 \right\rangle \tag{6-61}$$

where $\boldsymbol{k_{MCP}}$ is typically taken as 1.8 but varies with bias voltage.[33,34] The value $\boldsymbol{k_{MCP}}$ is also reported as an SNR degradation. The noise power is transformed into photoelectrons by the phosphor screen

$$\left\langle n_{SCREEN}^2 \right\rangle = \eta_{SCREEN}^2 \left\langle n_{MCP}^2 \right\rangle \tag{6-62}$$

It is attenuated by the fiberoptic bundle

$$\left\langle n_{CCD-PHOTON}^2 \right\rangle = T_{fo}^2 \left\langle n_{SCREEN}^2 \right\rangle \tag{6-63}$$

This value is converted to electrons by the CCD and the CCD noise is added

$$\left\langle n_{SYS}^2 \right\rangle = \eta_{CCD}^2 \left\langle n_{CCD-PHOTON}^2 \right\rangle + \left\langle n_{DARK}^2 \right\rangle + \left\langle n_{FLOOR}^2 \right\rangle + \left\langle n_{PATTERN}^2 \right\rangle \tag{6-64}$$

The total noise is

$$\left\langle n_{SYS}^2 \right\rangle = \eta_{CCD}^2 T_{fo}^2 \eta_{SCREEN}^2 k_{MCP}^2 G_{MCP}^2 \left\langle n_{PC-SHOT}^2 \right\rangle + \left\langle n_{CCD-DARK}^2 \right\rangle + \left\langle n_{FLOOR}^2 \right\rangle + \left\langle n_{PATTERBN}^2 \right\rangle \quad (6\text{-}65)$$

6.6. DYNAMIC RANGE

Dynamic range is defined as the maximum signal (peak) divided by the rms noise. If an antibloom drain is present, the knee value is used as the maximum (Figure 3-35). Expressed as a ratio, the dynamic range is

$$DR_{ARRAY} = \frac{SEE}{NEE} = \frac{V_{MAX}}{V_{NOISE}} \quad (6\text{-}66)$$

And when expressed in decibels,

$$DR_{ARRAY} = 20\log\left(\frac{SEE}{NEE}\right) \text{ dB} \quad (6\text{-}67)$$

When equated to electrons,

$$DR_{ARRAY} = \frac{N_{WELL} - n_{DARK}}{\left\langle n_{SYS} \right\rangle} \quad (6\text{-}68)$$

Often, only the noise floor is considered in $\langle n_{SYS} \rangle$. The noise is dependent upon amplifier design and is not related to charge well capacity.

Sometimes the manufacturer will specify dynamic range by the peak signal divided by the peak-to-peak noise. This peak-to-peak value includes any noise spikes and, as such, may be more useful in determining the lowest detectable signal in some applications. If the noise is purely Gaussian, the peak-to-peak is assumed to be six times the rms value. This multiplicative factor is author dependent.

If the system noise is less than the ADC's least significant bit (LSB), then the quantization noise limits the system dynamic range. The rms quantization noise, $\langle n_{ADC} \rangle$ is $V_{LSB}/\sqrt{12}$. The dynamic range is $2^N\sqrt{12}$. For a 12-bit ADC, the dynamic range is 14,189:1 or 83 dB. An 8-bit system cannot have a dynamic range greater than 887:1 or 59 dB.

The camera dynamic range is determined by the maximum signal, minimum signal, and the noise. This is not related to the scene dynamic range. When bright light is present, the scene dynamic range (intrascene dynamic range) can be much greater than the camera dynamic range. Here, darker scene components will fall into a gray level and not be perceptible in the image. The most difficult part of camera design is matching the camera dynamic range to the anticipated scene dynamic range. It the camera has more then 8 bits, it must be cleverly downsampled to match displays that are usually support only 8 bits.

Example 6-9
ARRAY PARAMETERS

An array has a dynamic range of 3000:1, average responsivity of 2 V/(μJ/cm^2), and a saturation equivalent exposure of 150 nJ/cm^2. These values apply only at λ_P. What are the remaining design parameters?

Maximum output:
The maximum output is $\boldsymbol{R_{AVE}}$ $\boldsymbol{SEE}$ = 300 mV.
Noise level:
The noise floor rms value is equal to the maximum output divided by the dynamic range or 100 μV rms.
Analog-to-digital Converter:
For a digital output, the ADC must have 12 bits (2^{12} = 4096 levels) to span the array's full dynamic range. The responsivity in DN is $2^{12}/\boldsymbol{SEE}$ = 27.3 DN/(nJ/cm^2) at λ_P.

Example 6-10
DIGITAL DYNAMIC RANGE

A device with a well capacity of 50,000 electrons has a noise floor of 60 electrons rms. What size analog-to-digital converter should be used?

The array dynamic range is 50,000/60 = 833 but the smallest available ADC has 1024 levels or 10 bits. The LSB is 48.8 electrons and the quantization noise is 14 e$^-$ rms. This ADC devotes 1.23 bits to array noise. The system noise is

$$\langle n_{SYS} \rangle = \sqrt{\langle n^2_{FLOOR} \rangle + \langle n^2_{ADC} \rangle} = 61.6\, e^- \; rms \qquad (6\text{-}69)$$

The ADC has minimal effect on system noise.

6.7. PHOTON TRANSFER and MEAN-VARIANCE

Either the rms noise or noise variance can be plotted as a function of signal level. The graphs are called the photon transfer curve and the mean-variance curve, respectively. Both graphs convey the same information. They provide[1,2,35,36] array noise and saturation level from which the dynamic range can be calculated. While the description that follows was developed for CCD chip characterization, it also applies to EMCCDs.[37]

For very low photon fluxes, the noise floor dominates. As the incident flux increases, the photoelectron shot noise dominates. Finally, for very high flux levels, the noise may be dominated by PRNU (assuming dark current and reset noise are negligible). As the signal approaches well saturation, the noise plateaus and then drops abruptly at saturation (Figure 6-24).

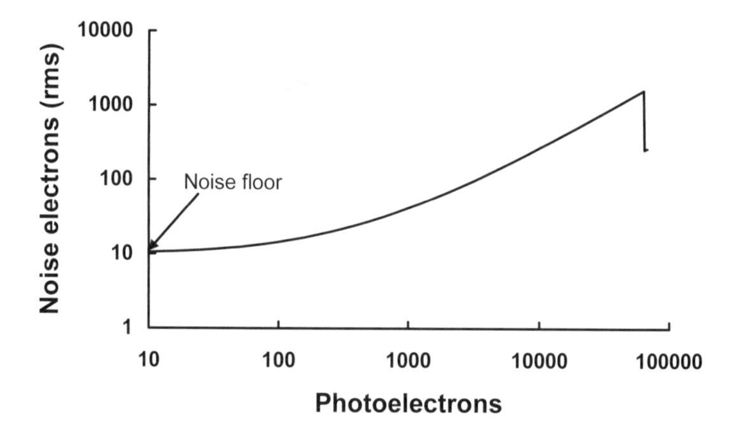

Figure 6-24. Photon transfer curve when $n_{DARK} = 0$.

As the charge well reaches saturation, electrons are more likely to spill into adjoining wells (blooming) and, if present, into the overflow drain. As a result, the number of noise electrons starts to decrease. Figure 6-25 illustrates the rms noise as a function of photoelectrons when the dynamic range is 60 dB. Here dynamic range is defined as $N_{WELL}/\langle n_{FLOOR} \rangle$. With large signals and small PRNU, the total noise is dominated by photoelectron shot noise. When PRNU is large, the array noise is dominated by U_{PRNU} at high signal levels.

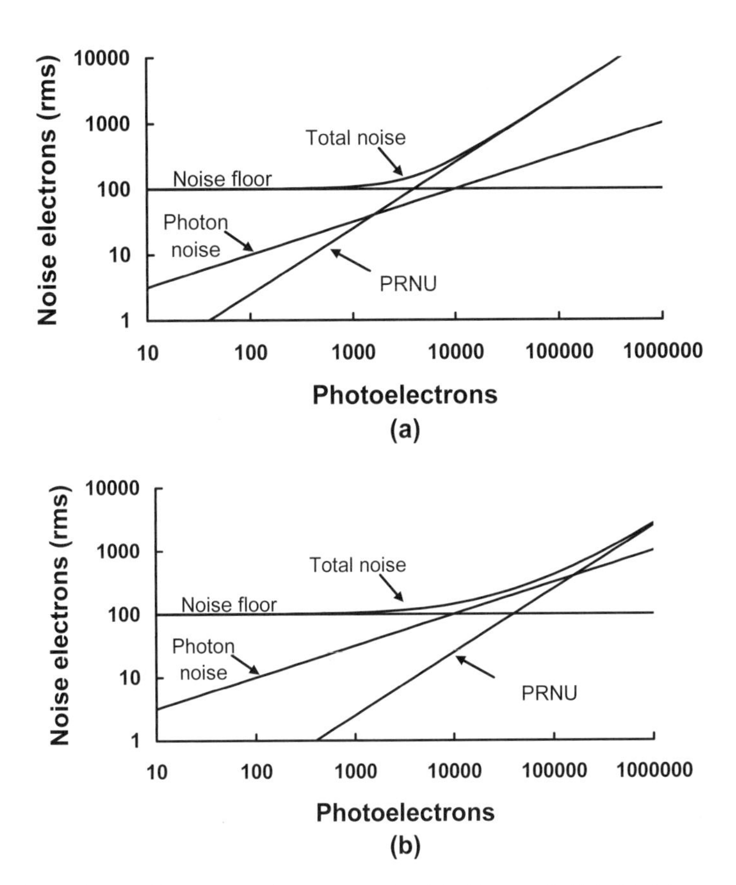

Figure 6-25. Photon transfer curves for (a) U_{PRNU} = 2.5% (b) U_{PRNU} = 0.25%. The charge well capacity is 100,000 electrons and the noise floor is 100 e^- rms to produce a dynamic range of 60 dB. The noise floor, photoelectron shot noise, and PRNU have slopes of 0, 0.5, and 1, respectively. Dark noise is considered negligible. The drop at saturation is not shown.

Dark current shot noise only affects those applications where the signal-to-noise ratio is low (Figure 6-26). At high illumination levels, either photoelectron shot noise or pattern noise dominates. Because many scientific applications operate in a low signal environment, cooling will improve performance. General video and industrial cameras tend to operate in high signal environments and cooling will have little effect on performance. A full SNR analysis is required before selecting a cooled camera.

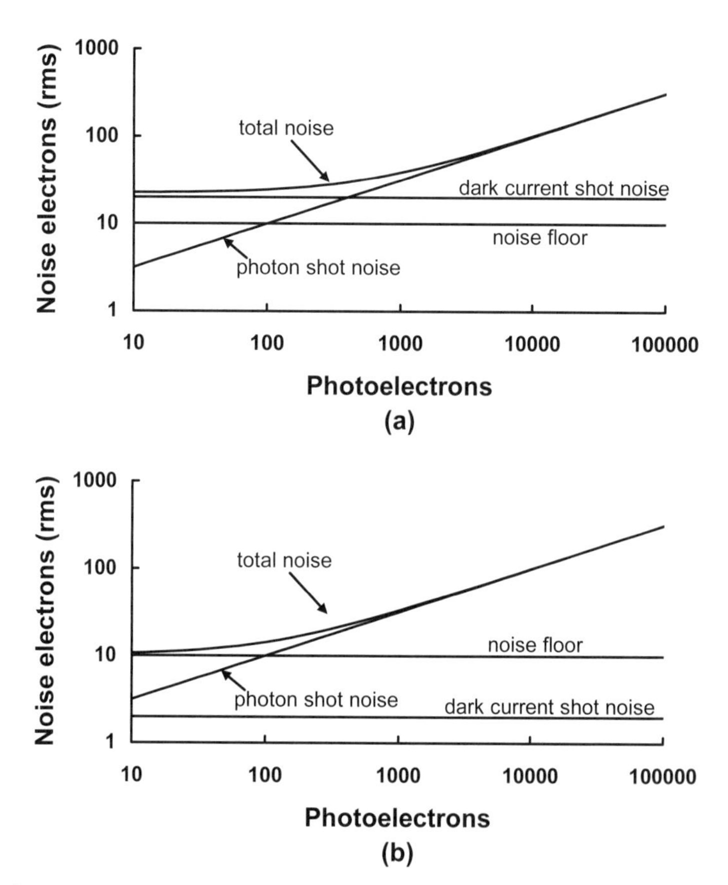

Figure 6-26. Dark current shot noise effects. The noise floor is 10 electrons rms, and PRNU and FPN is zero. (a) Dark current shot noise is 20 e^- rms and (b) dark current shot noise is 2 e^- rms. For large signals, photoelectron shot noise dominates the array noise. The drop at saturation is not shown.

Figure 6-27 illustrates the mean-variance technique. In the shot noise limited region

$$V_{SIGNAL} = \frac{Gq}{C} n_{PE} \tag{6-70}$$

and

$$\left\langle V_{NOISE}^2 \right\rangle = \left(\frac{Gq}{C} \right)^2 n_{PE} \tag{6-71}$$

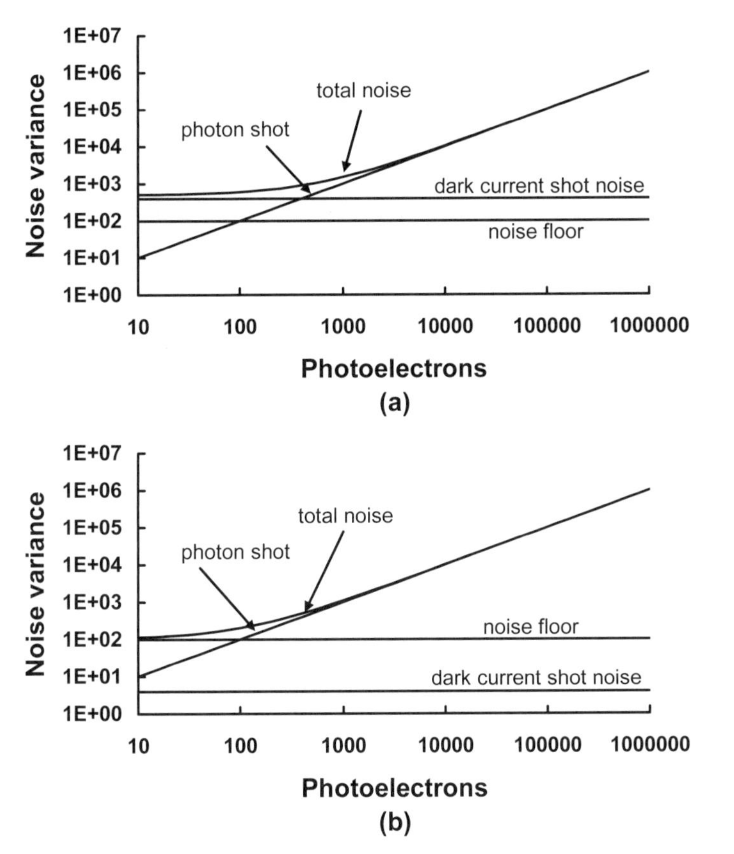

Figure 6-27. Mean-variance plot. (a) corresponds to Figure 6-26a. (b) corresponds to Figure 6-26b.

Then the output gain conversion is

$$\frac{Gq}{C} = \frac{\langle V_{NOISE}^{2} \rangle}{V_{SIGNAL}} \tag{6-72}$$

or the OGC is equal to the noise variance divided by the mean value. It is simply the slope of the photon shot noise in Figure 6-27. If both signal and noise are measured after the on-chip amplifier, the measured photogenerated signal is $V_{SIGNAL} = G\,q\,n_{PE}/C$ and the noise is

$$\langle V_{NOISE} \rangle = \frac{Gq}{C} \langle n_{SYS} \rangle \tag{6-73}$$

When $V_{NOISE} = 1$, the signal is C/Gq. The shot noise intercept on the photon transfer curve is 1/OGC (Figure 6-28). The OGC has units of μV/electrons. Similarly, if the values are collected after the ADC, when the shot noise is one, the shot noise signal provides the output gain conversion in units of DN/electrons.

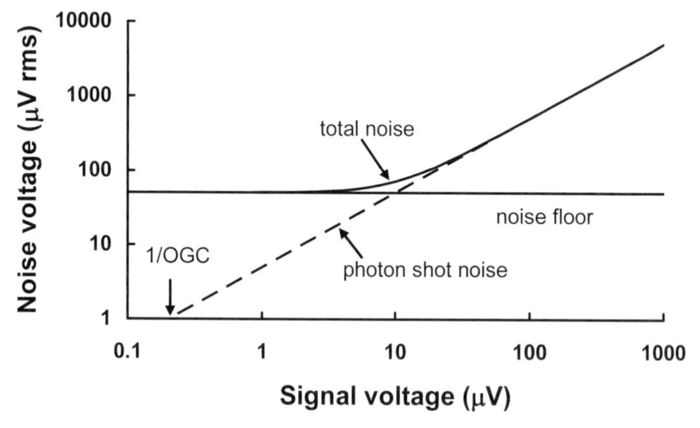

Figure 6-28. Photon transfer curve. The noise floor is 10 e⁻ and the OGC is 5 μV/e⁻. The drop at saturation is not shown.

6.8. ARRAY SIGNAL-TO-NOISE RATIO

SNR = 1 is an industry standard for defining system performance. It is also the value where a target can be detected 50% of the time. Machine vision systems require a higher SNR, and the value depends on image processing algorithm. For most machine vision applications, the SNR is very high. The human eye provides tremendous spatial and temporal integration. The eye can detect well-defined targets (e.g., alphanumerics, simple shapes) when the SNR is less than 0.1. Here the imagery is definitely noisy but detail can be discerned. Image intensifiers, ICCDs, EMCCDs, and EBCCDs are operated in this low light level region. For video and digital still cameras, the SNR must be greater than about 10 to produce good imagery with SNR = 40 producing excellent imagery. The SNR is for an extended source (covers many pixels). As the target size decreases, the MTF becomes important (discussed in Chapter 9, *MTF*).

The array signal-to-noise ratio is

$$SNR = \frac{n_{PE}}{\sqrt{\langle n^2_{SHOT}\rangle + \langle n^2_{FLOOR}\rangle + \langle n^2_{PATTERN}\rangle + \langle n^2_{ADC}\rangle + \langle n^2_{RESET}\rangle}} \tag{6-74}$$

6.8.1. CCD, CMOS, and SUPER CCD SNR

With negligible dark current shot noise, quantization, and reset noise

$$SNR = \frac{n_{PE}}{\sqrt{n_{PE} + \langle n^2_{FLOOR}\rangle + (U_{PRNU}\, n_{PE})^2}} \tag{6-75}$$

The SNR plots are similar to the photon transfer curves. With a fixed noise floor, at low flux levels, the SNR increases with the signal. At moderate levels, photoelectron shot noise limits the SNR and at high levels, the SNR approaches $1/U_{PRNU}$ (Figure 6-29).

The instantaneous dynamic range, which includes all noise sources, is the SNR. The theoretical maximum SNR is $\sqrt{N_{WELL}}$. While the actual SNR can never reach the value suggested by the dynamic range, dynamic range is used to select an appropriate analog-to-digital converter resolution (number of bits). This assures that low contrast targets can be seen.

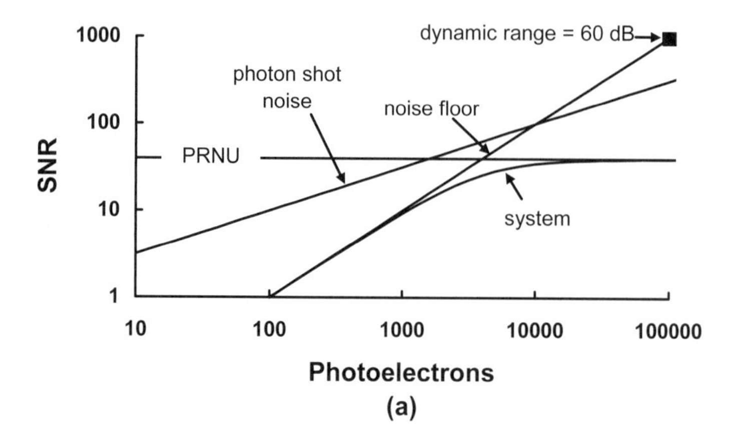

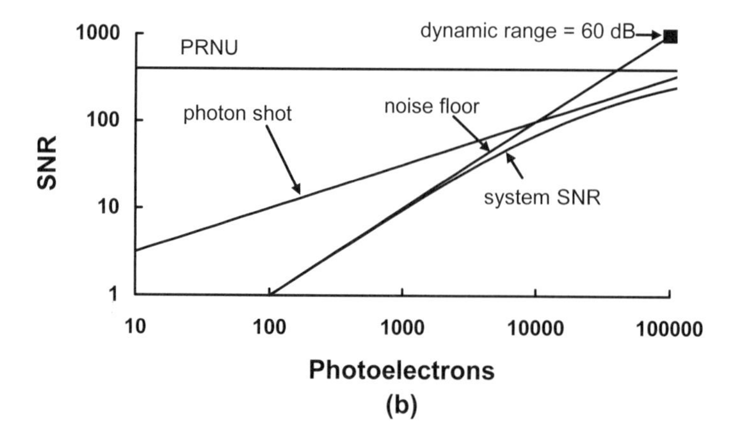

Figure 6-29. SNR as a function of noise for (a) $U = 2.5\%$ and (b) $U = 0.25\%$. $\langle n_{FLOOR} \rangle = 100\ e^-$. For most arrays, $U_{PRNU} < 1\%$. PRNU, photoelectron shot noise, and the noise floor have slopes of 0, 0.5, and 1, respectively. Dark current shot noise is considered negligible.

6.8.2. EMCCD SNR

With negligible reset and quantization noise, the EMCCD signal is amplified[37-39] by the gain to provide

$$SNR = \frac{Mn_{PE}}{\sqrt{k_{EMCCD}^2 M^2 \langle n_{SHOT}^2 \rangle + k_{EMCCD}^2 M^2 \langle n_{PATTERN}^2 \rangle + \langle n_{FLOOR}^2 \rangle}} \tag{6-76}$$

With high gain, no dark noise, no pattern noise and assuming that k_{EMCCD} = 1.4, the theoretical maximum SNR due to photoelectron shot noise only is less than a CCD by itself

$$SNR = 0.71\sqrt{n_{PE}} \tag{6-77}$$

The advantage of the EMCCD is that higher frame rates are possible. To increase the frame rate, the output bandwidth must increase and this increases the noise floor (Equation 6-43). The EMCCD gain diminishes the noise floor effect. Therefore, the EMCCD provides worse detection (lower SNR) but with much faster readout rates (trade between frame rate and image quality)[37]. While Figure 6-30 suggests that single photon counting may be possible, the ultimate limit depends on amplified clock-induced-charge (CIC).

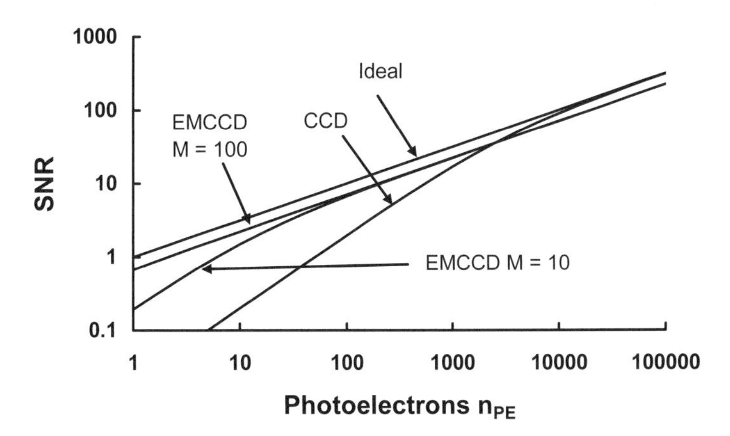

Figure 6-30. Comparison of CCD and EMCCD. $\langle n_{FLOOR} \rangle$ = 50 electrons.

CIC is the generation of spurious electrons during charge transfer. About one electron is generated for every 100 transfers. While the number of CIC electrons is small, the gain register amplifies the number and they appear as spikes in the output[14] thus mimicking single photon events. CCD, ICCD, and

EMCCD performance comparisons can be found in references 38 and 39. The EMCCD is a credible alternative to the ICCD when gating is not necessary.

6.8.3. ICCD SNR

While the image intensifier amplifies the image, the SNR may be reduced due to quantum losses and additive noise. For moderate light levels, the ICCD offers no advantage in SNR performance over the standard CCD camera. Its key attribute is the ability to gate the scene temporally.

The resultant signal-to-noise ratio is

$$SNR = \frac{\eta_{CCD}\eta_{SCREEN}T_{fo}G_{MCP}n_{CATHODE}}{\sqrt{\eta_{CCD}^2 T_{fo}^2 \eta_{SCREEN}^2 k_{MCP}^2 G_{MCP}^2 \langle n_{PC-SHOT}^2 \rangle + \langle n_{DARK}^2 \rangle + \langle n_{FLOOR}^2 \rangle + \langle n_{PRNU}^2 \rangle}} \tag{6-78}$$

Several limiting cases are of interest. When the gain is very high, the intensifier noise overwhelms the CCD noise:

$$SNR \approx \frac{n_{CATHODE}}{k_{MCP}\sqrt{\langle n_{CATHODE}^2 \rangle + \langle n_{PC-DARK}^2 \rangle}} \tag{6-79}$$

The SNR is independent of gain and CCD characteristics. Under high gain conditions, an expensive high quantum efficiency or thinned back-illuminated CCD is not necessary. Cooling the photocathode reduces the dark current

$$SNR \approx \sqrt{\frac{n_{CATHODE}}{k}} \tag{6-80}$$

When $M_{OPTICS} = 1$

$$SNR \approx \sqrt{\frac{\eta_{CATHODE} M_q(\lambda_o)\Delta\lambda}{4F^2 k_{MCP}^2} t_{INT} T_{OPTICS} A_{PIXEL}} \tag{6-81}$$

High gain is used for photon counting applications. SNR, in this case, is limited by the photocathode quantum efficiency, the excess noise factor, and the optical system.

Under moderate gain situations where the CCD noise is greater than the intensifier noise

$$SNR = \frac{\eta_{CCD}\eta_{SCREEN}T_{fo}G_{MCP}n_{CATHODE}}{\sqrt{\langle n_{SHOT}^2 \rangle + \langle n_{FLOOR}^2 \rangle + \langle n_{PRNU}^2 \rangle}} \tag{6-82}$$

Comparing this with Equation 6-75, the ICCD will have lower SNR than a comparable standard camera. The advantage of the ICCD under these conditions is the ability to gate the incoming photon flux temporally.

Example 6-11
CCD versus ICCD

Under moderate gain situations, what is the relative SNR of the ICCD to the CCD?

The ratio of SNRs is

$$\frac{SNR_{ICCD}}{SNR_{CCD}} = \eta_{SCREEN} T_{fo} G_{MCP} \eta_{PHOTOCATHODE} \tag{6-45}$$

The P20 phosphor screen has a peak emission at 0.56 μm with a conversion efficiency of 0.063. An S20 bialkali photocathode's efficiency is 0.2 at 0.53 μm. The value of T_{fo} is about 0.25. Combined, these factors give 0.00315. The MCP gain must be greater than 1/0.00315 = 317 for the SNR_{ICCD} to be greater than the SNR_{CCD}.

Example 6-12
IMAGE INTENSIFIER NOISE

Is there a relationship between the image intensifier noise equivalent input and the ICCD noise?

The noise in an image intensifier is due to random emissions (dark current noise) at the photocathode. When referred to the input (SNR = 1), it is called the equivalent background illumination (EBI). The EBI for an area A_{PIXEL} is

$$EBI = M_q(\lambda_o)\Delta\lambda\, t_{INT} A_{PIXEL} = \frac{4F^2 k_{MCP} \langle n_{PC-DARK} \rangle}{\eta_{CATHODE} T_{OPTICS}} \tag{6-83}$$

The ICCD will have the same equivalent background input under high gain situations. Cooling the image intensifier reduces the EBI and it may be negligible in gated applications. Under low gain situations, the CCD noise dominates.

6.9. NOISE EQUIVALENT INPUTS

The noise equivalent input occurs when the SNR is one. They represent the minimum input signal. The noise equivalent exposure (NEE) is an excellent diagnostic tool for production testing to verify noise performance. NEE is a poor array-to-array comparison parameter and should be used cautiously when comparing arrays with different architectures. This is so because it depends on array spectral responsivity and noise. The NEE is the exposure that produces a signal-to-noise ratio of one. If the measured rms noise on the analog output is V_{NOISE}, then the NEE is calculated from the radiometric calibration

$$NEE = \frac{V_{NOISE}}{R_{AVE}} \quad \frac{\text{J}}{\text{cm}^2} \text{ rms} \qquad (6\text{-}45)$$

When equated to electrons, NEE is simply the noise value in rms electrons. The absolute minimum noise level is the noise floor and this value is used most often for the NEE. Although noise is an rms value, the notation *rms* is often omitted. When divided by the average quantum efficiency, the noise equivalent signal (NES) is obtained

$$NES = \frac{NEE}{R_{q-AVE}} \quad \text{photons} \qquad (6\text{-}84)$$

Example 6-13
NOISE EQUIVALENT INPUT

The noise equivalent input (NEI) is that photon incidance that produces SNR = 1. Over a small spectral band,

$$NEI = \frac{4F^2}{A_D t_{INT} R_q \Delta\lambda T_{OPTICS}} \langle n_{SYS} \rangle \quad \frac{photons}{s - m^2 - \mu m} \qquad (6\text{-}85)$$

While valid for paraxial rays (1st order approximation), when $F < 3$, the equation must be modified (see Appendix). When the SNR is unity, the number of photons is small and photon noise is negligible If the detector is cooled, dark current is minimized. With PRNU and FPN correction, the minimum NEI is

$$NEI = \frac{4F^2}{A_D t_{INT} R_q \Delta\lambda T_{OPTICS}} \langle n_{FLOOR} \rangle \quad \frac{photons}{s - m^2 - \mu m} \qquad (6\text{-}86)$$

If TDI is used, then NTDI is multiplied in the denominator. With binning or super pixeling, the number of pixels combined is multiplied in the denominator.

Example 6 -14
NOISE EQUIVALENT REFLECTANCE

The NEI assumes that the target is against a black background. For real scenes, both the target and background create signals. The target is only visible when a sufficient reflectance difference exists. For detecting objects at long ranges, the atmospheric transmittance loss must be included. The signal difference between the target and its background is

$$\Delta n_{PE} = \int_{\lambda_1}^{\lambda_2} R_q(\lambda) \frac{[\rho_T(\lambda) - \rho_B(\lambda)] M_q(\lambda) t_{INT} A_D}{4F^2} T_{OPTICS}(\lambda) T_{ATM}(\lambda)\, d\lambda \qquad (6\text{-}87)$$

Both the target and background are illuminated by the same source. The variable $M_q(\lambda)[\rho_T(\lambda) - \rho_B(\lambda)]$ is a complicated function that depends on the cloud cover, atmospheric conditions, sun elevation angle, surface properties of the target, and the angle of the target with respect to the line-of-sight and sun angle. As a back-of-the-envelope approximation, the spectral reflectances are considered constants. Using the same simplifying assumptions as before, solving for ρ_T - ρ_B and calling this value the noise equivalent reflectance difference (NEΔρ) provides

$$NE\Delta\rho = \frac{4F^2}{A_D t_{INT} R_q M_q(\lambda_o) \Delta\lambda T_{OPTICS} T_{ATM}} \langle n_{SYS} \rangle \qquad (6\text{-}88)$$

This back-of-the-envelope approximation assumes that the total received energy is from the target only. In fact, over long path lengths, path radiance contributes to the energy detected. Path radiance, specified by visibility, reduces contrast and is quantified by the sky-to-background ratio, SGR (discussed in Section 12.3. *Contrast transmittance*). If the SGR is one, T_{ATM} = exp($-\sigma_{ATM}R$). Then

$$NE\Delta\rho = \frac{4F^2 \exp(\alpha_{ATM} R)}{A_D t_{INT} R_q M_q(\lambda_o) \Delta\lambda T_{OPTICS}} \langle n_{SYS} \rangle = \frac{NEI}{M_q(\lambda_o)} e^{\alpha_{ATM} R} \qquad (6\text{-}89)$$

Thus, $NE\Delta\rho$ depends on the source illumination, system noise, and the atmospheric conditions. As the range increases, the target reflectance must increase to keep the SNR at one. The maximum range occurs when the NEΔρ is one. At this maximum range,

$$NEI = M_q(\lambda_o)\; e^{-\sigma_{ATM} R_{MAX}} \qquad (6\text{-}90)$$

6.10. LUX TRANSFER

The literature is rich with natural and artificial luminance incidance values (see, for example, Tables 2-5 and 2-6). Therefore, consumer camera responsivity is often provided in photometric terms. By using Equation 6-16, selecting a minimum SNR, the minimum photometric signal is determined. This value is unique to the camera's spectral response and source spectral characteristics (discussed in Section 2.8. *Normalization issues*). In spite of these difficulties, the lux transfer concept has been developed[2] to relate SNR (based on electrons) to input lux. The SNR depends upon conversion factor of lux to electrons. Since this requires some assumptions, the lux transfer should be viewed as a design tool to identify those noise sources that limit performance.

The eye is sensitive from 0.38 to 0.75 μm whereas consumer cameras tend to have "high" quantum efficiency from 0.45 to 0.65 μm. When viewing a source of "infinite" size (***r*** » ***R*** in Figure 2-12), the lux value is

$$N_{LUX} = k_{CAL} 683 \int_{0.38}^{0.75} V(\lambda) M_e(\lambda, T)\, d\lambda \qquad (6\text{-}91)$$

where k_{CAL} is a scaling value that makes N_{LUX} = 1. The number of photoelectrons is approximated by

$$n_{PE} \approx k_{CAL} R_{q-AVE} \int_{0.45}^{0.65} \frac{M_q(\lambda, T) A_D t_{INT}}{F^2 (1 + M_{OPTICS})^2} \tau_{OPTICS}(\lambda)\, d\lambda \qquad (6\text{-}92)$$

While valid for paraxial rays (1st order approximation), when $F < 3$, the equation must be modified (see Appendix).

$$n_{PE} \approx \frac{N_{LUX} R_{q-AVE} \int_{0.45}^{0.65} \frac{M_q(\lambda, T) A_D t_{INT}}{F^2 (1 + M_{OPTICS})^2} \tau_{OPTICS}(\lambda)\, d\lambda}{683 \int_{0.38}^{0.75} V(\lambda) M_e(\lambda, T)\, d\lambda} \qquad (6\text{-}93)$$

Assuming the magnification approaches zero,

$$n_{PE} \approx \frac{R_{q-AVE} A_D t_{INT} \tau_{OPTICS}}{F^2} \left[\frac{\int_{0.45}^{0.65} M_q(\lambda, T)\, d\lambda}{683 \int_{0.38}^{0.75} V(\lambda) M_e(\lambda, T)\, d\lambda} \right] N_{LUX} \qquad (6\text{-}94)$$

This the basis for the lux transfer curves: the number of photoelectrons (and hence the SNR) is proportional to the detector area, integration time and focal ratio. The bracketed term is the band averaged photons/(m^2-s)/lux. Its value depends on the chosen limits of integration. Changing the limits on the denominator has a minimal effect since the eye's sensitivity is less than 1% when $\lambda < 0.43$ μm and $\lambda > 0.69$ μm. Changing the numerator can have a large effect. As illustrated in Figure 6-31, it is nearly constant for all practical source temperatures [$\approx 7.646\times10^{15}$ photons/(m^2-s)/lux]. Reference 2 considered "green" lux values and used 4×10^{15} photons/(m^2-s)/lux. This difference does not significantly affect the general conclusions.

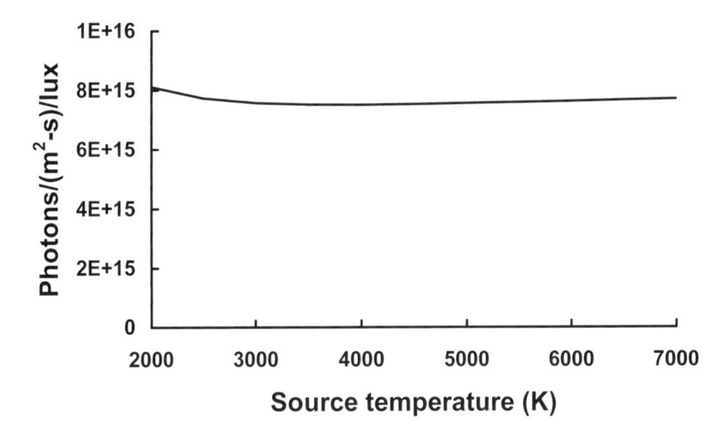

Figure 6-31. Proportionality constant.

The reset noise is assumed to be negligible. Lux transfer curves also include quantization noise, dark current noise, pattern noise, and the noise floor. The relative magnitude depends upon the assumed parameters (Table 6-5 and Figure 6-32). Consistent with Figure 6-20, FPN and PRNU have minimal effect on the overall noise. Figure 6-33 illustrates the noise components on the SNR. For the values selected, quantization noise dominates at low integration times and dark current shot noise dominates at long integration times. The dark current partially fills the charge wells and this limits the total integration time. Figure 6-34 provides the SNR for three different luminance incidances. Assuming a scene reflectivity of 10%, the incidances represent full daylight, overcast sky, and a very dark day, respectively (Table 2-5). With digital still cameras, the maximum shutter time is typically 1/30 s to avoid motion blur. Figures 6-35 through 6-37 illustrate the SNR for 3 different sized pixels (5 μm, 3 μm, and 1 μm). The charge well capability is 1500 e$^-$/μm^2. The well size is 37,500, 13,500, and 1,500 electrons, respectively. The horizontal lines at SNR = 10 and SNR = 40 represent “good” and “excellent” imagery.

Table 6-5
ARRAY PARAMETERS
(Typical for consumer digital still cameras)

Array parameters	Value
Input lux	Variable
Pixel size	Variable
Well capability	1500 $e^-/\mu m^2$
Dark current	30 pA/cm^2
Quantum efficiency	0.5
f-number	2
Noise floor	5 e^-
Optical transmission (CFA)	0.5
PRNU	0.7%
FPN	1.0%
ADC	8 bits

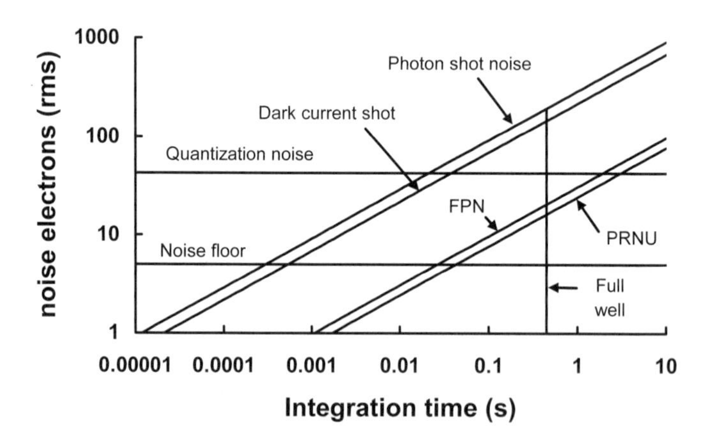

Figure 6-32. Noise electrons versus integration time for 10 lux. The 5 μm square pixel’s well capacity is (5)(5)(1500) = 37,500 e^- Signal photons fill the well in 0.448 s. With an 8-bit ADC, quantization noise dominates at low integration times. For high integration times both photon shot noise and dark current shot noise dominate.

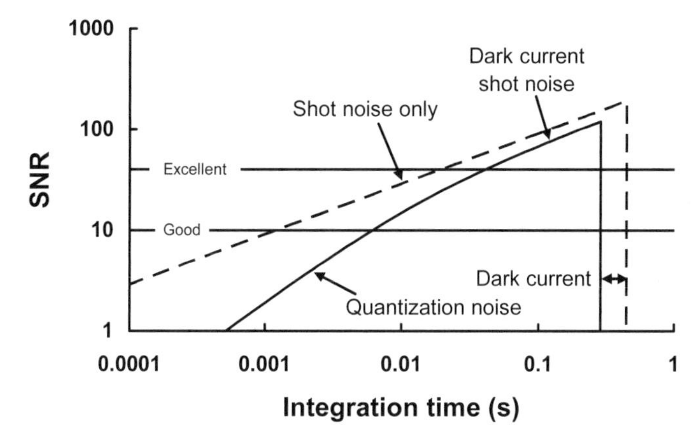

Figure 6-33. SNR versus integration time when the luminance exitance is 10 lux for a 5 μm square pixel. Dark current partially fills the charge well and it saturates in 0.287 s. When SNR > 10, the imagery is considered good and when SNR > 40, the imagery is considered excellent.

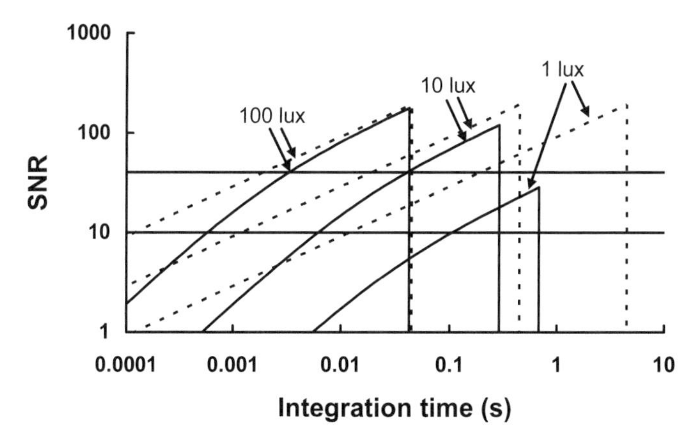

Figure 6-34. SNR versus integration time for three different scene luminance incidance for 5μm square pixel. The dashed curves are for shot noise only and are the theoretical maximum. Dark current significantly affects the SNR for long integration times. The horizontal lines represent "good" and "excellent" imagery.

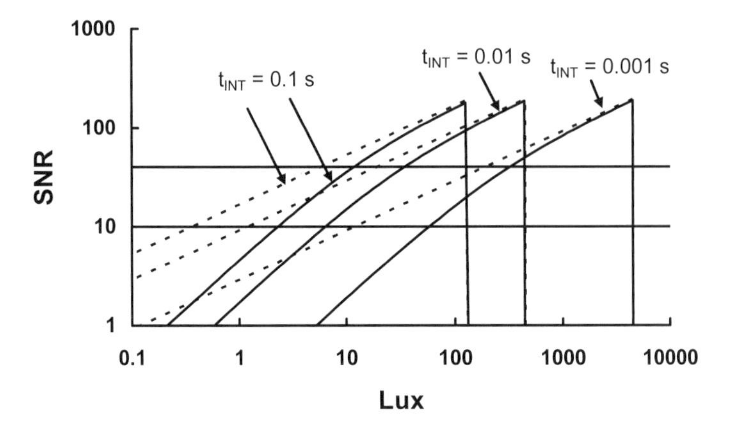

Figure 6-35. SNR versus luminance incidance for a 5 μm square pixel.

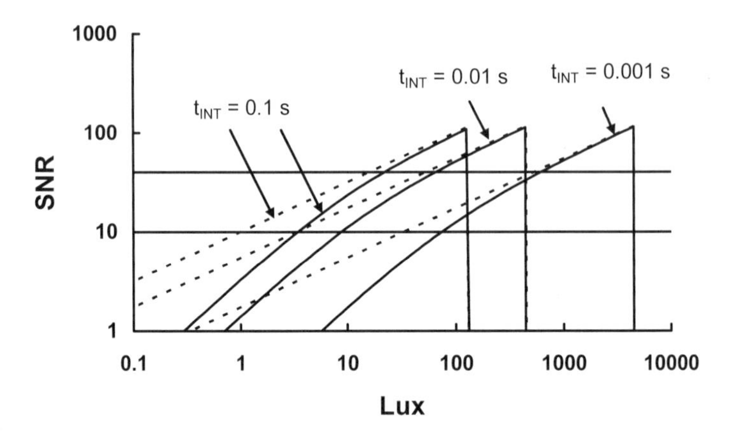

Figure 6-36. SNR versus luminance incidance for a 3 μm square pixel.

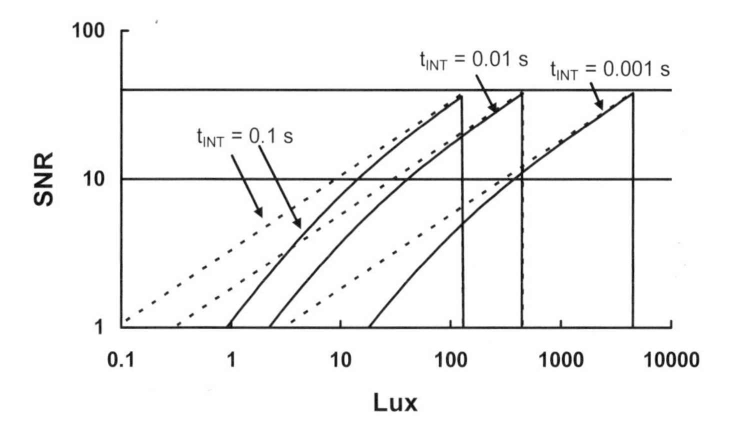

Figure 6-37. SNR versus luminance incidance for a 1 μm square pixel.

The signal dynamic range is defined as the maximum lux value divided by the minimum lux value at fixed SNR

$$DR_{SIGNAL} = \left.\frac{Max\ lux}{Min\ lux}\right|_{FIXED\ SNR} \tag{6-95}$$

The dynamic range for a "good" image (SNR = 10) and an excellent image (SNR = 40) is plotted in Figure 6-38. For 1 μm square pixel, the SNR never exceeds 40 (Figure 6-37) so that the dynamic range is 0. Based on the assumptions in Table 6-4, the minimum desirable detector size is about 5 for "excellent" imagery. For "good" imagery, the minimum size is about 3 μm. The signal dynamic range will increase by reducing the dark current, increasing the size of the ADC (10 or 12 bits), increasing the well capacity, and reducing the focal ratio. Obviously, the SNR can be improved by increasing the integration time, or equivalently, averaging multiple frames. This reduces shot noise but not FPN or PRNU. It also makes image quality very susceptible to image motion. The conversion from lux to photoelectrons assumed that the array quantum efficiency is uniform from 0.45 to 0.65 μm. If the spectral response is over a narrower region, the signal decreases. A conservative 50% quantum efficiency was used. Higher quantum efficiencies will increase the SNR. The maximum increase is minimal since it depends upon the square root of the ratios. The CFA transmission was optimistic when considering its spectral characteristics. A lower transmission will reduce the SNR.

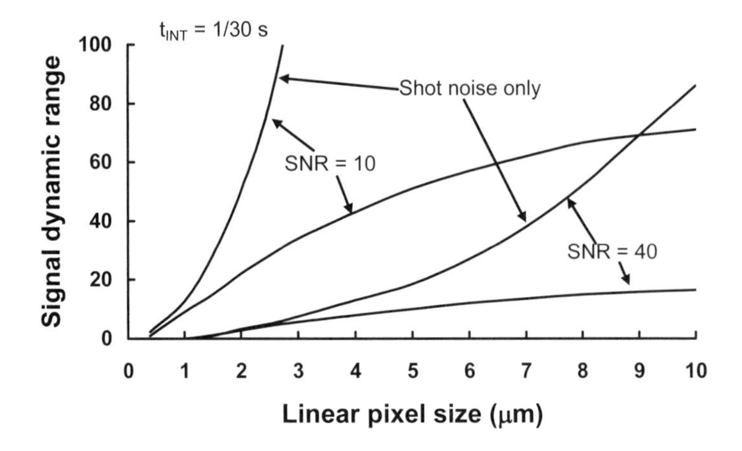

Figure 6-38. Estimated signal dynamic range for t_{int} = 1/30 s as a function of linear pixel dimension (pixels are square). The theoretical maximum signal dynamic range is larger (i.e., photon shot noise is the only noise present).

The "minimum" signal depends upon the SNR (usually unity), integration time, and pixel size. Considering Figures 6-32 through 6-37, the value selected is unique to the assumptions made. Camera manufacturers publish a "minimum" signal but do not specify the operating conditions or the SNR. This makes it very difficult to compare cameras. Some cameras have an IR filter that can be switched into the field of view. While the lux value does not change, the signal increases without the IR filter. This makes the minimum signal appear lower. Finally, some digital still cameras have a built-in LED operating in the near IR that acts a flash. It cannot be seen, but it illuminates the scene and creates an image. Here, the manufacturer claims "zero lux" minimum signal.

SNR analysis assumes very large targets with no structure. Image quality depends on scene content (edges and gradients) and target contrast (signal with respect to its background). The MTF (discussed in Chapter 9. *MTF*) is a measure of edge reproduction. Image motion effects are described by an MTF. For long integration times, the MTF decreases. An important consideration is the ratio of the optical blur diameter (2.44λF) to the detector size. Smaller detectors drive the optical design to lower focal ratios (discussed in Chapter 11, *Image quality*).

6.11. SPEED - ISO

Film speed (ISO rating) is related to the energy necessary to increase film opacity. The energy is the incidence photon flux multiplied by the integration speed (shutter speed). Higher ISO ratings indicate more sensitive film. For fixed integration time, it is possible to take pictures in lower light level conditions. For fixed light levels, the integration time can be reduced to avoid image blur. When digital still cameras replaced film, the photography community wanted an equivalent ISO rating. This was somewhat difficult since DSCs are noisy at low light levels.

The International Organization for Standardization issued ISO-12232 which describes[40] methods to determine camera "speed." Base ISO (saturation base ISO) depends upon the focal plane flux level in lux-seconds. The exposure index is

$$EI = \frac{10}{E_v t_{INT}} \tag{6-96}$$

The focal flux is generally not available and can be computed from[41,42]

$$ISO_{BASE} = EI_{BASE} = 10\frac{(1+M_{OPTICS})^2 F^2}{k_1\, L_v\, t_{INT}} \tag{6-97}$$

The variable L_v is the average luminance of the scene and k_1 is a camera constant that includes optical transmission, $\cos^4(\theta)$ roll-off, etc. To be consistent with photographic terminology, t_{INT} is the exposure time. Since EI_{BASE} does not include noise, it does not indicate image quality.

Reference 42 defines the SNR as

$$SNR = \frac{\frac{\Delta n}{\Delta L} L_v}{\langle n_{SYS} \rangle} \tag{6-98}$$

where $\Delta n/\Delta L$ is the incremental system response in electrons/(cd/m^2) at scene luminance L_v. If nonlinear image processing is present, $\Delta n/\Delta L$ varies with L_v and if the system is linear, $\Delta n/\Delta L$ is a constant. Let L_{SNR} be the scene luminance that provides the desired SNR. The noise-based exposure index is

$$ISO_{BASE} = EI_{BASE} = 10\frac{(1+M_{OPTICS})^2 F^2}{k_1\, L_{SNR} t_{INT}} \frac{1}{SNR\, \langle n_{SYS} \rangle} \frac{\Delta n}{\Delta L} \tag{6-99}$$

As the SNR increases, the exposure index decreases. There is no standard for acceptability in terms of SNR. Generally, SNR = 10 is an "acceptable" image and that SNR = 40 is an "excellent" image. Reference 43 illustrates representative imagery for ISO values ranging from 160 to 5120.

6.12. REFERENCES

1. J. R. Janesick, *Scientific Charge-Coupled Devices,* Chap 2, SPIE Press, Bellingham, WA (2001).
2. J. R. Janesick, "Lux transfer: CMOS versus CCD," in *Sensors and Camera Systems for Scientific, Industrial, and Digital Photography Applications* III; M. M. Blouke, J. Canosa, and N. Sampat, eds., SPIE Proceedings Vol. 4669, pp. 232-249 (2002).
3. R. H. Dyck, "VLSI Imagers," in *VLSI Electronics: Microstructure Science, Volume 5*, N. G. Einspruch, ed., pp. 594, Academic Press, New York (1985).
4. R. Lyon and P. Hubel, "Eyeing the Camera: into the Next Century," in *Tenth Color Imaging Conference: Color Science and Engineering Systems, Technologies, and Applications,* IS&T/SID Proceedings, pp. 349-355 (2002).
5. P. M. Hubel, J. Liu, and R. J. Guttosch, "Spatial frequency response of color image sensors: Bayer color filters and Foveon X3," in *Sensors and Camera Systems for Scientific, Industrial, and Digital Photography Applications* V; M. M. Blouke, N. Sampat, and R. J. Motta, eds., SPIE Proceedings Vol. 5301, pp. 402-407 (2004).
6 .R. H. Dyck, "Design, Fabrication, and Performance of CCD Imagers," in *VLSI Electronics: Microstructure Science, Volume 3*, N. G. Einspruch, ed., pp. 70-71, Academic Press, New York (1982).
7. A. Ciccarelli, B. Davis, W. Des Jardin, H. Doan, E. Meisenzahl, G. Putnam, J. Shepherd, E. Stevens, J. Suma, and K. Wetzel, "Front-illuminated Full-frame Charge-coupled Device Image Sensor Achieves 85% Peak Quantum Efficiency," in *Sensors and Camera Systems for Scientific, Industrial, and Digital Photography Applications* III, M. M. Blouke, J. Canosa, and N. Sampat, eds., SPIE Proceedings Vol. 4669, pp. 153-160 (2002).
8. W. A. R. Franks, M. J. Kiik, and A. Nathan, "UV-responsive CCD Image Sensors with Enhanced Inorganic Phosphor Coatings," IEEE *Trans on Electron Devices*, Vol. 50, pp. 352-358 (2003).
9. J. Janesick and G. Putnam, "Developments and Applications of High-Performance CCD and CMOS Imaging Arrays," *Annual Review of Nuclear and Particle Science*, Vol. 53, pp. 263–300 (2003).
10. See Princeton Instruments "Etaloning in CCDs – NIR" at http://www.piacton.com/library
11. T. E. Dutton, J. Kang, T. S. Lomheim, R. Boucher, R. M. Shima, M. D. Nelson, C. Wrigley, X. Zheng, and B. Pain, "Measurement and Analysis of Pixel Geometric and Diffusion Modulation Transfer Function (MTF) Components in Photodiode Active Pixel Sensors", 2003 *IEEE Workshop on CCDs and Advanced Image Sensors*, Elmau, Germany, May 16, 2003.
12. P. B Catrysse, X. Liu, and A. Gamal, "QE Reduction due to Pixel Vignetting in CMOS Image Sensors", in *Sensors and Camera Systems for Scientific, Industrial, and Digital Photography Applications*, M. M. Blouke, N, Sampat, G. M. Williams, Jr., and T. Yeh, eds., SPIE Proceedings Vol. 3965, pp. 420 – 430 (2000).
13. I. Csorba, *Image Tubes*, p. 209, Howard Sams, Indianapolis, Indiana (1985).
14. D. J. Denvir and E. Conroy, "Electron multiplying CCDs," in *Opto-Ireland 2002: Optical Metrology, Imaging, and Machine Vision,* A. Shearer, F. D. Murtagh, J. Mahon, and P. F. Whelan, eds., SPIE Proceedings Vol. 4877 pp. 55-68 (2003).

15. T. Ashida, H. Yamashita, M. Yoshida, O. Saito, T. Nishimura, and K. Iwabe, "Signal Processing and Automatic Camera Control for Digital Still Cameras Equipped with a New Type CCD': in *Sensors and Camera Systems for Scientific, Industrial, and Digital Photography Applications*, IV, M. M. Blouke, N. Sampat, and R, J. Motta, eds., Proceedings of SPIE Vol. 5310, pp. 42-50 (2004).
16. Mark Sartor, private communication.
17. Y. Talmi, "Intensified Array Detectors," in *Charge-Transfer Devices in Spectroscopy*, J. V. Sweedler, K. L. Ratzlaff, and M. B. Denton, eds., Chapter 5, VCH Publishers, New York (1994).
18. J. Janesick, "Open Pinned-Phase CCD Technology," in *EUV, X-ray, and Gamma Ray Instruments for Astronomy and Atomic Physics,* C. J. Hailey and O. H. Siegmund, eds., SPIE Proceedings Vol. 1159, pp 363-373 (1989).
19. J. Janesick, T. Elliot, G. Fraschetti, S. Collins, M. Blouke, and B. Corey, "CCD Pinning Technologies," in *Optical Sensors and Electronic Photography*, M. M. Blouke, ed., SPIE Proceedings Vol. 1071, pp. 153-169 (1989).
20. J. S. Campbell, "TC271 Characterization Report" in Area Array Image Sensor Products, Texas Instruments Data Catalog, pp. B-43 to B-46 (1996).
21. J. R. Janesick, T. Elliott, S. Collins, M. M. Blouke, and J. Freeman, "Scientific Charge-coupled Devices," *Optical Engineering*, Vol. 26(8), pp. 692-714 (1987).
22. C. H. Sequin and M. F. Tompsett, *Charge Transfer Devices*, Academic Press, New York, NY (1975).
23. E. S. Yang, *Microelectronic Devices*, McGraw-Hill, NY (1988).
24. E. L. Dereniak and D. G. Crowe, *Optical Radiation Detectors*, John Wiley and Sons, New York, NY (1984).
25. D. G. Crowe, P. R. Norton, T. Limperis, and J. Mudar, "Detectors," in *Electro-Optical Components*, W. D. Rogatto, pp. 175-283. This is Volume 3 of *The Infrared & Electro-Optical Systems Handbook*, J. S. Accetta and D. L. Shumaker, eds., copublished by Environmental Research Institute of Michigan, Ann Arbor, MI, and SPIE Press, Bellingham, WA (1993).
26. J. R. Janesick, T. Elliott, S. Collins, M. M. Blouke, and J. Freeman, "Scientific Charge-coupled Devices," *Optical Engineering*, Vol. 26(8), pp. 692-714 (1987).
27. T. W. McCurnin, L. C. Schooley, and G. R. Sims, "Charge-coupled Device Signal Processing Models and Comparisons," *Journal of Electronic Imaging*, Vol. 2(2), pp. 100-107 (1994).
28. M. D. Nelson, J. F. Johnson, and T. S. Lomheim, "General Noise Process in Hybrid Infrared Focal Plane Arrays," *Optical Engineering*, Vol. 30(11), pp. 1682-1700 (1991).
29. T. S. Lomheim and L. S. Kalman, "Analytical Modeling and Digital Simulation of Scanning Charge-Coupled Device Imaging Systems," in *Electro-Optical Displays*, M. A. Karim, ed., pp. 551-560, Marcel Dekker, New York (1992).
30. J. M. Mooney, "Effect of Spatial Noise on the Minimum Resolvable Temperature of a Staring Array," *Applied Optics*, Vol. 30(23), pp. 3324-3332, (1991).
31. J. M. Mooney, F. D. Shepherd, W. S. Ewing, J. E. Murguia, and J. Silverman, "Responsivity Nonuniformity Limited Performance of Infrared Staring Cameras," *Optical Engineering*, Vol. 28(11), pp. 1151-1161 (1989).
32. M. S. Robbins and B. J. Hadwen, "The Noise Performance of Electron Multiplying Charge-coupled Devices," *IEEE Trans Electron Devices*, Vol. 50, pp 1227-1232 (2003).
33. I. Csorba, *Image Tubes*, pp. 120-124, Howard Sams, Indianapolis, Indiana (1985).
34. R. J. Hertel, "Signal and Noise Properties of Proximity Focused Image Tubes," in *Ultrahigh Speed and High Speed Photography, Photonics, and Videography*, G. L. Stradling, ed., SPIE Proceedings Vol. 1155, pp. 332-343 (1989).
35. J. E. Murguia, J. M. Mooney, and W. S. Ewing, "Diagnostics on a PtSi Infrared Imaging Array" in *Infrared Technology XIV*, I. Spiro, ed., SPIE Proceedings Vol. 972, pp. 15-25 (1988).
36. B. Stark, B. Nolting, H. Jahn, and K. Andert, "Method for Determining the Electron Number in Charge-coupled Measurement Devices," *Optical Engineering*, Vol. 31(4), pp. 852-856 (1992).

37. M. J. DeWeert, J. B. Cole, A. W. Sparks, and A. Acker, "Photon Transfer Methods and Results for Electron Multiplication CCDs," in *Applications of Digital Image Processing* XXVII, A. G. Tescher, ed., SPIE Proceedings Vol. 5558, pp. 248-259 (2004).
38. A. O'Grady, "A Comparison of EMCCD, CD, and Emerging Technologies Optimized for Low Light Spectroscopy Applications," in *Biomedical Vibrational Spectroscopy* III, A. Mahadevan-Jansen and W. H. Petrich, eds., SPIE Proceedings Vol. 6093, paper 60930S (2006).
39. D. Dussault and P. Hoess, "Noise Performance Comparison of ICCD with CCD and EMCCD Cameras," in *Infrared Systems and Photolectronic Technology*, E. L. Dereniak, R. E. Sampson, C. B. Johnson, eds., SPIE Proceeding Vol. 5563, pp. 195-204 (2004).
40. International Organization for Standardization, ISO-12232, Technical Committee 42, Photography, Digital Still Cameras, Determination of Exposure Index, ISO, Speed Ratings, Standard Output Sensitivity, and Recommended Exposure Index. Available at WWW.ISO.CH.
41. G. Putnam, "Understanding Sensitivity," *OEMagazine*, pg 56, January 2002
42. G.Putnam, "Noise-based ISO Gives Measure of Camera Performance in Low-light, High-noise Conditions," *OEMagazine*, pg. 56, January 2003
43. G. Putnam, S. Kelly, S. Wang, W. Davis, E. Nelson, and D. Carpenter, "Photography with an 11-megapixle 35-mm format CCD," in *Sensors and Camera Systems for Scientific, Industrial, and Digital Photography Applications* IV, M. M. Blouke, N. Sampat, and R. J. Motta, eds., SPIE Proceedings Vol. 5017, pp. 371-384 (2003).

7
CAMERA DESIGN

This chapter is an overview of solid state camera design. It highlights those features that may affect performance. Together with the preceding chapters, it provides the necessary information to compare and analyze camera systems. In principle, any size of array can be manufactured; the problem is how to create a viewable image. This forces array sizes to be consistent with standard displays. If an array has more pixels than are supported by the display, only a portion of the field of view can be displayed. Downsampling is always possible, but this reduces resolution.

Both monochrome and color cameras are quite complex as evidenced by the 1622-page *Television Engineering Handbook*, Revised edition, K.B. Benson, ed., revised by J. Whitaker, McGraw-Hill, New York (1992) and the 692-page *Digital Video and HDTV* by C. Poynton, Morgan Kaufmann (2003). The major display performance requirements are listed in Table 7-1. This table is not meant to be all-inclusive. It just indicates that different users have different requirements. As displays are built for each user, they are specified in units familiar to that user. For example, consumer displays will not have the MTF or spot size listed. These parameters are provided for militarized displays.

Table 7-1
DISPLAY REQUIREMENTS

APPLICATION	REQUIREMENTS
Consumer TV	"Good" image quality
Computer	Alphanumeric character legibility "Good" graphical capability
Scientific and military	Wide dynamic range High resolution Wide color gamut

Television receivers have been available for over 50 years, and the technology has matured. Cathode ray tubes (CRTs) are considered quite goo, and little improvement in technology is expected. Advances in television displays currently focus on high-definition television, flat-panel displays, the legibility of alphanumeric characters, and computer graphics capability. As a result, display specifications are a mix of CRT terminology, video transmission standards, alphanumeric character legibility, and graphics terminology. CRTs are low cost with high resolution, wide color gamut, high luminance, and long life. Flat panels do not have all these attributes. Nevertheless, it appears that all CRTs will eventually be replaced by flat-panel displays.

The symbols used in this book are summarized in the *Symbol List*, which appears after the *Table of Contents*.

7.1. CAMERA OPERATION

Figure 7-1 illustrates a generic solid state camera. The lens assembly images the light onto the detector array, and the analog detector output is digitized. Extensive image processing exists in all cameras. Image processing reformats the data into signals and timing that are consistent with transmission and display requirements. The digital output can be directly fed into a machine vision system or digital data capture system (e.g., a computer). The digital-to-analog converter (DAC) and sample-and-hold circuitry converts the digital data into an analog stair-step signal. The post-reconstruction filter removes the stair-step appearance to create a smooth analog signal (discussed in Section 9.9.4. *Post-reconstruction filter*). This signal must be re-digitized for computer analysis. These multiple conversions will modify the imagery. Thus, staying within the digital domain ensures maximum image fidelity.

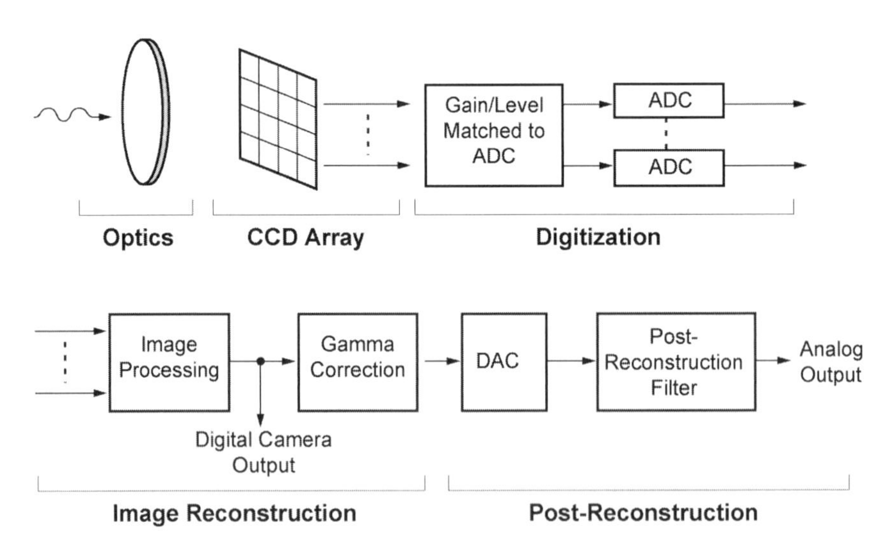

Figure 7-1. Generic functional block diagram. The gamma correction circuitry is found in both video and digital still cameras.

While the ADC can provide up to 16 bits, the display typically can only accept 8 bits. The gamma correction algorithm efficiently encodes the signal into perceptually linear scale. If displays could handle 16 bits, gamma correction would not be needed.

7.2. OPTICAL DESIGN

Camera optical design can be quite complex. It usually consists of several different lenses to control chromatic aberrations, coma, and distortion. On-axis image quality is usually better than off-axis imagery. There seems to be an infinite number of designs; many optical design considerations are listed in Reference 1. For this text, the lens system is treated as a simple lens where paraxial ray-tracing formulas are valid. This simple lens has the same focal length and diameter as the actual lens system. Its performance is specified by its optical transfer function (discussed in Section 9.2. *Optics MTF*).

7.3. ANALOG-TO-DIGITAL CONVERTERS

An analog-to-digital converter is an essential component of all cameras. For consumer applications, an 8-bit converter is used. Sometimes seemingly slight deviations from perfection in an ADC can modify image fidelity[2] when examined by a trained photointerpreter or by an automatic target detection algorithm. While a nonlinear ADC may not be visually obvious with a single readout, it may become obvious if the array has multiple outputs with each output having its own ADC. Here, any object which spans the two areas with different ADCs will have a discontinuity in intensity.

ADC linearity is specified by differential nonlinearity (DNL) and integral nonlinearity (INL) (Figure 7-2). The differential nonlinearity is a test of missing codes or misplaced codes. Two adjacent codes should be exactly one LSB apart. Any positive or negative deviation from this is the DNL. If the DNL exceeds one LSB, there is a missing code. Many nonlinearity specifications simply state "no missing codes." This means that the DNL is less than one LSB. However, 1.01 LSB DNL (fail specification) and 0.99 LSB DNL (pass specification) probably look the same visually. Poor DNL can affect the cosmetic quality of images.

Integral nonlinearity is the deviation of the transfer function from an ideal straight line. Most ADCs are specified with endpoint INL. That is, INL is specified in terms of the deviation from a straight line between the end points. Good INL is important for radiometric applications. Imagery with DNL and INL are provided in Reference 3.

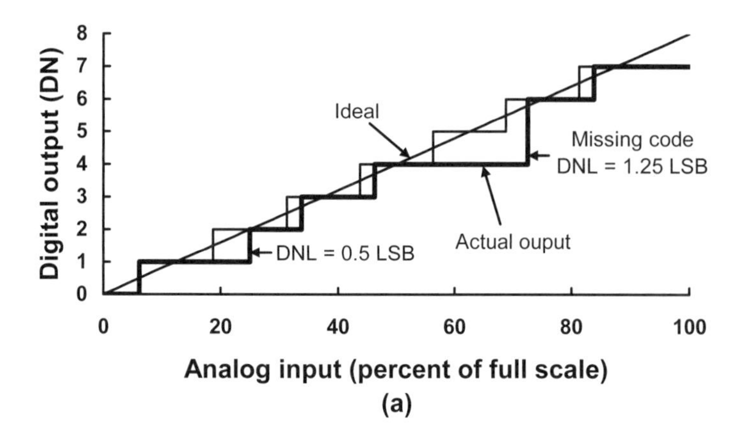

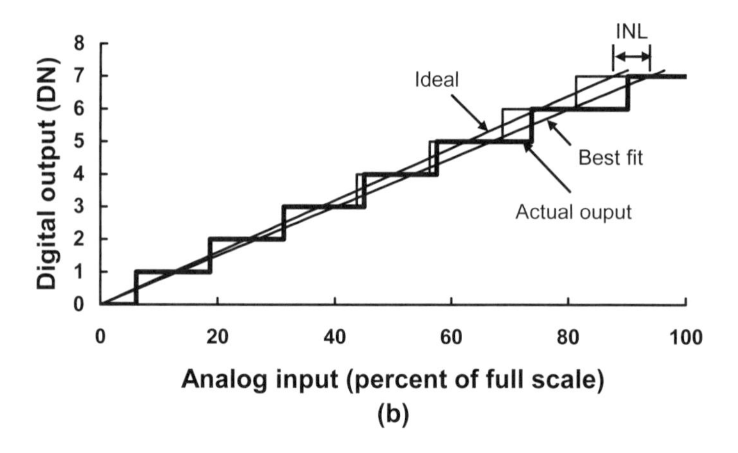

Figure 7-2. ADC nonlinearity for 3-bit ADC. (a) Differential nonlinearity and (b) integral nonlinearity. Missing codes can be caused by the accumulation of DNLs. Accumulation can also cause INL. The light stair-step line is the ideal ADC output. The heavy line is the actual output.

With commercially available cameras, the manufacturer has selected the ADC. To ensure that the ADC is linear, the user must experimentally characterize the performance in a manner consistent with his application.[4] This is rather difficult if the camera output is an analog signal. The analog output represents a smoothed output where the smoothing can hide ADC nonlinearities.

7.4. IMAGE PROCESSING

Extensive image processing occurs within a camera. This includes PRNU correction algorithms (called nonuniformity correction of NUC), dark current subtraction, color correction, aperture correction, gamma, and a myriad of others that are deemed to improve image quality. Monochrome cameras may also have these algorithms, with the obvious exception of color correction. Figure 7-3 illustrates a typical color camera functional block diagram.

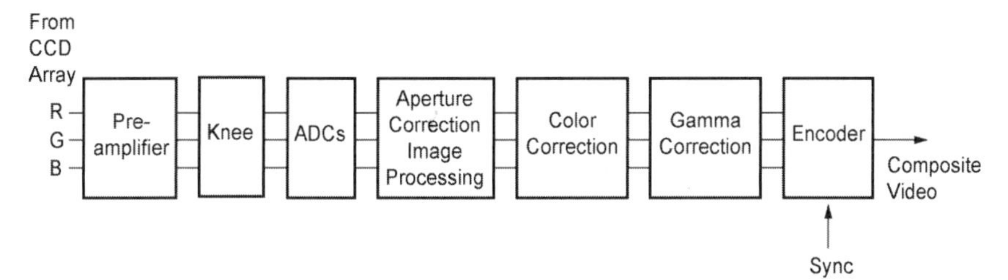

Figure 7-3. Typical color camera functional block diagram. The subsystem location and design vary by manufacturer. Consumer cameras typically have a composite video output that is created by the encoder. Scientific and industrial cameras may have a component output which is the signal before the encoder. To increase versatility, many cameras can provide NTSC, RGB, and complementary color output.

7.4.1. THE KNEE

The antibloom drain (Figure 3-35) creates a knee in the linear response to brightness. The knee slope and location can be any value. Arrays without an anti-bloom drain may have an off-chip knee to extend the dynamic range (Figure 7-4). While scientific cameras operate in a linear fashion, there is no requirement to do so with consumer cameras. Consumer camera manufacturers provide a mapping that they think will provide a cosmetically pleasing image. The knee also simplifies requirements in that an 8-bit ADC can provide aesthetically pleasing imagery over a wide input dynamic range.

Scientific and industrial applications often require a linear input-output system. For example, linearity is required for radiometric analysis. Interestingly, as some manufacturers add more knee control, others are actively designing algorithms to "de-knee" the signal before image processing.

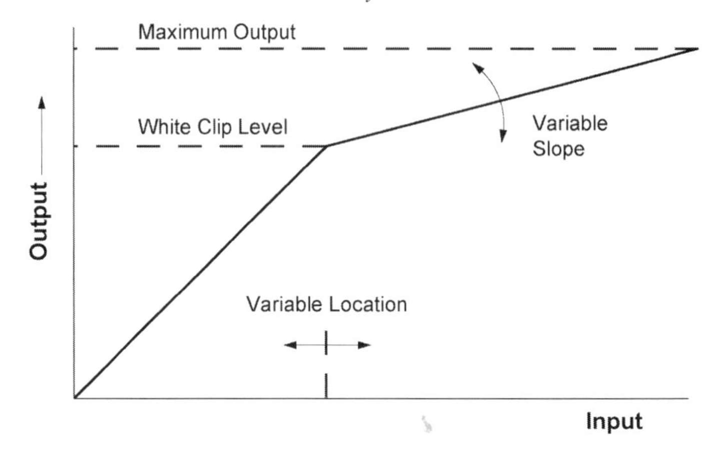

Figure 7-4. The knee extends the dynamic range by compressing high-intensity signals. By decreasing the knee slope, the camera can handle a larger scene dynamic range.

7.4.2. APERTURE CORRECTION

Image sharpness is related to the modulation transfer function (discussed in Chapter 11. *Image quality*). By increasing the MTF at high spatial frequencies, the image usually appears "sharper."

For analysis purposes, a camera is separated into a series of subsystems: optics, detector, electronics, etc. MTF degradation in any one subsystem will soften imagery. The degradation can be partially compensated with electronic boost filters (discussed in Section 9.9.5. *Boost*). Historically, each subsystem was called an "aperture." Therefore boost provides "aperture correction." This should not be confused with the optical system, where the optical diameter is the entrance aperture.

7.4.3. COLOR CORRECTION

Figure 7-3 illustrated a typical color camera block diagram. The "color" signals may have originated from a single CFA (see Section 5.6. *Color filter arrays*) or three separate arrays as illustrated in Figure 7-5. Because of high cost, three-chip cameras were used almost exclusively for high-quality broadcast applications. Consumer camera designs, driven by cost, tend to use only one chip. This is changing, however.

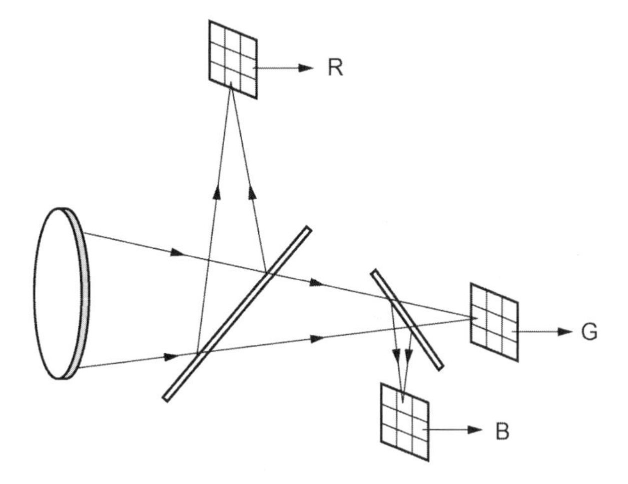

Figure 7-5. A high-quality color camera has three separate arrays: one for each primary color or its complement. Light is spectrally separated by coatings on the beam splitters and additional filters. The three-chip camera is listed as CCD×3 or 3CCD in sales brochures. Each color samples the scene equally; this minimizes color moiré patterns when compared to a CFA. Although beam splitters are shown, most 3CCD cameras use prisms for compactness.

The color correction circuitry is an algorithm that can be represented as a matrix

$$\begin{pmatrix} R_2 \\ G_2 \\ B_2 \end{pmatrix} = \begin{pmatrix} k_{11} & k_{12} & k_{13} \\ k_{21} & k_{22} & k_{23} \\ k_{31} & k_{32} & k_{33} \end{pmatrix} \begin{pmatrix} R_1 \\ G_1 \\ B_1 \end{pmatrix} \tag{7-1}$$

When used to improve color reproducibility (i.e., matching the camera spectral response to the CIE standard observer), the process is called masking.[5,6] When manipulating signals into the NTSC standard, it is called matrixing.

Matrixing can be used to alter the spectral response and intensities of the primaries to compensate for changes in scene color temperature. Intensity changes are achieved simply by multiplying a primary by a constant. Spectral changes are approximated by adding various ratios of the primaries together. For example, if an object is viewed at twilight, the matrix can be adjusted so that the object appears to be in full sun light (neglecting shadows). A scene illuminated with a 2600 K color temperature source may appear somewhat yellow compared to a scene illuminated with a 6500 K source. With white balancing (matrixing) a white object can appear to be illuminated by different blackbodies with color temperatures ranging from 3200 to 5600 K.

A CFA creates unequal sampling where monochrome edges may appear colored and small objects may change color. Sophisticated image processing is required to overcome the unequal sampling of the colors and to format the output into the NTSC standard. The algorithm selected depends on the CFA design (Figures 5-8 and 5-9). Algorithms may manipulate the ratios of blue-to-green and red-to-green,[7] separate the signal into high and low frequency components,[8] use a pixel shift processor,[9] or add two alternating rows together[10,11] to minimize color artifacts. The decoding algorithm is unique for each CFA configuration.

Figure 7-6 illustrates the Bayer pattern with a blue pixel in the center. For the color display, there must also be a green and red value at the same location. The green value may be an interpolation of the immediately surrounding greens, and the weightings may be a function of the red values. Likewise, the reds must be interpolated and the weightings may depend upon the immediate green values. The weighting may include more pixels (the lighter cells in Figure 7-6).

B	G	B	G	B
G	R	G	R	G
B	G	B	G	B
G	R	G	R	G
B	G	B	G	B

Figure 7-6. Bayer pattern

Algorithm design is more of an art than a science because the acceptability of color aliasing is not known until after the camera is built. There are numerous algorithms available; each claims to provide the "best" imagery[12-18].

7.4.4. GAMMA CORRECTION

The human visual system can just detect a just noticeable difference (JND) when the contrast is $\Delta L/L > 0.01$. If the signal is linearly digitized, then the contrast for each digital step is 1/DN, where DN is the digital number (ranges from 0 to 255 for an 8-bit system). For very small DNs, contrast is large and the imagery will appear contoured. For large DNs, the contrast is less than a JDN indicating inefficient coding (Figure 7-7).

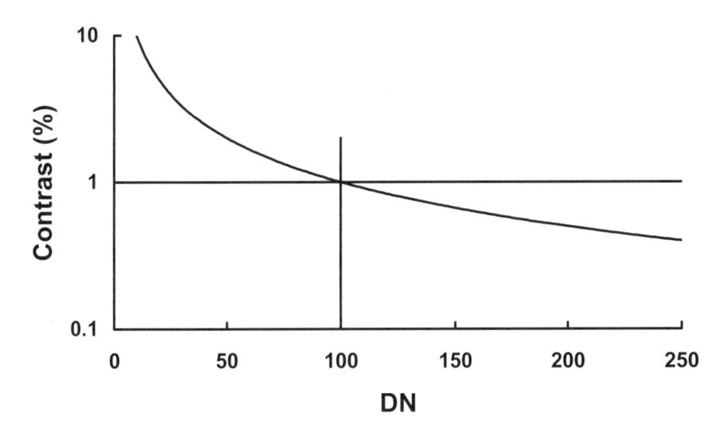

Figure 7-7. Percent contrast as a function of digital values for linear conversion of signal to DN

ITU (International Telecommunications Union) Recommendation 709 stretches the blacks and compresses the whites to create a perceptually uniform signal

$$
\begin{aligned}
V_{VIDEO} &= 4.5V_{SCENE} & \quad when \quad 0 \le V_{SCENE} < 0.018 \\
V_{VIDEO} &= 1.099V_{SCENE}^{0.45} - 0.099 & \quad when \quad 0.018 \le V_{SCENE} \le 1
\end{aligned}
\tag{7-2}
$$

The variable V_{SCENE} is the normalized voltage ($0 \le V_{SCENE} \le 1$) before the gamma correction and V_{VIDEO} is the voltage after (Figure 7-8). The exponent 0.45 is the gamma correction, but is simply called gamma. Some cameras may have a continuous variable gamma from 0.4 to 1. The user selects a value that seems to provide an aesthetically pleasing image without regard to its true purpose. This setting will vary from individual to individual.

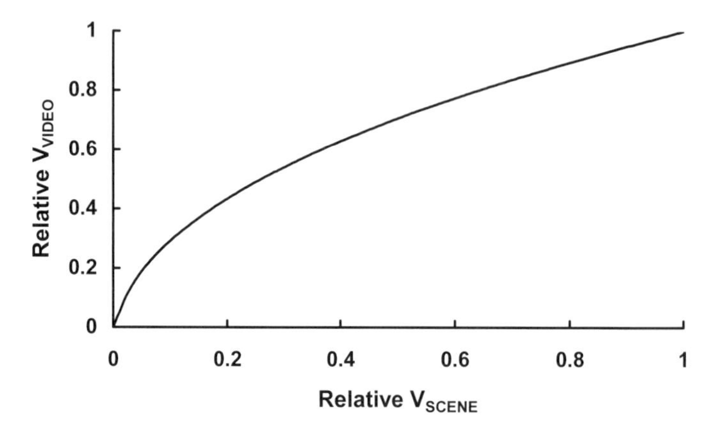

Figure 7-8. Recommended gamma correction algorithm.

The luminous output, $L_{DISPLAY}$, of a CRT is related to its grid voltage by a power law. The grid voltage is related to the input voltage (called the command level)

$$L_{DISPLAY} = aV_{GRID}^{\gamma} + L_B \tag{7-3}$$

The exponent is "gamma." If the camera has the inverse gamma (Figure 7-9), then radiometric fidelity is approximately preserved. Letting V_{GRID} = V_{VIDEO},

$$L_{DISPLAY} = aV_{VIDEO}^{\gamma} + L_B \approx a\left(V_{SCENE}^{1/\gamma}\right)^{\gamma} + L_B = aV_{SCENE} + L_B \tag{7-4}$$

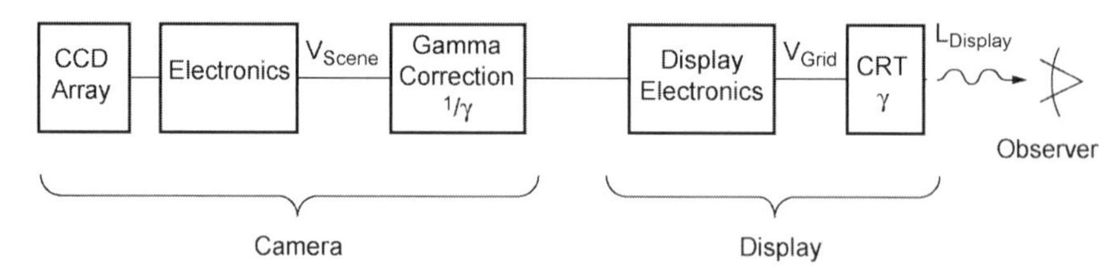

Figure 7-9. Camera and display. V_{SCENE} is the "electronic" version of the scene.

It has been reported that the CRT gamma can range[19] from one to five. This variation was probably caused by erroneously assuming the background luminance L_B = 0 in Equation 7-3. For design purposes, NTSC standardized gamma at 2.2. PAL and SECAM standardized gamma at 2.8. Computers may have different gammas. Its value depends upon look-up tables that the manufacturer feels provides the "best" imagery. The combination of the camera gamma and computer gamma often results in a power law, $L_{DISPLAY} = (V_{SCENE})^k$. Reference 20 provides examples of different computer designs.

Image processing should be performed on linear data. If gamma is present, it should be removed. Figure 7-10 illustrates a poorly designed machine vision system. Obviously, image processing is simplified when gamma is removed (i.e., use a camera without a gamma correction). Image fidelity is improved with a digital output camera. Although a machine vision system does not require a monitor, it adds convenience.

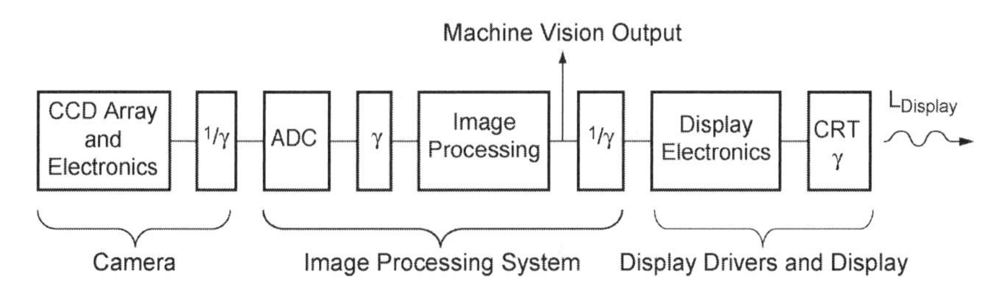

Figure 7-10. An analog camera system with multiple gammas. A digital camera without gamma correction is ideally suited for machine vision applications.

7.5. VIDEO FORMATS

Video standards were originally developed for television applications. The purpose of a standard is to ensure that devices offered by different manufacturers will operate together. The standards apply only to the transmission of video signals. It is the camera manufacturer's responsibility to create an output signal that is compatible with the standards. Similarly, it is the display manufacturer's responsibility to ensure that the display can recreate an adequate image.

In 1941, the Federal Trade Commission (FCC) adopted the 525-line, 30 Hz frame standard for broadcast television that was proposed by the NTSC (National Television Systems Committee). The standard still exists today with only minor modifications. Although originally a monochrome standard, it has been modified for color transmission. Amazingly, color video fits within the video electronic bandwidth originally set aside for black-and-white television.

NTSC initially selected a 60 Hz field rate. It was later changed to 59.94 Hz for color systems to avoid interactions (beat frequencies) between the audio and color subcarrier frequencies. The monochrome standard is often called EIA 170 (originally called RS 170) and the color format is simply known as NTSC (originally called EIA 170A or RS 170A). The timing for these two standards is nearly the same.[21]

Worldwide, three color broadcast standards exist: NTSC, PAL, and SECAM. NTSC is used in the United States, Canada, Central America, some of South America, and Japan. The phase alteration line (PAL) system was standardized in most European countries in 1967 and is also found in China. Many of the PAL standards were written by the CCIR. As a result, PAL is sometimes called CCIR. In 1967, France introduced SECAM (Sequentiel colour avec mémoire). SECAM is used in France, Russia, and the former Soviet bloc nations. Reference 22 provides a complete worldwide listing. These standards are incompatible for a variety of reasons.

In the NTSC and EIA 170 standards, a frame is composed of 525 scan lines separated into two sequential fields (even and odd). Because time is required for vertical retrace, only 485 lines are displayed. There are 242.5 active lines per field. Because field 1 ends in a half-line, it is called the odd field. Figure 7-11 illustrates how the information is presented on a display. For commercial television, the aspect ratio (horizontal to vertical extent) is 4:3. Solid state arrays do not have "half-lines". The half-line is blanked out. Since monitors usually operate in an overscan mode, the half-lines are never seen.

When a camera is designed to be compatible with a transmission standard, the manufacturer uses the broadcast terminology as part of the camera specifications. Similarly, consumer display manufacturers use the same terminology.

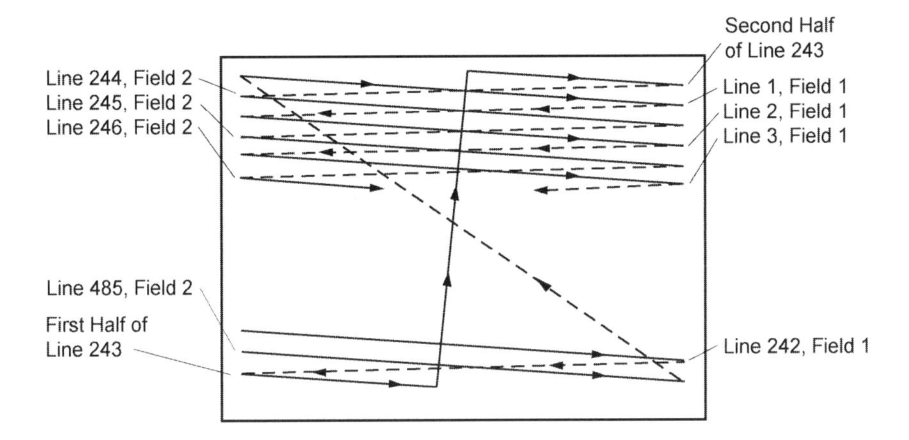

Figure 7-11. Presentation of EIA 170 video on a display. Standard timing is defined for video transmission only. Solid state arrays may not read out data in this precise order. The camera image processing algorithm formats the data into a serial stream that is consistent with the video standard.

7.5.1. "CONVENTIONAL" VIDEO TIMING

Because EIA 170 and NTSC are so pervasive, nearly all cameras are built to these video standards. This permits saving imagery on off-the-shelf video recorders and displaying it on just about any monitor. With progressive-scan cameras, the camera's internal electronics must reformat the data into an interlace mode. If the camera's output is fed directly into a computer, timing is no longer a concern. If digital data are transferred, only a sync signal is necessary to indicate the start of the data. If analog data are transferred, then the frame capture board should have a resolution that is greater than the camera's. That is, there should be more datels than pixels.

Table 7-2 provides the relevant parameters for the monochrome and color video standards. The standards specify the line frequency and blanking period. The scanning frequency is the frame rate multiplied by the number of lines. The total line time is calculated from the line frequency (e.g., 1/15,750 = 63.492 μs) and the active line time is the total line time minus the horizontal blanking period. The blanking period has a tolerance of about ±0.25% so that the active time line also varies slightly. Table 7-2 provides the minimum active time line. It is assumed that the camera field-of-view exactly fills the active line time.

However, there is no requirement to do so. If the FOV is less, then vertical black lines will appear on either side of the display. Since displays typically operate in the overscan mode, the black lines will not be seen. Everyone uses slightly different values for the active lime time.

Table 7-2
STANDARD VIDEO TIMING

FORMAT	FRAME RATE (Hz)	LINES PER FRAME	ACTIVE LINES	TOTAL LINE TIME (μs)	MINIMUM ACTIVE LINE TIME (μs)
EIA 170	30	525	485	63.492	52.092
NTSC	29.97	525	485	63.555	52.456
PAL	25	625	575	64.0	51.7
SECAM	25	625	575	64.0	51.7

Current camera design philosophy matches the number of detectors to the number of pixels supported by the video bandwidth. The actual number of detector elements depends on the manufacturer's philosophy. Image processing algorithm complexity is minimized when the spatial sampling frequency is equal in the horizontal and vertical directions. The 4:3 aspect ratio suggests that there should be 4/3 times more detectors in the horizontal direction compared to the vertical direction. If the number of elements in the vertical direction is simply equal to the number of active scan lines, then the number of horizontal detectors should be about 646 for NTSC, and 766 for PAL and SECAM. The number of pixels per line is often chosen so that the horizontal clock rate is an integer multiple of the color subcarrier. For NTSC, 768 pixels equate to four times the NTSC color subcarrier.

For NTSC and EIA 170 compatible arrays, it is common to have about 480 detectors in the vertical direction. The image formatting algorithm inserts blank lines (all black) to meet video standards (485 lines). Because many displays operate in an overscan mode, the blank (black) lines are not noticed.

For higher vertical resolution, more lines are required. EIA 343A (originally RS 343A) was created[23] for high-resolution closed-circuit television cameras (CCTVs). CCTVs have the same formats as TVs but do not broadcast to the public. Usually a cable links the camera to a recorder and/or monitor. Although the standard encompasses equipment that operates from 675 to 1023 lines, the recommended values are 675, 729, 875, 945, and 1023 lines per frame (Table 7-3). The military primarily uses 875 lines/frame but requires "standards conversion" boxes that can record and present imagery on EIA 170 compatible equipment. EIA 343A is still a monochrome standard.

Table 7-3
RECOMMENDED EIA 343A VIDEO FORMATS

LINES/FRAME	ACTIVE LINES	SCANNING FREQUENCY (kHz)	LINE TIME μs	ACTIVE LINE TIME μs
675	624	20.25	49.38	42.38
729	674	21.87	45.72	38.72
875	809	26.25	38.09	31.09
945	874	28.35	35.27	28.27
1023	946	30.69	32.58	25.58

7.5.2. DIGITAL TELEVISION

Digital image processing is already present in most transmitters and receivers. It seems natural to transmit digitally. The major advantage is that digital signals are relatively immune to noise. Either a bit is present or it is not. A digital voltage signal, which may have been reduced in value and had noise added, can be restored to its full value with no noise in the receiver. An all-digital system avoids the multiple analog-to-digital and digital-to-analog conversions that degrade image quality. Digital signal "transmission" was first introduced into tape recorders. Because a bit is either present or not, multiple-generation copies retain high image quality. In comparison, analog recorders, such as VHS, provide very poor quality after just a few generations.

The current accepted meaning of high-definition television (HDTV) is a television system providing approximately twice the horizontal and vertical resolution of present NTSC, PAL, and SECAM systems. While the term HDTV implies a receiver, the industry is equally concerned about transmission standards. The video format must be standardized before building cameras and receivers. The acronym HDTV is slowly being replaced with the advanced television system, or ATS. The formats are provided in Table 7-4.

The goal is to have worldwide compatibility so that HDTV receivers can display NTSC, PAL, and SECAM transmitted imagery. With today's multimedia approach, any new standard must be compatible with a variety of imaging systems ranging from 35 mm film to the various motion picture formats. The desire is to create composite imagery[24,25] and special effects from a mixture of film clips, motion pictures, and 35 mm film photos. Compatible frame rates allow the broadcasting of motion pictures and the filming of television. Reference 26 provides a historical overview of HDTV development.

Table 7-4
HDTV FORMATS
Aspect ratio 16:9

SYSTEM	PIXELS	SCAN PATTERN	TOTAL LINES	FRAME RATE
S1	1280×720	Progressive	750	50 Hz
S2	1920×1080	Interlaced	1125	25 Hz
S3	1920×1080	Progressive	1125	25 Hz
S4	1920×1080	Progressive	1125	50 Hz

7.6. CRT OVERVIEW

Figure 7-12 illustrates a typical color cathode ray tube display. Although a color display is illustrated, the block diagram applies to all CRT displays. The analog input video signal is decoded into its components and then digitized. While digital electronics may be used to process the signal, the CRT is an analog device. Similarly, displays accepting digital inputs must convert the signal into an analog voltage to drive the CRT. Numerous transformations occur to create a visible image. The display matrix and display gain controls allow linear mapping from input image to visible image.[27]

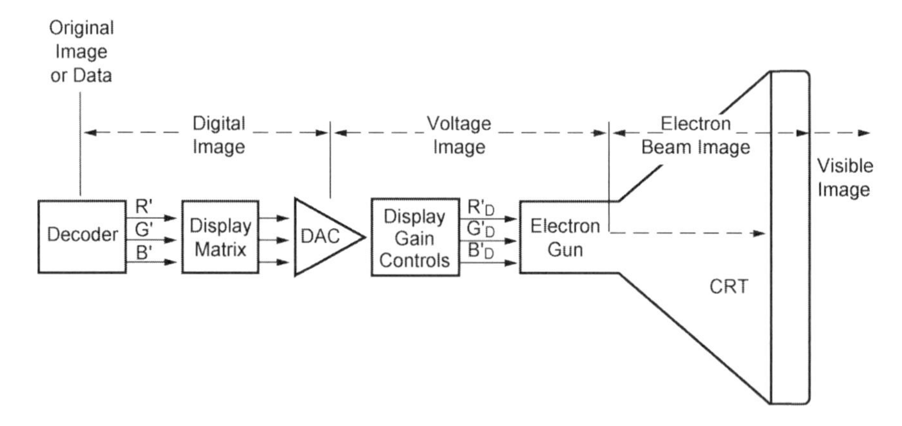

Figure 7-12. Although a color display is illustrated, the block diagram applies to all CRT displays. The digital-to-analog converter (DAC) provides the analog voltages that modulate the electron beam.

The display matrix provides the mapping between the standard video color components and the actual colors emitted by the phosphors on the screen. This color correction circuitry is an algorithm that can be represented as a matrix (Equation 7-1).

With the NTSC format, the electron beams paint an image on the CRT phosphor in the scan format shown in Figure 7-11. The color of the visible image depends on the specific phosphors used. Electron beam density variations represent scene intensity differences. Monochrome displays have a single electron gun and a single phosphor. A color CRT must produce three images (three primary colors) to give a full range of color. To do so requires three electron guns and three phosphors. The three electron beams scan the CRT screen in the same way as a monochrome beam, but arrive at the screen at slightly different angles (discussed in Section 9.10. *CRT display*.) The beams strike three different phosphors' dots. When the dots (*sub*disels) are beyond the eye's limiting resolution, the eye blends them to create the illusion of a full color spectrum (discussed in Section 9.13. *The observer*).

Digital displays provide excellent alphanumeric and graphic imagery. There is no distortion, and the disels are accurately located because there is no jitter in timing signals compared to what might be expected with EIA 170. "Digital" seems to imply high resolution, but the displayed resolution is limited by the electron beam spot size. Although called "digital," the CRT is an analog device.

7.6.1. MONOCHROME DISPLAYS

Many industrial and scientific applications use monochrome cameras. The military uses monochrome monitors extensively to display imagery from non-visible scenes (e.g., thermal and radar imagery). Monochrome displays are found in cockpits and medical imaging instruments. For television receivers, these are simply called black-and-white TVs. Depending on the phosphor used, monochrome displays can provide nearly any color desired. Certain colors (e.g., orange) tend to reduce eye strain. Others (e.g., blue) tend not to disrupt scoptic vision adaptation. Although computers have color monitors, word processing is often performed in a monochrome format.

Black-and-white scenes can be imaged on a color monitor by assigning a color palette (false color or pseudo color) to various intensity levels. This does not increase the resolution, but draws the observer's attention to selected features. The eye can easily differentiate color differences but may have trouble distinguishing small intensity differences.

7.6.2. COLOR DISPLAYS

A color monitor has three electron beams that scan the CRT faceplate. Near the faceplate is a metal mask that has a regular pattern of holes (Figure 7-13). With the dot arrangement, it is called as shadow mask. With slots, it is called an aperture grille. The three guns are focused onto the same hole, and the emerging beams strike the CRT faceplate at different locations. At each location there is a different phosphor. One fluoresces in the blue, one in the red, and one in the green region of the spectrum. The arrangement is such that the beam corresponding to the desired color will strike only the phosphor dot producing that color. The electron guns emit only electrons and have no color associated with them. Color is only created when the beam hits the phosphor even though the guns are called red, green, and blue guns.

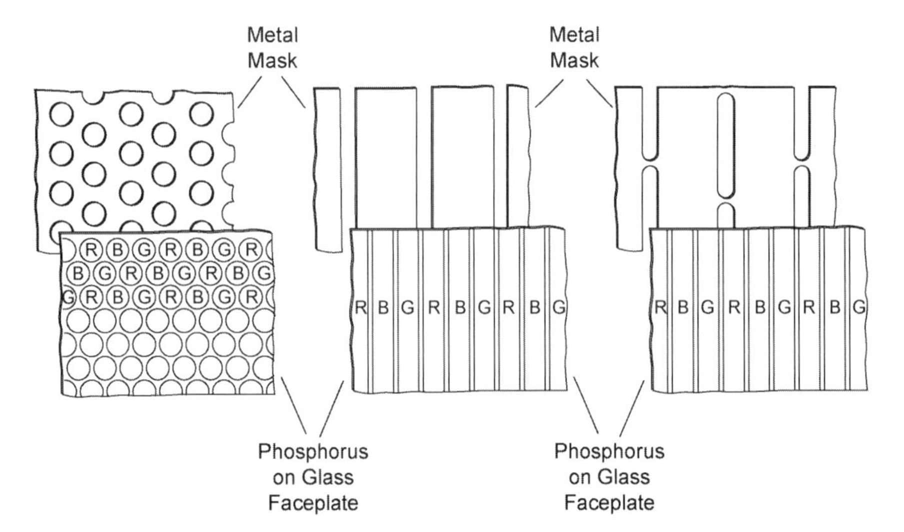

Figure 7-13. Color CRT mask patterns. (a) Dotted, (b) stripes, and (c) slotted. Most consumer television receivers use slots whereas computer displays use dots. The mask design varies with manufacturer.

Hexagonal structure masks (the dots) are inherently more rigid than slots. However, slots are easier to manufacture and therefore are usually found in television receivers. The slots have a larger area and therefore aperture-grille CRTs tend to be brighter than shadow-mask CRTs. With this brightness it is said that the colors are richer. The distance between dots is less and more difficult to perceive. Shadow-mask CRTs tend to provide sharper images.

Because computer monitors are viewed at a close distance, the dots are used most often. The selection of slots over dots depends upon the application.

Dots and stripes cannot be easily compared because the distance between adjacent spots is different and depends upon orientation and the way measured. Because there is $\sqrt{3}/2$ difference, a 0.24 mm aperture-grille pitch is roughly equal to a 0.28 mm shadow mask. Figure 7-14 illustrates one possible arrangement of color dots. The dot pitch is the center-to-center spacing between adjacent like-colored dots. The three dots create a triad. Because each electron beam passes through the same hole, the mask pitch is the same as the triad pitch. Table 7-5 relates the triad dot pitch to the more common definitions of CRT resolution.

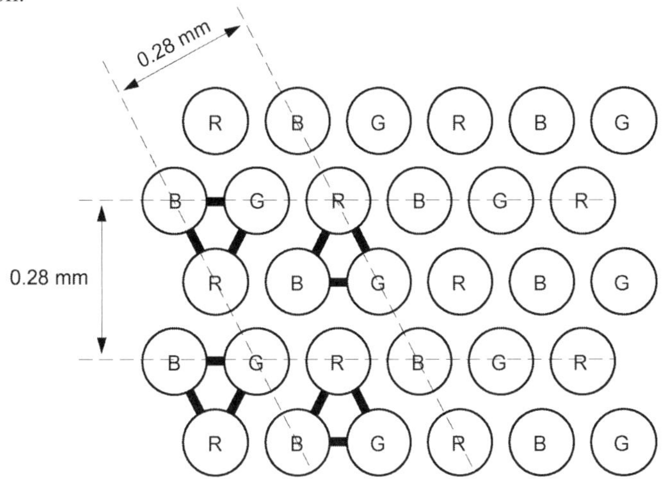

Figure 7-14. Arrangement of phosphor dots on a CRT. There is one shadow-mask hole for each triad (shown with heavy lines). Most computer monitors have a triad pitch of 0.28 mm.

Table 7-5
RESOLUTION OF COLOR DISPLAYS

RESOLUTION	TRIAD DOT PITCH
Ultra-high	< 0.27 mm
High	0.27 - 0.32 mm
Medium	0.32 - 0.48 mm
Low	> 0.48 mm

7.7. FLAT PANEL DISPLAYS

Flat-panel displays consist of rectangular elements where the horizontal and vertical disel (triad) pitches are equal (Figure 7-15).

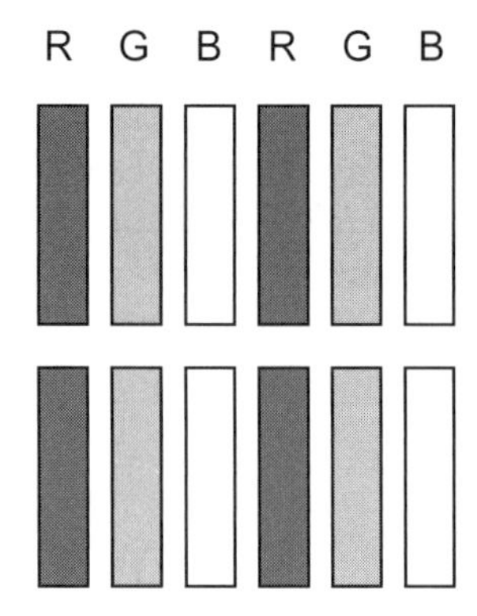

Figure 7-15. Flat-panel display elements. The triad is square.

Flat panels do not offer the color gamut nor brightness dynamic range that is available with CRTs. This suggests that in critical applications, CRT should be used. However, flat panels dominate the display market because of power, size, and weight advantages. They are still more expensive than CRTs, but not significantly. The output luminance versus input[28] tends to be S-shaped (Figure 7-16).

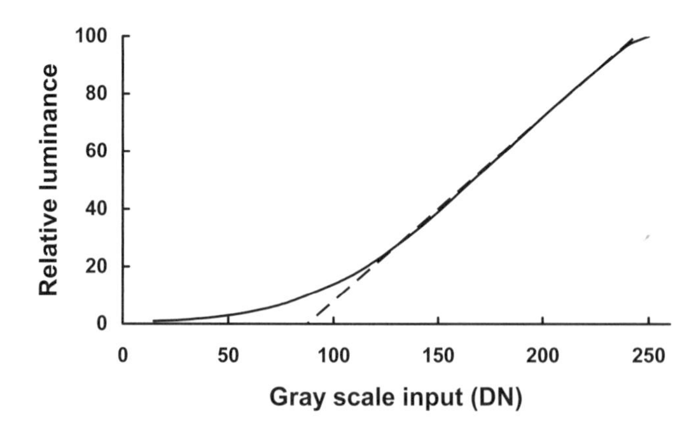

Figure 7-16. Typical flat-panel display response. The linear fit (dashed line) suggests that the display's dynamic range is about 6 bits.

7.8. COMPUTER INTERFACE

There is no requirement that the camera's analog output be precisely matched to the active time line. If the FOV is less, vertical black lines will appear on either side of the display. Since displays typically operate in the overscan mode, the black lines will not be seen. Similarly, frame grabbers (frame capture board) are designed to operate at the standard video formats. Its internal ADC clock rate depends upon the manufacturer's assumptions about the active line time. As a result, a mismatch[29] results in edge ambiguity and may create moiré patterns in periodic structures. DNL and INL problems may exist within a frame capture board; he user must test the board for any nonlinearities. Because the frame capture board usually resides within a computer, testing is relatively easy.

A frame capture board is still necessary for a digital output camera. It acts as a buffer between the camera and the computer. A sych signal identifies valid data for the board.

7.9. REFERENCES

1. N. Schuster, "Optical Systems for High-resolution Digital Still Cameras," in *EUROTO Conference on Design and Engineering of Optical Systems*, SPIE Proceedings Vol. 3737, pp. 202-213 (1999).
2. C. Sabolis "Seeing is Believing," *Photonics Spectra*, pp. 119-124, October 1993.
3. T. S. Lomheim, J. D. Kwok, T. E. Dutton, R. M. Shima, J. F. Johnson, R. H. Boucher, and C. Wrigley, "Imaging Artifacts due to Pixel Spatial Sampling Smear and Amplitude Quantization in Two-dimensional Visible Imaging Arrays," in *Infrared Imaging Systems: Design, Analysis, Modeling, and Testing* X, G. C. Holst ed., SPIE Vol. 3701, pp 36-60 (1999).
4. D. H. Sheingold, *Analog-Digital Conversion Handbook*, pp. 317-337, Prentice-Hall, Englewood Cliffs, NJ (1986).
5. K. B. Benson, ed., revised by J. Whitaker, *Television Engineering Handbook*, revised edition, Chapter 2 and pp. 22.29 - 22.30, McGraw-Hill, New York, NY (1992).
6. H. R. Kang, *Color Technology for Electronic Imaging Devices*, pp. 3-101, SPIE Press, Bellingham, WA (1997).
7. K. A. Parulski, L. J. D'Luna, and R. H. Hubbard, "A Digital Color CCD Imaging Systems using Custom VLSI Circuits," *IEEE Transactions on Consumer Electronics*, Vol. CE-35(8), pp. 382-388 (1989).
8. S. Nishikawa, H. Toyoda, V. Miyakawa, R. Asada, Y. Kitamura, M. Watanabe, T. Kiguchi, and M. Taniguchi, "Broadcast-Quality TV Camera with Digital Signal Processor," *SMPTE Journal*, Vol. 99(9), pp. 727-733 (1990).
9. Y. Takemura, K. Ooi, M. Kimura, K. Sanda, A. Kuboto, and M. Amano, "New Field Integration Frequency Interleaving Color Television Pickup System for Single Chip CCD Camera," *IEEE Transactions on Electron Devices*, Vol. ED-32(8), pp. 1402-1406 (1985).
10. H. Sugiura, K. Asakawa, and J. Fujino, "False Color Signal Reduction Method for Single Chip Color Video Cameras," *IEEE Transactions on Consumer Electronics*, Vol. 40(2), pp. 100-106 (1994).

11. J.-C. Wang, D.-S. Su, D.-J. Hwung, and J.-C. Lee, "A Single Chip CCD Signal Processor for Digital Still Cameras," *IEEE Transactions on Electron Devices*, Vol. 40(3), pp. 476-483 (1994).
12. T. Kuno, H. Sugiura, M. Asamura, and Y. Hatano, "Aliasing Reduction Method for Color Digital Still Cameras with a Single-chip Charged-coupled Device," *Journal of Electronic Imaging*, Vol. 8, pp 457-466 (1999).
13. M. R. Gupta and T. Chen, "Vector Color Filter Array Demosaicing," in *Sensors and Camera Systems for Scientific, Industrial, and Digital Applications* II, M. M. Blouke, J. Canosa, and N. Sampat, eds., SPIE Proceedings Vol. 4306, pp. 374-382 (2001).
14. O. Kalevo and H. Rantanen, "Sharpening Methods for Images Captured Through Bayer Matrix," in *Sensors and Camera Systems for Scientific, Industrial, and Digital Applications* IV, M. M. Blouke, N. Sampat, and R. J. Motta, eds., SPIE Proceedings Vol. 5017, pp. 286-297 (2003).
15. N. Kehtarnavaz, H-j.Oh, and Y. Yoo, "Color Filter Array Interpolation Using Correlations and Directional Derivatives," *Journal of Electronic Imaging*, Vol. 12, pp 621-632 (2003).
16. I. Tsubaki and K. Aizawa, "A Restoration and Demosaicing Method for a Pixel Mixture Image," in *Sensors and Camera Systems for Scientific, Industrial, and Digital Applications* V, M. M. Blouke, N. Sampat, and R. J. Motta, eds., SPIE Proceedings Vol. 5301, pp. 346-355 (2004).
17. M. Shao, K. E. Barner, R. C. Hardie, "Partition-based Interpolation for Color Filter Array Demosaicking and Super-resolution Reconstruction," *Optical Engineering*, Vol. 44, paper 107003 (2005).
18. W. Feng, and S. J. Reeves, "Estimation of Color Filter Array Data from JPEG Images for Improved Demosiacking," in *Computation Imaging IV*, C. A. Bouman. E. L. Miller, and I. Pollack, eds., SPIE Proceedings Vol. 60065, paper 606510 (2006).
19. T. Olson, "Behind Gamma's Disguise," *SMPTE Journal*, Vol. 104(7), pp. 452-458 (1995).
20. C. Poynton, *Digital Video and HDTV: Algorithms and Interfaces*, Chapter 23, Morgan Kaufmann (2003).
21. The timing requirements can be found in many books. See, for example, D. H. Pritchard, "Standards and Recommended Practices," in *Television Engineering Handbook*, revised edition, K. B. Benson, ed., revised by J. Whitaker, pp. 21.40 - 21.72, McGraw-Hill, New York, NY (1992).
22. D. H. Pritchard, "Standards and Recommended Practices," in *Television Engineering Handbook*, revised edition, K. B. Benson, ed., revised by J. Whitaker, pp. 21.73 - 21.74, McGraw-Hill, New York, NY (1992).
23. "EIA Standard EIA-343A, Electrical Performance Standards for High Resolution Monochrome Closed Circuit Television Camera," Electronic Industries Association, 2001 Eye Street, NW Washington, D.C. 20006 (1969).
24. L. J. Thorpe, "HDTV and Film - Issues of Video Signal Dynamic Range," *SMPTE Journal*, Vol. 100(10), pp. 780-795 (1991).
25. L. J. Thorpe, "HDTV and Film - Digitization and Extended Dynamic Range," *SMPTE Journal*, Vol. 102(6), pp. 486-497 (1993).
26. Y. Wu, S. Hirakawa, U. H. Reimers, and J. Whitaker, "Overview of Digital Television Development Worldwide" *Proceeding of the IEEE* Vol. 94, pp.8 – 21 (2006). This January 2006 issue is the "Special Issue On Global Digital Television: Technology And Emerging Services"
27. W. Cowan, "Displays for Vision Research," in *Handbook of Optics*, M. Bass, ed., Second edition, Vol. 1, Chapter 27, McGraw-Hill, New York, NY (1995).
28. Flat Panel Display Measurements Standard, Version 2, Manual FPDM Video Electronics Standards Association (VESA), (June 1, 2001)
29. P. Cencik, "Matching Solid State Camera with Frame Grabber – A Must for Accurate Gauging," in *Optics, Illumination, and Image Sensing for Machine Vision* VI, SPIE Proceedings Vol. 1614, pp 112-120 (1991).

8
LINEAR SYSTEM THEORY

Linear system theory was developed for electronic circuitry and has been extended to optical, electro-optical, and mechanical systems. It forms an indispensable part of system analysis. For modeling purposes, electronic imaging systems are characterized as linear spatial-temporal systems that are shift-invariant with respect to both time and two spatial dimensions. Although space is three-dimensional, an imaging system displays only two dimensions.

Electrical filters are different from spatial filters in two ways. They are single-sided in time and must satisfy the causality requirement that no change in the output may occur before the application of an input. Optical filters are double-sided in space. Electrical signals may be either positive or negative, whereas optical intensities are always positive. As a result, optical designers and circuit designers often use different terminology.

With a linear-shift-invariant (LSI) system, signal processing is linear. A shift of the input causes a corresponding shift in the output. For electronic circuits, the shift is in time, whereas for optical systems, the shift is in space. In additional, the input-to-output mapping is single-valued. For optical systems, one additional condition exists: the radiation must be incoherent if the system is to be described in terms of radiance.

A sampled-data system may be considered "globally" shift-invariant on a macro scale. For example, with an electronic imaging system, as a target moves from the top of the field of view to the bottom, the image also moves from the top to the bottom. On a micro scale, moving a point source across a single detector does not change the detector output. That is, the system is not shift-invariant on a micro scale. Similarly, with electronic circuits, large time shifts of the input produce corresponding large time shifts in output. A shift during a sample time, does not change the output. Here also, the system is not shift-invariant on a micro scale.

Single-valued mapping only occurs with non-noisy and non-quantized systems. No system is truly noiseless, but can be approximated as one when the signal has sufficient amplitude. With digital systems, the output is quantized. The smallest output is the least significant bit, and the analog-to-digital converter generally limits the largest signal. If the signal level is large compared to the least significant bit, then the system can be treated as quasi-linear over a restricted region.

In spite of the disclaimers mentioned, systems are treated as quasi-linear and quasi-shift-invariant over a restricted operating region to take advantage of the wealth of mathematical tools available. An LSI system merely modifies the amplitude and phase of the input. These are specified by the modulation transfer function and phase transfer function.

The symbols used in this book are summarized in the *Symbol List*, which appears after the *Table of Contents*.

8.1. LINEAR SYSTEM THEORY

A linear system provides single mapping from an input to an output. Superposition allows the addition of individual inputs to create a unique output. Let $h\{\ \}$ be a linear operator that maps one function, $f(t)$, into another function, $g(t)$

$$h\{f(t)\} = g(t) \tag{8-1}$$

Let the response to two inputs, $f_1(t)$ and $f_2(t)$, be $g_1(t)$ and $g_2(t)$:

$$h\{f_1(t)\} = g_1(t) \quad and \quad h\{f_2(t)\} = g_2(t) \tag{8-2}$$

For a linear system, the response to a sum of inputs is equal to the sum of responses to each input acting separately. For any arbitrary scale factors, the superposition principle states

$$h\{a_1 f_1(t) + a_2 f_2(t)\} = a_1 g_1(t) + a_2 g_2(t) \tag{8-3}$$

When passing the signal through another linear system, the new operator provides

$$h_2\{g(t)\} = h_2\{h_1\{f_1(t)\}\} \tag{8-4}$$

8.1.1. TIME VARYING SIGNALS

An electrical signal can be thought of as the sum of an infinite number of impulses (Dirac delta functions) located inside the signal boundaries. Thus, the signal can be decomposed into a series of weighted Dirac delta functions (Figure 8-1)

$$v_{IN}(t) = \sum_{t'=-\infty}^{\infty} v_{IN}(t')\delta(t-t') \tag{8-5}$$

where $\delta(t - t')$ is the Dirac delta or impulse function at $t = t'$ and $v_{IN}(t')$ is the signal evaluated at t'.

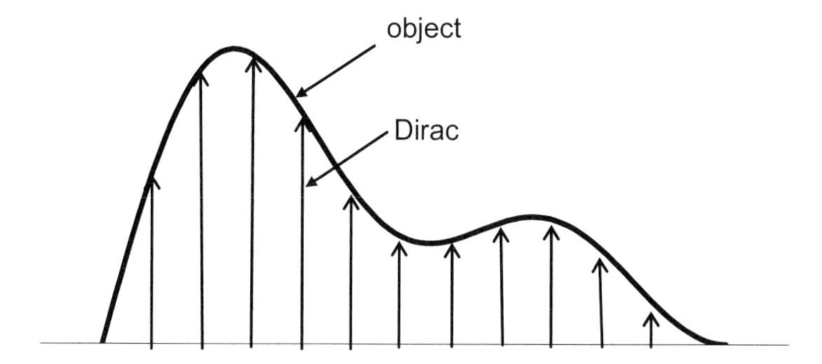

Figure 8-1. A time-varying signal can be decomposed into a series of closely spaced impulses with amplitude equal to the signal at that value. They are shown widely separated for clarity and should not be confused with sampling.

If $v_{IN}(t)$ is applied to a linear circuit, using $h\{\ \}$ will provide the output

$$v_{OUT}(t) = \sum_{t'=-\infty}^{\infty} h\{v_{IN}(t')\delta(t - t')\Delta t'\} \qquad (8\text{-}6)$$

Comparing Equation 8-5 to Equation 8-3, for each t', $v_{IN}(t')$ replaced a_i and $\delta(t - t')$ replaced $f_i(t)$. That is, the input has been separated into a series of functions $a_1 f_1(t) + a_2 f_2(t) + \cdots$.

As $\Delta t' \to 0$, this becomes the convolution integral

$$v_{OUT}(t) = \int_{t'=-\infty}^{\infty} v_{IN}(t') h\{\delta(t - t')\} dt' \qquad (8\text{-}7)$$

and is symbolically represented by the convolution operator * (other texts may use ★ or ⊗)

$$v_{OUT}(t) = v_{IN}(t) * h(t) \qquad (8\text{-}8)$$

The system impulse response is $h\{\delta(t)\}$. In Figure 8-2, each impulse response is given for each t'. The individual impulse responses are added to produce the output. This addition is the definition of superposition.

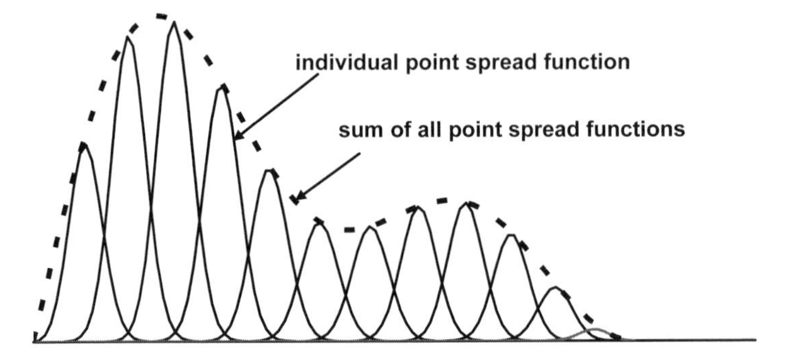

Figure 8-2. The linear operator, $h\{\ \}$, transforms each input Dirac delta into an output impulse response. The sum of the impulse responses creates the output. Although shown separated for clarity, the impulse responses have infinitesimal separations.

The convolution theorem of Fourier analysis states that the Fourier transform of the convolution of two functions equals the product of the transform of the two functions. That is, multiple convolutions are equivalent to multiplication in the frequency domain. Using frequency domain analysis is convenient because multiplications are easier to perform and easier to visualize than convolutions. Then

$$V_{OUT}(f_e) = V_{IN}(f_e)H(f_e) \tag{8-9}$$

The variable f_e denotes electrical frequency (Hz). Upper-case variables represent the Fourier transform of the signal (lower-case variables). Let the input be an impulse. The Fourier transform of an impulse response is

$$V_{IN}(f_e) = 1 \tag{8-10}$$

That is, all frequencies are present. Because the input to the system contains all frequencies, the output must be a function of the system only. Any variation in phase or amplitude must be due to the system itself. The output is

$$V_{OUT}(f_e) = H(f_e) \tag{8-11}$$

8.1.2. SPATIALLY VARYING SIGNALS

The one-dimensional electronic signal decomposition can be easily extended to two-dimensional imagery. An object can be thought of as the sum of an infinite array of impulses located inside the target boundaries. Thus, an object can be decomposed into a two-dimensional array of weighted Dirac delta functions, $\delta(x - x')$, $\delta(y - y')$

$$o(x, y) = \sum_{x'=-\infty}^{\infty} \sum_{y'=-\infty}^{\infty} o(x', y')\delta(x - x')\delta(y - y')\Delta x' \Delta y' \tag{8-12}$$

An optical system with operator $h\{\ \}$ produces an image

$$i(x, y) = \sum_{x'=-\infty}^{\infty} \sum_{y'=-\infty}^{\infty} h\{o(x', y')\delta(x - x')\delta(y - y')\Delta x' \Delta y'\} \tag{8-13}$$

For small increments, this becomes the convolution integral

$$i(x, y) = \int_{-\infty}^{\infty} \int_{-\infty}^{\infty} o(x', y')h\{\delta(x - x')\delta(y - y')\}\, dx' dy' \tag{8-14}$$

It is symbolically represented by the two-dimensional convolution operator **

$$i(x, y) = o(x, y) ** h(x, y) \tag{8-15}$$

The function $h(x,y)$ is the optical system's response to an input impulse. The resulting image is the point spread function (PSF). For an ideal circular aperture, the central portion of the PSF is the Airy disk from which the Rayleigh criterion and other resolution metrics are derived. In one dimension $h(x)$ is the line spread function, LSF. The LSF is the resultant image produced by the imaging system when viewing an ideal line. There is no equivalent interpretation for $h(t)$ for electrical circuits. The function $h(t)$ is simply the impulse response.

The transform of the convolution becomes a multiplication in the frequency domain

$$I(u, v) = O(u, v)H(u, v) \tag{8-16}$$

If all frequencies are present in the image (i.e., an impulse), $O(u,v) = 1$, then $I(u,v) = H(u,v)$. The relationship between spatial frequency (u,v) and electrical frequency (f_e) is further discussed in Section 9.1. *Frequency Domains*.

8.2. ELECTRONIC IMAGING SYSTEM

The electronic imaging system response consists of both the optical response and the electronic response: $h(x,y,t)$. Time and spatial coordinates are treated separately. For example, optical elements do not generally change with time and therefore are characterized only by spatial coordinates. Similarly, electronic circuitry exhibits only temporal responses. The detector provides the interface between the spatial and temporal components, and its response depends on both temporal and spatial quantities. The conversion of two-dimensional optical information to a one-dimensional electrical response assumes a linear photodetection process. Implicit in the detector response is the conversion from input photon flux to output voltage.

The optical transfer function (OTF) plays a key role in the theoretical evaluation and optimization of an optical system. The modulation transfer function (MTF) is the magnitude and the phase transfer function (PTF) is the phase of the complex-valued OTF. In many applications, the OTF is real-valued and positive so that the OTF and MTF are equal. When focus errors or aberrations are present, the OTF may become negative or even complex valued.

Electronic circuitry also can be described by an MTF and PTF. The combination of the optical MTF and the electronic MTF creates the electronic imaging system MTF. The MTF is the primary parameter used for system design, analysis, and specifications.

Symbolically

$$OTF(u,v) = H_{SPATIAL}(u,v) = MTF(u,v)\, e^{j\,PTF(u,v)} \tag{8-17}$$

and

$$H_{ELECTRONICS}(f_e) = MTF(f_e)\, e^{j\,PTF(f_e)} \tag{8-18}$$

With appropriate scaling, the electronic frequencies can be converted into spatial frequencies. This is symbolically represented by $f_e \rightarrow u$. The electronic circuitry is assumed to modify the horizontal signal only (although this depends on the system design). The combination of spatial and electronic responses is sometimes called the system OTF

$$OTF_{SYSTEM}(u,v) = MTF(u,v)\, MTF(f \rightarrow u)\, e^{j[PTF(u,v)+PTF(f\rightarrow u)]} \tag{8-19}$$

For mathematical convenience, the horizontal and vertical responses are considered separable (usually coincident with the detector array axes). The electrical response is considered to affect only the horizontal response

$$H_{SYSTEM}(u) = H_{SPATIAL}(u) H_{ELECTRONICS}(f_e \to u) \tag{8-20}$$

and

$$H_{SYSTEM}(v) = H_{SPATIAL}(v) \tag{8-21}$$

A system is composed of many components that respond to either spatial or temporal signals. For independent MTFs

$$MTF_{SYSTEM}(u,v) = \prod_{i=1}^{N} \prod_{j=1}^{m} MTF_i(u,v) MTF_j(f_e \to u) \tag{8-22}$$

While individual lens elements each have their own MTF, the MTF of the lens system is not usually the product of the individual MTFs. This occurs because one lens may minimize the aberrations created by another.

8.3. MTF and PTF INTERPRETATION

The system MTF and PTF alter the image as it passes through the circuitry. For linear-shift-invariant systems, the PTF simply indicates a spatial or temporal shift with respect to an arbitrarily selected origin. An image where the MTF is drastically altered is still recognizable, whereas large PTF nonlinearities can destroy recognizability. Modest PTF nonlinearity may not be noticed visually except those applications where target geometric properties must be preserved (i.e., mapping or photogrammetry). Generally, PTF nonlinearity increases as the spatial frequency increases. Because the MTF is small at high frequencies, the nonlinear phase shift effect is diminished.

The MTF is the ratio of output modulation to input modulation normalized to unity at zero frequency. While the modulation changes with system gain, the MTF does not. The input can be as small as desired (assuming a noiseless system with high gain) or it can be as large as desired because the system is assumed not to saturate.

Modulation is the variation of a sinusoidal signal about its average value (Figure 8-3). It can be considered as the AC amplitude divided by the DC level. The modulation is

$$MODULATION = M = \frac{V_{MAX} - V_{MIN}}{V_{MAX} + V_{MIN}} = \frac{AC}{DC} \tag{8-23}$$

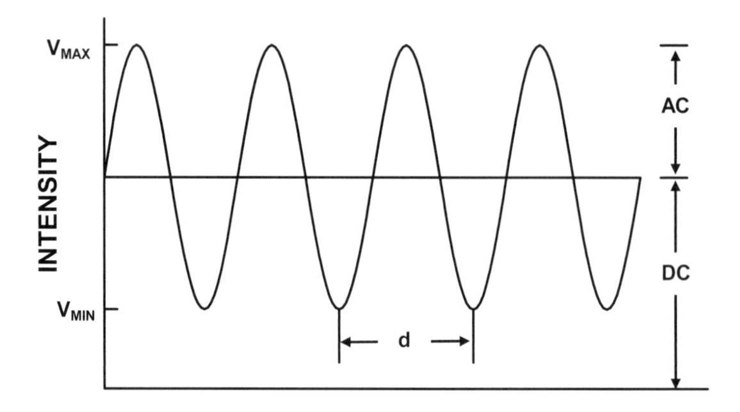

Figure 8-3. Definition of target modulation. The value $\boldsymbol{d_o}$ is the extent of one cycle.

The variables V_{MAX} and V_{MIN} are the maximum and minimum signal levels, respectively. The modulation transfer function is the output modulation produced by the system divided by the input modulation as a function of frequency

$$MTF(f_e) = \frac{M_{OUTPUT}(f_e)}{M_{INPUT}(f_e)} \qquad (8\text{-}24)$$

The concept is presented in Figure 8-4. Three input and output signals are plotted in Figures 8-4a and 8-4b, respectively, and the resultant MTF is shown in Figure 8-4c. As a ratio, the MTF is a relative measure with values ranging from zero to one.

The MTF is a measure of how well a system will faithfully reproduce the input. The highest frequency that can be faithfully reproduced is the system cutoff frequency. If the input frequency is above the cutoff, the output will be proportional to the signal's average value with no modulation.

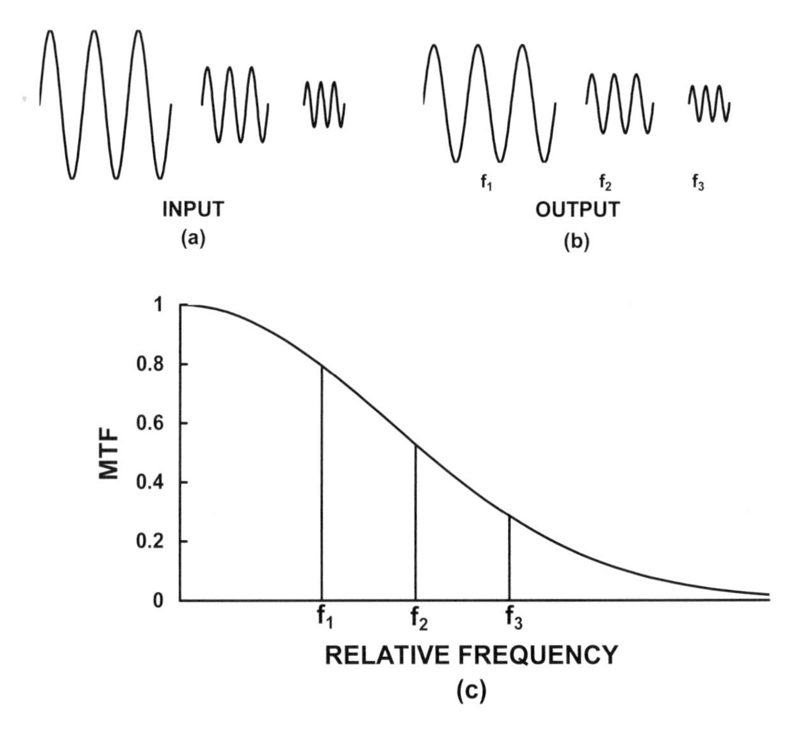

Figure 8-4. Modulation transfer function. (a) Input signal for three different spatial frequencies, (b) output for the three frequencies, and (c) the MTF is the ratio of output-to-input modulation.

8.4. SUPERPOSITION APPLIED to OPTICAL SYSTEMS

If the system MTF is known, the image for any arbitrary object can be computed. First, the object is dissected into its constituent spatial frequencies (i.e., the Fourier transform of the object is obtained). Next, each of these frequencies is multiplied by the system MTF at that frequency. Then, the inverse Fourier transform provides the image.

To illustrate the superposition principle and MTF approach, we will show how an ideal optical system modifies an image. An ideal optical system is, by definition, a linear phase shift system. The most popular test target consists of a series of bars: typically three or four bars, although more may be used. For illustrative purposes, the periodic bars are assumed of infinite extent. A one-dimensional square wave, when expanded into a Fourier series about the origin, contains only odd harmonics

$$o(x) = \frac{1}{2} + \frac{2}{\pi}\sum \frac{1}{n}\sin\left(\frac{2\pi\, n\, x}{d_o}\right) \quad n = 1, 3, 5, \cdots \tag{8-25}$$

The variable d_o is the square wave period. The fundamental frequency u_o is $1/d_o$. Note that the peak-to-peak amplitude of the fundamental is $4/\pi$ times the square wave amplitude. In the frequency domain, the square wave provides discrete spatial frequencies of $u_n = 1/d_o, 3/d_o, 5/d_o, \cdots$, with amplitudes $2/\pi, 2/3\pi, 2/5\pi, \cdots$, respectively.

Let a circular optical system image the square wave. Consider a circular, clear aperture, whose cutoff is u_C. The diffraction-limited MTF is

$$MTF_{OPTICS}(u) = \frac{2}{\pi}\left[\cos^{-1}\left(\frac{u}{u_C}\right) - \left(\frac{u}{u_C}\right)\sqrt{1 - \left(\frac{u}{u_C}\right)^2}\right] \quad when\ u < u_C$$
$$= 0 \quad elsewhere \tag{8-26}$$

By superposition, the diffraction-limited optical MTF and square wave Fourier series amplitudes are multiplied together at each spatial frequency:

$$I(u_n) = MTF_{OPTICS}(u_n)O(u_n) \tag{8-27}$$

Taking the inverse Fourier transform provides the resultant image. Equivalently,

$$i(x) = \frac{1}{2} + MTF_{OPTICS}(u_o)\left[\frac{2}{\pi}\sin(2\pi\, u_o x)\right]$$
$$+ MTF_{OPTICS}(3u_o)\left[\frac{2}{3\pi}\sin(6\pi\, u_o x)\right] + \ldots. \tag{8-28}$$

If u_o is greater than $u_C/3$, only the fundamental of the square wave will be faithfully reproduced by the optical system. Here, the square wave will appear as a sine wave. Note that the optical MTF will reduce the image amplitude. As u_o decreases, the image will look more like a square wave (Figure 8-5). Because the optics does not pass any frequencies above u_C, the resultant wave form is a truncation of the original series modified by the optical MTF. As the input frequency approaches u_C (Figure 8-5d), the modulation is barely present in the image. However, the average scene intensity approaches 0.5. That is, the scene detail can no longer be perceived at or above u_C. *The scene did not disappear.*

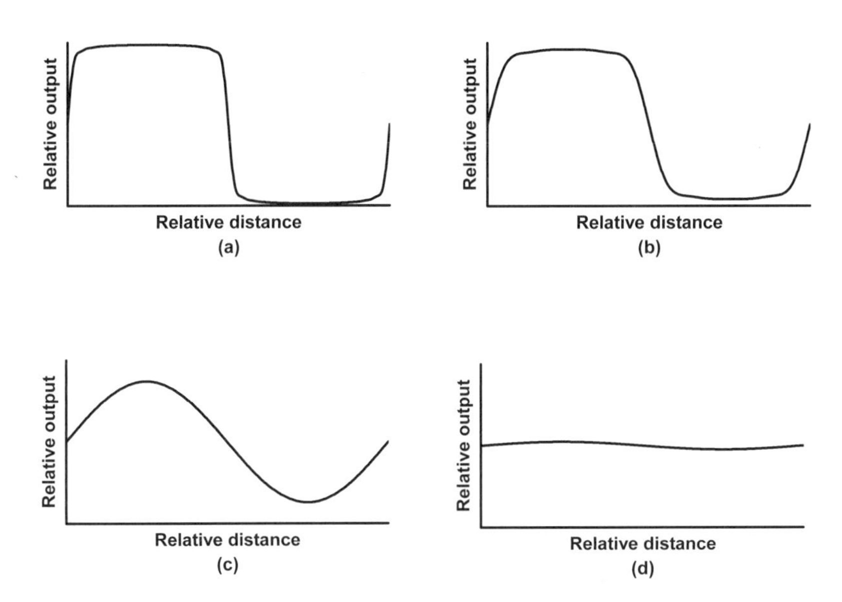

Figure 8-5. A square wave modified by a circular aperture as a function of x/d_o. As the square-wave fundamental frequency increases, the edges become rounded. It will appear as a sinusoid when $u_o \geq u_C/3$. (a) $u_o = u_C/30$, (b) $u_o = u_C/10$, (c) $u_o = u_C/3$, and (d) $u_o = 0.9\ u_C$.

9

MTF

The equations describing array performance (Chapter 6) assume that the image is very large and illuminates many detector elements. Equivalently, the object contained only very low spatial frequencies and that the system MTF is essentially unity over these spatial frequencies. The image consists of many different frequencies, each of which has different amplitude (frequency spectrum). The optics and detector MTFs modify the spectral amplitudes. These amplitudes may also be affected by electronic filter MTFs, although these filters are typically designed to pass the entire image frequency spectrum without attenuation.

A solid state camera consists of optics, an array of pixels, and a signal chain with electronic frequency response characteristics. The camera MTF, as presented in this chapter, is

$$MTF_{SYS} = MTF_{MOTION} MTF_{OPTICS} MTF_{DETECTOR} MTF_{SIGNAL-CHAIN} \quad (9\text{-}1)$$

Camera manufacturers have no control over how an observer will process the imagery. The perceived MTF depends upon display characteristics and the human visual system (HVS) interpretation

$$MTF_{PERCEIVED} = MTF_{SYS} MTF_{DISPLAY} MTF_{HVS} \quad (9\text{-}2)$$

Although not truly separable, for convenience, the horizontal MTF, MTF($\boldsymbol{u}$), and vertical MTF, MTF($\boldsymbol{v}$), are analyzed separately. Electrical filters are causal, one-dimensional, and considered to operate in the direction of the readout clocking only. This chapter presents all the MTFs associated with an analog system. Spatial sampling, present in all imaging systems, corrupts imagery. The system is not longer shift invariant with respect to the target location. Sampling defines the highest spatial frequency (Nyquist frequency) that can be faithfully reproduced. Its MTF is ill-defined but can be represented by average-type MTF. Sampling issues are provided in the next chapter.

As the MTF increases, edge sharpness increases. MTF requirements tend to be more important in high-performance applications (e.g., scientific, military, and NASA applications). Because of diffusion, CMOS tends to have a lower MTF than an equivalent CCD. This suggests that CCDs should be used in the most demanding applications.

The symbols used in this book are summarized in the *Symbol List,* which appears after the *Table of Contents*.

9.1. FREQUENCY DOMAINS

There are four different locations where spatial frequency domain analysis is appropriate. They are object space (before the camera optics), image space (after the camera optics), display, and observer (at the eye) spatial frequencies. Simple equations relate the spatial frequencies in these domains.

Figure 9-1 illustrates the spatial frequency associated with a bar target. Bar patterns are the most common test targets and are characterized by their fundamental spatial frequency. Using the small angle approximation, the angle subtended by one cycle (one bar and one space) is $d_o/\mathbf{R}_1$, where d_o is the spatial extent of one cycle and $\mathbf{R}_1$ is the distance from the imaging system entrance aperture to the target. When using a collimator to project the targets at apparently long ranges, the collimator focal length replaces $\mathbf{R}_1$.

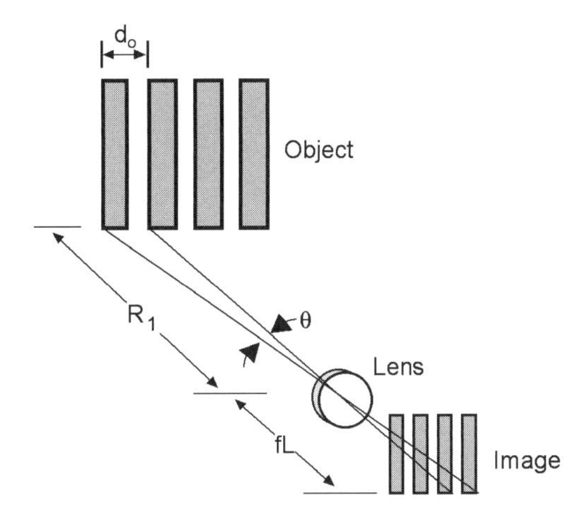

Figure 9-1. Correspondence of spatial frequencies in object and image space. Although the MTF is defined for sinusoidal signals, the bar target is the most popular test target.

The horizontal object-space spatial frequency, u, is the inverse of the horizontal target angular subtense and is usually expressed in cycles/mrad

$$u = \frac{1}{1000}\left(\frac{R_1}{d_o}\right) \quad \frac{\text{cycles}}{\text{mrad}} \tag{9-3}$$

A similar equation exists for the vertical object-space spatial frequency. Optical designers typically quote spatial frequencies in image space to specify the resolving capability of lens systems. Image-space spatial frequency is

$$u_i = \frac{u}{fl} \quad \frac{\textbf{line pairs}}{\textbf{mm}} \quad or \quad \frac{\textbf{cycles}}{\textbf{mm}} \tag{9-4}$$

The variable u_i is the inverse of one cycle in the focal plane of the lens system. Although the nomenclature is used interchangeably, line-pairs suggest square wave targets and cycles suggest sinusoidal targets. To maintain dimensionality, if u_i is measured in cycles/mrad then the focal length must be measured in meters to obtain cycles/mm.

The field of view and monitor size link image spatial frequency to the horizontal and vertical display spatial frequencies:

$$u_d = \frac{HFOV \bullet fl}{W_{MONITOR}} u_i \quad \frac{\textbf{cycles}}{\textbf{mm}} \tag{9-5}$$

and

$$v_d = \frac{HFOV \bullet fl}{H_{MONITOR}} v_i \quad \frac{\textbf{cycles}}{\textbf{mm}} \tag{9-6}$$

Both the monitor width, $W_{MONITOR}$, and height, $H_{MONITOR}$, are usually measured in millimeters. When the monitor aspect ratio is the same as the FOV ratio (the usual case)

$$\frac{HFOV}{W_{MONITOR}} = \frac{VFOV}{H_{MONITOR}} \tag{9-7}$$

The spatial frequency presented to the observer depends on the observer's viewing distance, D, and the image size on the display (Figure 9-2). Observer response is assumed to be rotationally symmetric so that the horizontal and vertical responses are the same. The usual units are cycles/deg

$$u_{eye} = \frac{1}{2\tan^{-1}\left(\frac{1}{2Du_d}\right)} \quad \frac{cycles}{\textbf{deg}} \tag{9-8}$$

Here, the arc tangent is expressed in degrees. Because image detail is important, the small angle approximation provides

$$u_{eye} = 0.01745 D u_d \quad \frac{cycles}{\deg} \tag{9-9}$$

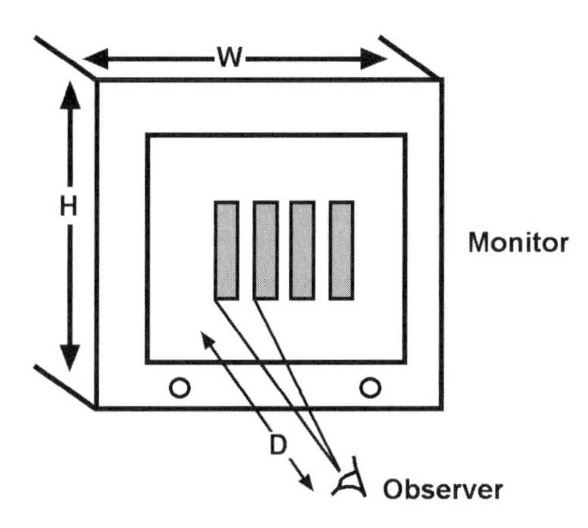

Figure 9-2. Correspondence of spatial frequencies for the display and for the eye.

Analog electronic filters are one-dimensional and modify a serial data stream. The electrical frequency is related to the horizontal field of view (HFOV) and the time it takes to read out one line. For analog filters immediately following the detector array

$$f_e = \frac{HFOV \bullet fl}{t_{H-LINE}} u_i = \frac{(N_H - 1) d_{CCH} + d_H}{t_{H-LINE}} u_i \qquad Hz. \tag{9-10}$$

The readout time is $t_{H\text{-}LINE}$.

After the digital-to-analog converter, the serial stream data rate (and therefore the filter design) is linked to the video timing. The active line time, $t_{VIDEO\text{-}LINE}$, (Table 7-2) and array size provide the link between image space and video frequencies:

$$f_v = \frac{HFOV \bullet fl}{t_{VIDEO-LINE}} u_i = \frac{(N_H - 1) d_{CCH} + d_H}{t_{VIDEO-LINE}} u_i \quad Hz \tag{9-11}$$

Table 9-1 provides the transform pairs.

Table 9-1
TRANSFORM PAIR UNITS

SUBSYSTEM	SIGNAL VARIABLES	FREQUENCY VARIABLES
Optics, detector	x denotes distance (mm) y denotes distance (mm)	u_i expressed as cycles/mm v_i expressed as cycles/mm
Optics, detector	θ_x denotes angle (mrad) θ_y denotes angle (mrad)	u expressed as cycles/mrad v expressed as cycles/mrad
Analog electronics	t denotes time (sec)	f_e expressed as cycles/s (Hz)
Video electronics	t denotes time (sec)	f_v expressed as cycles/s (Hz)
Display	x denotes distance (mm) y denotes distance (mm)	u_d expressed as cycles/mm v_d expressed as cycles/mm
Observer	θ_x denotes angle (deg) θ_y denotes angle (deg)	u_{EYE} expressed as cycles/deg v_{EYE} expressed as cycles/deg

9.2. OPTICS MTF

Optical systems consist of several lenses or mirror elements with varying focal lengths and varying indices of refraction. Multiple elements are used to minimize aberrations. While individual MTFs are used for design and fabrication, these individual MTFs cannot be cascaded to obtain the optical system MTF. For modeling purposes, the optical system is treated as a single lens that has the same effective focal length as the lens system.

Optical spatial frequency is two-dimensional, with the frequency ranging from -∞ to +∞. The diffraction-limited MTF for a circular aperture was first introduced in Section 8.4. *Superposition Applied to Optical Systems*. For an aberration-free and radially symmetric optical system, OTF_{OPTICS} is the same in the horizontal and vertical directions. Since the OTF is positive, it is labeled as the MTF. In the horizontal direction, the diffraction-limited MTF is

$$MTF_{DIFFRACTION}(u_i) = \frac{2}{\pi}\left[\cos^{-1}\left(\frac{u_i}{u_{iC}}\right) - \frac{u_i}{u_{iC}}\sqrt{1-\left(\frac{u_i}{u_{iC}}\right)^2}\right] \qquad (9\text{-}12)$$

The image-space optics cutoff is $u_{iC} = D_o/(\lambda fl) = 1/(F\lambda)$, where D_o is the aperture diameter and fl is the effective focal length. Figure 9-3 illustrates MTF-$_{\text{OPTICS}}$ as a function of u_i/u_{iC} and imagery is provided in Figure 9-4.

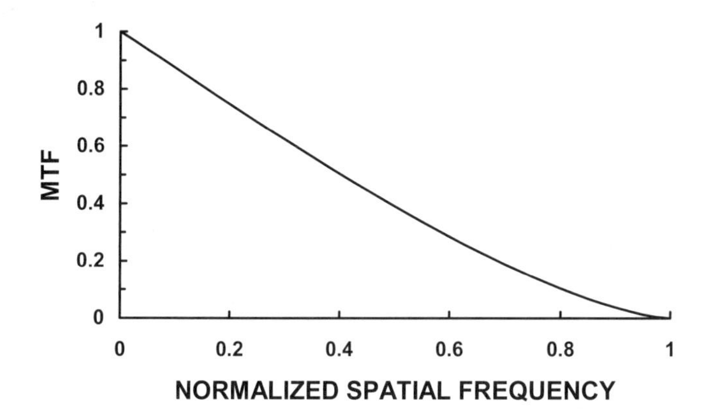

Figure 9-3. MTF_{OPTICS} for a circular aperture normalized to u_i/u_{iC}.

Because the cutoff frequency is wavelength dependent, Equation 9-12 and Figure 9-3 are only valid for noncoherent monochromatic light. The extension to polychromatic light is lens-specific. Most lens systems are color corrected (achromatized) and therefore there is no simple way to apply this simple formula to predict the MTF. As an approximation to the polychromatic MTF, the average wavelength is used to calculate the cutoff frequency

$$\lambda_{AVE} \approx \frac{\lambda_{MAX} + \lambda_{MIN}}{2} \tag{9-13}$$

For example, if the system spectral response ranges from 0.4 to 0.7 μm, then $\lambda_{AVE} = 0.55$ μm.

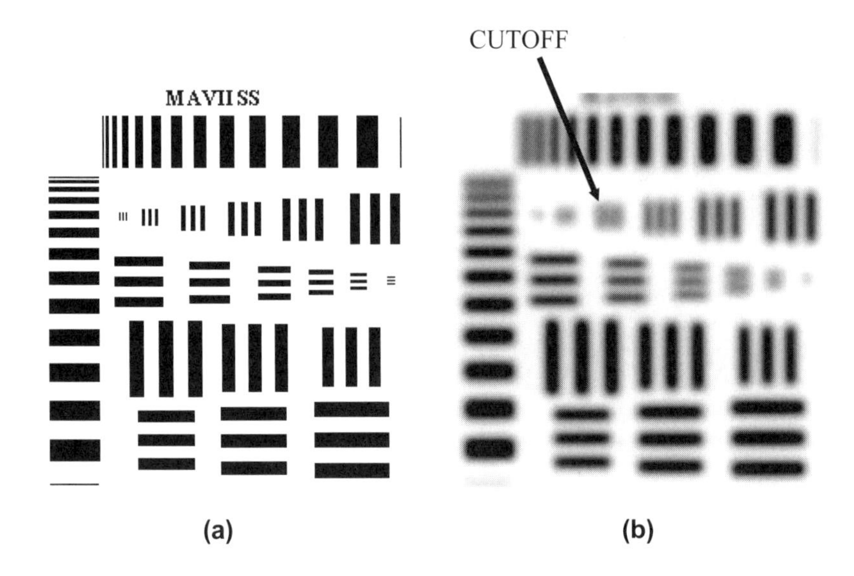

Figure 9-4. The bar target is viewed though diffraction-limited optics. (a) Very high optical cutoff frequency (compared to the target frequency) and (b) low optical cutoff frequency. The smallest bars are above the optical cutoff. Although the modulation is zero, the targets did not disappear but appear as blobs with no detail. This is the image that appears at the detector plane. The detector response further degrades the image. Imagery created[1] by MAVIISS. ***NOTE***: Imagery is enlarged so that your eye MTF and the printing do not significantly affect image quality. This allows you to easily see the image degradation created by the system MTF.

It is mathematically convenient to represent an aberrated optical system by two MTFs. This assumes that the blur is symmetrical about the center. Aberrations such as coma are not symmetrical and cannot be modeled in closed form.

$$MTF_{OPTICS} \approx MTF_{DIFFRACTION} MTF_{ABERRATION} \tag{9-14}$$

Shannon developed an empirical relationship that encompasses most aberrations of real lens systems

$$MTF_{ABERRATION} \approx 1 - \left(\frac{W_{rms}}{A}\right)^2 \left[1 - 4\left(\frac{u_i}{u_{iC}} - \frac{1}{2}\right)^2\right] \quad when\ u_i \le u_{iC} \tag{9-15}$$

where W_{rms} is the rms wave-front error expressed as a fraction of waves and A = 0.18. Maréchal suggested that rms wave-front error is related to the peak-to-peak wave-front error by $W_{rms} = W_{PP}/3.5$. Although $W_{PP} = 0$ could be used for diffraction limited optics, a practical choice would be $W_{PP} = 0.25$ or, equivalently, $W_{rms} = 1/14$. This simulates the wave-front error that typically occurs during manufacturing.

For most systems operating in the visible spectral region, the optical system may be considered near diffraction-limited and in focus. Approximations for defocused, rectangular aperture, and telescopes with a central obscuration (Cassegrainian optics) can be found in Reference 2. The optical anti-aliasing filter (optical low-pass filter or OLPF), which is used to blur the image on a single-chip color filter array, is also part of the lens system. Its design depends on the detector pitch (sampling lattice) and is discussed in Section 10.4. *Optical low pass filter*.

9.3. DETECTORS

Figure 9-5 illustrates the spatial integration afforded by a staring array. In Figure 9-5a, the detector width is one-half of the center-to-center spacing (detector pitch). That is, only 50% of the input sinusoid is detected. The detector averages the signal over the range indicated and the heavy lines are the detector outputs. As the detector size increases relative to the input frequency, it integrates over a larger region and the signal decreases. The heavy lines represent the detector OTF. The detector MTF is sometimes called the detector aperture MTF or pixel aperture MTF. Recall that with a finite fill factor, the detector is smaller than a pixel.

The detector OTF cannot exist by itself. Rather, the detector OTF must also have the optical MTF to make a complete imaging system. The optical system cutoff limits the absolute highest spatial frequency that can be faithfully imaged. The detector cutoff may be either higher (optics-limited system) or lower (detector-limited system). Most imaging systems sensitive in the visible spectrum are detector-limited (discussed in Chapter 11, *Image quality*).

CCD detectors are typically rectangular shaped with the exception of the super CCD, which is approximated as a circular detector. CMOS detectors may be L-shaped or notched rectangles. The MTF of virtual phase detectors is a complex expression.[3]

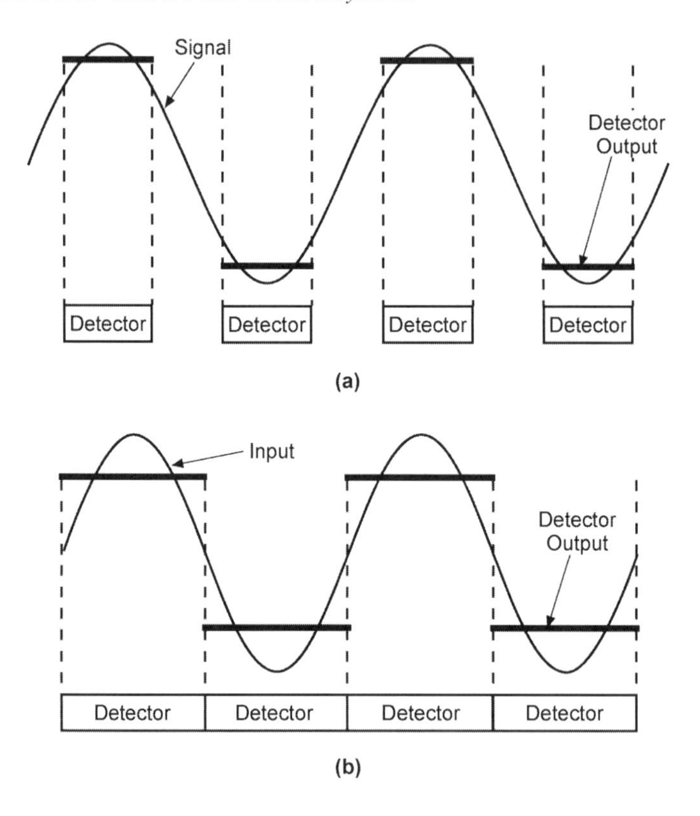

Figure 9-5. A detector spatially integrates the signal. (a) Detector width is one-half of the center-to-center spacing and (b) detector width is equal to the center-to-center spacing. (a) Typifies an interline transfer CCD array in the horizontal direction and (b) is representative of frame transfer arrays. The heavy lines are the detector output voltages.

9.3.1. RECTANGULAR

In the horizontal direction, the OTF of a single rectangular detector is

$$\boldsymbol{OTF}_{DETECTOR}(\boldsymbol{u}_i) = \frac{\sin(\pi\, \boldsymbol{d}_H \boldsymbol{u}_i)}{\pi\, \boldsymbol{d}_H \boldsymbol{u}_i} \tag{9-16}$$

The variable $\boldsymbol{d_H}$ is the physical extent of the photosensitive surface. The OTF is equal to zero when $\boldsymbol{u_i} = \boldsymbol{k}/\boldsymbol{d_H}$ (Figure 9-6). The first zero ($\boldsymbol{k} = 1$) is considered the detector cutoff, u_{iD}. It is customary to plot the OTF only up to the first zero (Figure 9-7). These figures illustrate the OTF without regard to the array Nyquist frequency. The analyst must add the array Nyquist frequency to the curve (discussed in Section 10.3. *Array Nyquist frequency*).

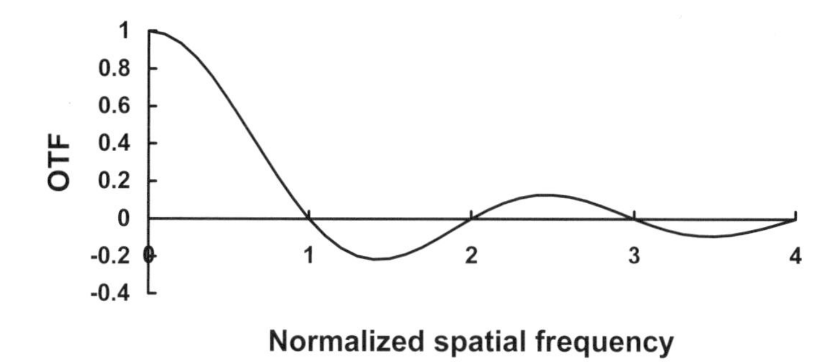

Figure 9-6. Detector OTF as a function of normalized spatial frequency u_i/u_{iD}. Detector cutoff is $u_{iD} = 1/d_H$.

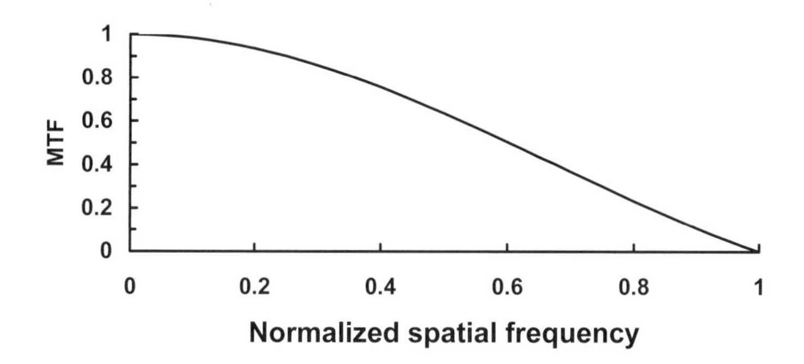

Figure 9-7. The OTF is usually plotted up to the first zero ($d_H u_i = 1$). Since the OTF is positive, it is called the MTF. This representation erroneously suggests that the detector does not respond to frequencies greater than u_{iD}.

Most arrays have rectangular detectors. The two-dimensional spatial response of a rectangular detector is

$$MTF_{DETECTOR}(u_i, v_i) = \left| \frac{\sin(\pi\, d_H\, u_i)}{(\pi\, d_H\, u_i)} \right| \left| \frac{\sin(\pi\, d_V\, v_i)}{(\pi\, d_V\, v_i)} \right| \qquad (9\text{-}17)$$

Note that arrays are often specified by the pixel size. With 100% fill factor arrays, the pixel size is equal to the detector size. With finite fill factor arrays, the photosensitive detector dimensions are less than the pixel dimensions. A microlens will optically increase the effective size (see Section 5.5. *Microlenses*).

9.3.2. CIRCULAR

Fujifilm's octagonal detector cannot be described by equations that are separable in Cartesian coordinates. As an approximation, the active area can be considered a circle with radial symmetry[4]. The OTF is

$$OTF_{SUPER\ CCD}(u_r, \theta) = 2\frac{J_1(\pi\, d_{CIRCLE}\, u_r)}{\pi\, d_{CIRCLE}\, u_r} \tag{9-18}$$

where $u_r = (u_i^2 + v_i^2)^{0.5}$, d_{CIRCLE} is the detector's equivalent diameter, and $J_1(\ )$ is the first-order Bessel function. As illustrated in Figure 9-8, the first zero occurs at $d_{CIRCLE}u_r = 1.22$. When compared to a rectangular detector the circular detector has a higher cutoff frequency.

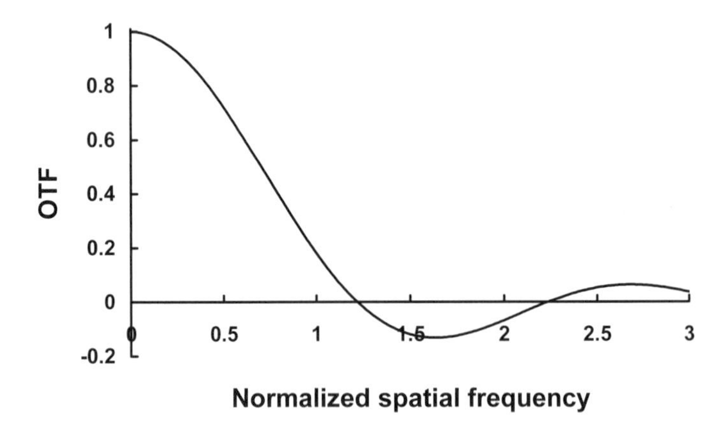

Figure 9-8. Circular detector OTF as a function of normalized spatial frequency, $d_{CIRCLE}u_r$.

9.3.3. L-SHAPED

The active area of a CMOS detector tends to be L-shaped where transistors and connectors occupy the open area (Figure 9-9). The OTF equations are complex[5-9]. As an approximation, it is treated as an equivalent square or rectangle (Equation 9-17).

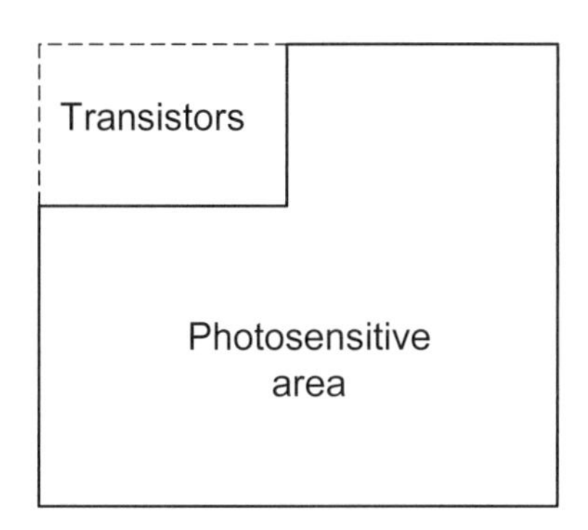

Figure 9-9. Typical L-shaped active area of a CMOS detector.

9.3.4. NOTCHED RECTANGLE

CMOS detectors may also be a notched rectangle. However, the spatial response is not limited by the detector shape (Figure 6-10). As such, the MTF is unique to the pixel architecture and fabrication process. Performing a two-dimensional Fourier transform of the pixel response (Figure 6-9) yields the two-dimensional MTF (Figure 9-10). While the MTF is not separable, for convenience, two orthogonal axes are selected corresponding to the array axes to provide representative horizontal and vertical MTFs (Figure 9-11).

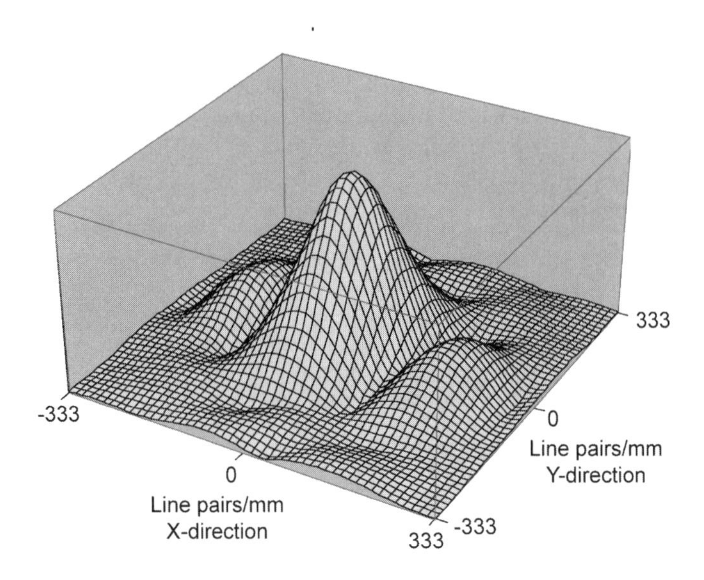

Figure 9-10. MTF of the 9 μm square CMOS pixel calculated from the photodiode response shown in Figure 6-9.

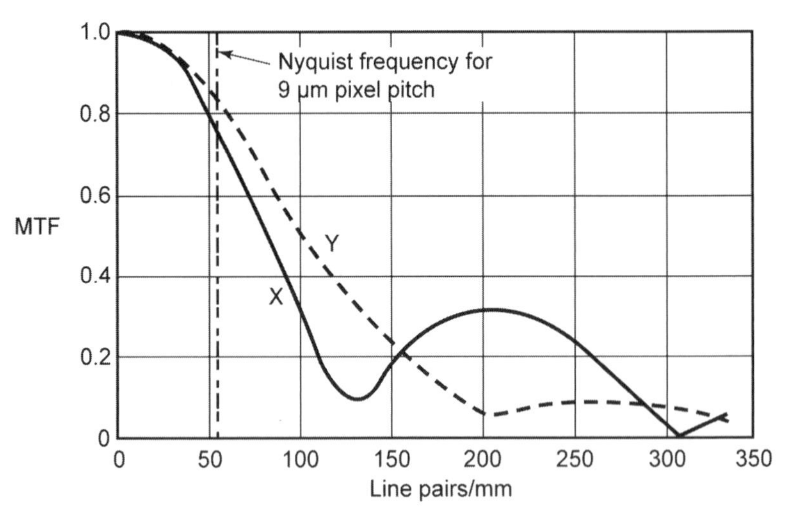

Figure 9-11. Representative horizontal and vertical MTFs for a notched rectangular CMOS detector. Nyquist frequency is described in Chapter 10.

9.4. DIFFUSION MTF

Diffusion can take place deep within the silicon substrate, where photoelectrons randomly walk in three dimensions. The photoelectron travel may be limited by channel blocks (lateral diffusion). Longer-wavelength photons are absorbed deeply, and these diffusions only affect the longer-wave MTF.

9.4.1. BULK DIFFUSION

As the wavelength increases, photon absorption occurs at increasing depths in the detector material (see Section 6.1. *Quantum efficiency*). A photoelectron generated deep within the substrate will experience a three-dimensional random walk until it recombines or reaches the edge of a depletion region where the pixel electric field exists. A photoelectron generated under one well may eventually land in an adjoining well. This blurs the image and its effect is described by an MTF. Beyond 0.8 μm, diffusion (random walk) affects the detector MTF and the response is expressed by $MTF_{DETECTOR}MTF_{DIFFUSION}$. Imagery produced by near-IR sources will appear slightly blurry compared to imagery produced by a visible source.

The diffusion MTF[10] for a front-illuminated device with a thick, bulk absorbing layer is

$$MTF_{DIFFUSION}(u_i,\lambda) = \frac{1 - \dfrac{\exp[-\alpha_{ABS}(\lambda)L_D]}{1+\alpha_{ABS}(\lambda)L(u_i)}}{1 - \dfrac{\exp[-\alpha_{ABS}(\lambda)L_D]}{1+\alpha_{ABS}(\lambda)L_{DIFF}}} \qquad (9\text{-}19)$$

The variable $\alpha_{ABS}(\lambda)$ is the spectral absorption coefficient. Figure 6-1 illustrated the absorption coefficient in bulk p-type silicon. The factor L_D is the depletion width and $L(u_i)$ is the frequency-dependent component of the diffusion length (L_{DIFF}) given by

$$L(u_i) = \frac{L_{DIFF}}{\sqrt{1+(2\pi\ L_{DIFF}u_i)^2}} \qquad (9\text{-}20)$$

The diffusion length typically ranges between 50 μm and 200 μm. The depletion width is approximately equal to the gate width. For short wavelengths

(λ < 0.6 μm), $\alpha_{ABS}(\lambda)$ is large and lateral diffusion is negligible. Here, $MTF_{DIFFUSION}$ approaches unity. For the near IR (λ > 0.8 μm), the diffusion MTF may dominate the detector MTF (Figure 9-12). By symmetry, $MTF_{DIFFUSION}(u_i) = MTF_{DIFFUSION}(v_i)$ and $L(u_i) = L(v_i)$.

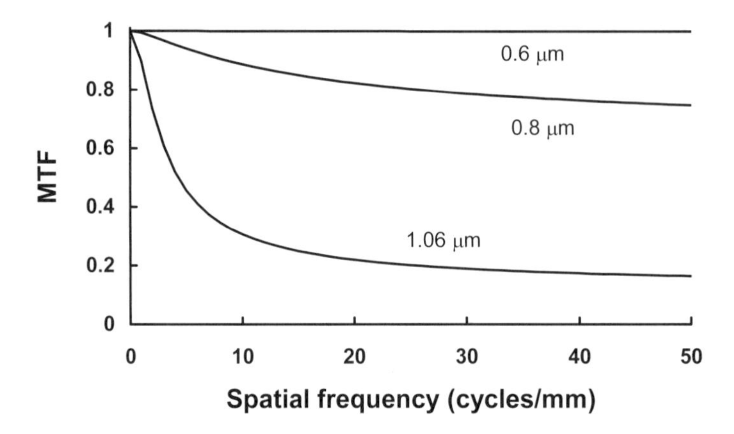

Figure 9-12. $MTF_{DIFFUSION}$ as a function of wavelength when $L_D = 10$ μm and $L_{DIFF} = 100$ μm. As the diffusion length increases, the MTF decreases.

9.4.2. SURFACE LATERAL DIFFUSION

CCD imager pixels are isolated from each other by clocked gates in the direction of charge motion, and by channel stops in the direction perpendicular to the charge motion. The terminology "channel stop" derives from action of avoiding charge flow from one pixel to another in this direction. However, photoelectrons generated near the channel stops can diffuse towards either pixel well. To account for the effects of lateral crosstalk, Schumann and Lomheim[11] introduced a trapezoidal response approximation. This MTF only applies in the direction perpendicular to charge motion

$$MTF_{DIFFUSION}(u_i) \approx \frac{\sin(d_{CCH}u_i)}{d_{CCH}u_i}\frac{\sin[(d_{CCH}-\beta)u_i]}{(d_{CCH}-\beta)u_i} \qquad (9\text{-}21)$$

where β is the region over which the detector responsivity is flat (Figure 9-13). At the boundary between pixels, the trapezoidal response is 0.5, indicating equal probability of an electron going to either pixel's charge well.

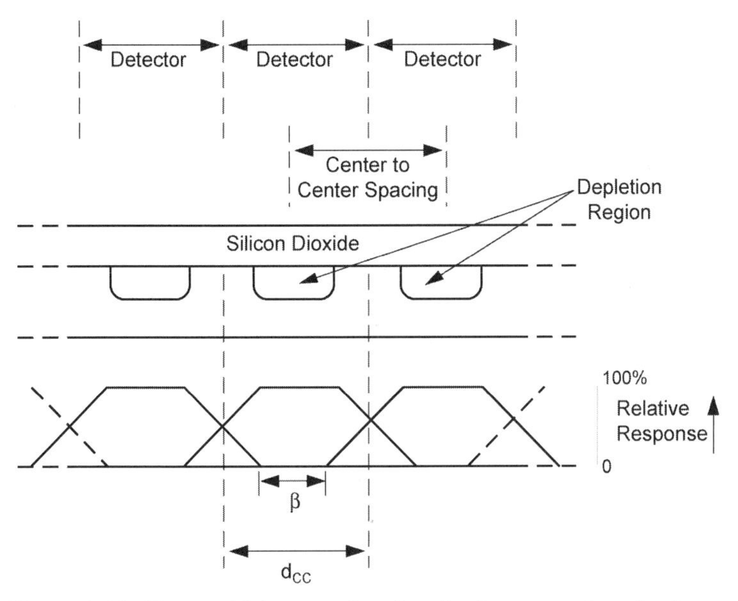

Figure 9-13. Trapezoidal approximation for long-wavelength photon absorption. Most of the charge goes to the charge well nearest to the photon absorption site (see also Figure 6-4).

9.4.3. EPITAXIAL LAYER DIFFUSION

For front-illuminated arrays that are fabricated using epitaxially doped silicon wafers (often called epi-wafers), the layer structure consists of a thin lightly doped epitaxial layer, which starts at the imager surface and is typically 7 to 15 μm deep. Beneath this thin layer is a thick layer (typically 300 μm) that is an extremely heavily doped layer; the heavy doping makes it electrically conductive. Photon absorption and carrier diffusion occur within the epitaxial layer. However, the highly doped layer underneath it has no photoresponse (carriers instantly recombine) and it acts essentially as an electrical contact and also serves to provide structural support for the array. The thin epitaxial layer is broken down two regions: a depletion region near the surface and a field-free region occupying the remainder of the layer thickness. For standard CMOS processing, the epitaxial layer is thinner (usually closer to 7 μm) and the depletion region can be very thin (less than 1 μm) unless deliberate steps are taken during the manufacturing process. This impacts both the spectral quantum efficiency and diffusion MTF in CMOS in a way that is distinctly different than that typically observed in CCD pixels.

Using the tilted edge technique[12], the MTF of an 11.9 μm square CMOS detector was measured as a function of wavelength (Figures 9-14 and 9-15). The rectangular detector [11.5 μm × 6 μm (H×V)] was located in the upper half of the pixel area. The 256×256 pixel CMOS array contained an n+-p-substrate silicon photodiode. The variation in the MTF as a function of wavelength is somewhat similar to optical chromatic aberration. Diffusion will make blue edges appear sharper than red edges. This impact image quality.

Several MTF models[3,13-15] account for the epitaxial layer structure found in modern CCD and CMOS arrays. The simplist[3] model assumes that the rectangular detector and diffusion MTFs are multiplicative. The parameters selected to match the data were: a shallow photodiode depletion width (< 1 μm), a diffusion layer thickness of 7 μm, and a rectangular photodiode dimension of 11 μm × 6 μm (H×V). These parameters are physically plausible for the CMOS imager design with the exception of a larger detector vertical dimension (8 μm rather than 6 μm). As illustrated in Figures 9-16 and 9-17, the MTF at 400 nm is similar to the rectangular detector MTF.

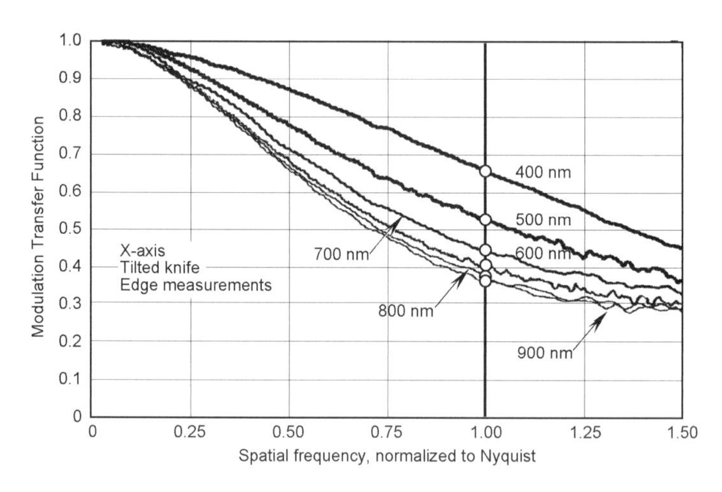

Figure 9-14. Detector horizontal MTF (includes diffusion and rectangular detector MTF).

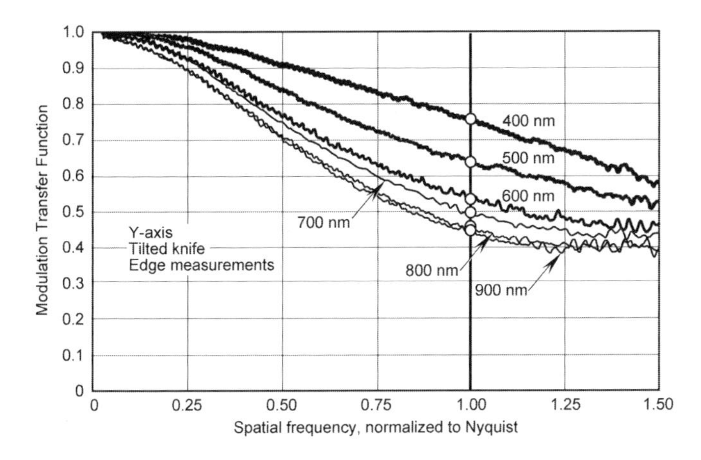

Figure 9-15. Detector vertical MTF (includes diffusion and rectangular detector MTF).

The fact that a pixel aperture of 8 μm rather than 6 μm best fits the data is an indication of lateral diffusion current from the pixel area adjacent to the photodiode; namely the area occupied by the CMOS pixel circuitry. For wavelengths greater than 500 nm, light is absorbed below the rather shallow depletion layer boundary and the onset of diffusion is evident in the subsequent increasing wavelength MTF curves. For wavelengths greater than about 800 nm the MTF curves coalesce, indicating that the light is penetrating deeper than the diffusion layer thickness. This also leads to poor red quantum efficiency.

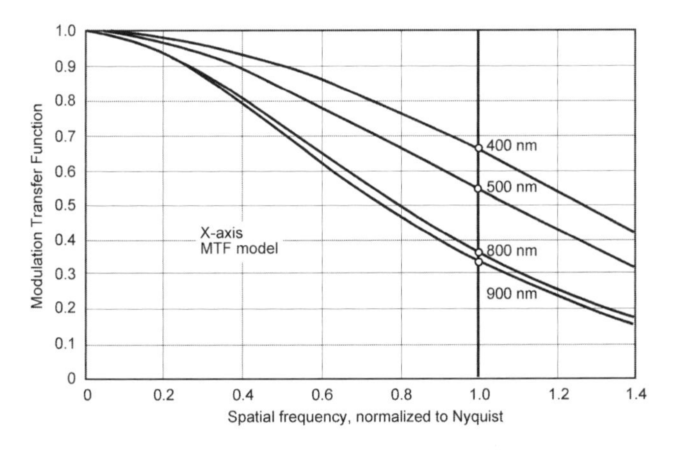

Figure 9-16. Predicted horizontal MTF.

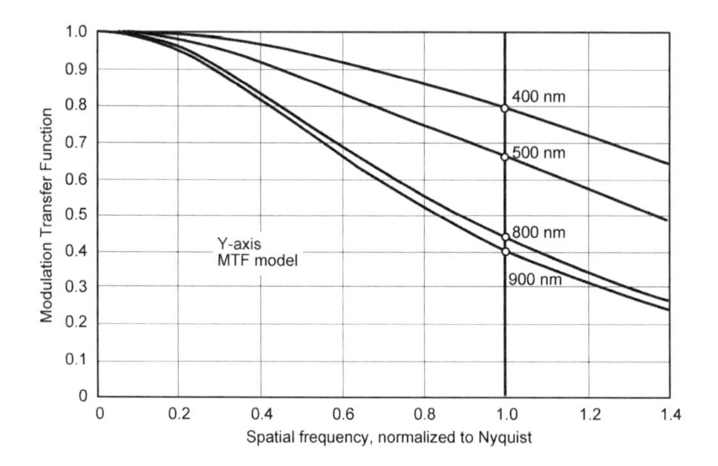

Figure 9-17. Predicted vertical MTF.

9.6. CHARGE TRANSFER EFFICIENCY

The early CCDs transferred electrons in channels that were at the surface. Surface defects interfere with the transfer and create poor charge transfer efficiency (CTE). These were SCCD or surface channel CCDs. By burying the channel (BCCD), the charge was no longer affected by surface defects; this improved the CTE significantly. Nearly all CCDs use buried channel technology.

As the charge packet moves from storage site to storage site, a few electrons are left behind. The ability to transfer all the charge is given by the CTE. As the number of transfers increases, the charge transfer efficiency must also increase to maintain signal strength.

A fraction of the charge, as specified by the charge inefficiency, is left behind. After the first transfer, ε (ε is the CTE) are in the leading well and $(1-\varepsilon)$ remain the first trailing well. After the second transfer, the leading well transfers ε of the previous transfer and loses $(1-\varepsilon)$ to the trailing well. The trailing well gains the loss from the first well and adds it to the charge transferred into the trailing well in the last go-around. Charge is not lost, just rearranged. The fractional values follow a binomial probability distribution

$$P_R = \binom{N}{R} \varepsilon^R (1-\varepsilon)^{N-R} \tag{9-22}$$

where N is the number of transfers. For the first well, $R = N$. For the second well, $R = N - 1$ and so on.

Consider a very small spot of light that illuminates just one pixel and creates n_o electrons. A well, N transfers away, will contain n_N electrons. The fractional amount of charge left is

$$\frac{n_N}{n_o} = \varepsilon^N \tag{9-23}$$

The fractional amount of charge appearing in the first trailing well after N transfers is

$$\frac{n_{(FIRST\ TRAILING\ WELL)-N}}{n_o} = N\varepsilon^{N-1}(1-\varepsilon) \tag{9-24}$$

The fractional signal strength, n_N/n_o, is plotted in Figure 9-18. The first trailing well fractional strength, $n_{FIRST\ TRAILING\ WELL\text{-}N}/n_o$, is plotted in Figure 9-19. Most of the electrons lost from the first well appear in the second and third wells. For high CTE arrays, a negligible amount appears in successive wells.

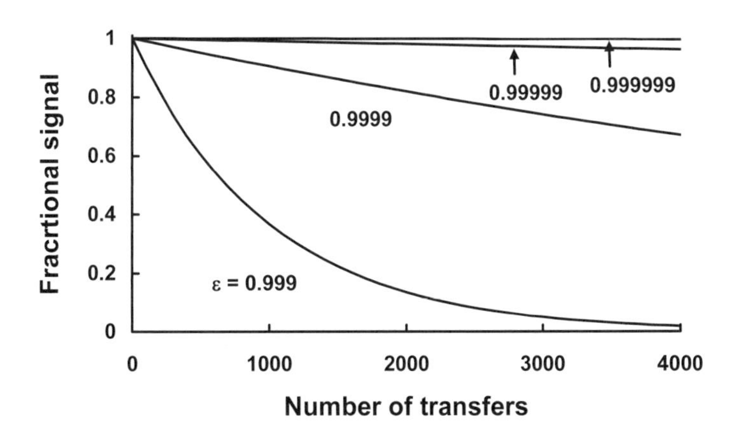

Figure 9-18. Fractional signal as a function of transfer efficiency and number of transfers.

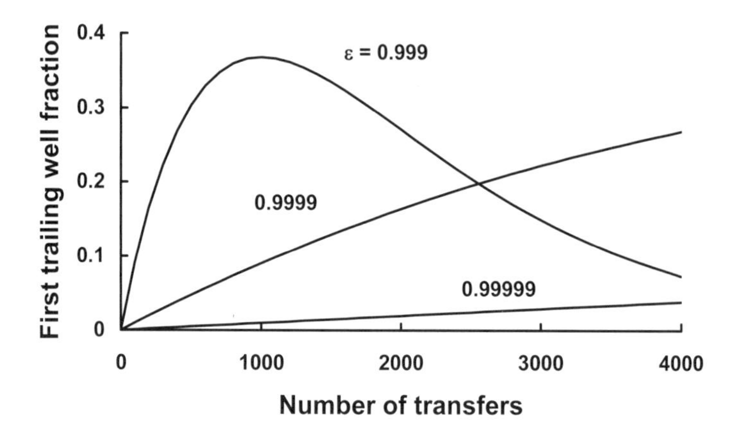

Figure 9-19. Fractional signal for the first trailing charge well. With low CTEs, as the number of transfers increases, more charge spills into the second trailing well.

CTE is not a constant value, but depends upon the charge packet size. For very low-charge packets, the CTE decreases due to surface state interactions. This is particularly bothersome when the signal-to-noise ratio is low and the array is very large (typical of astronomical applications). The CTE also decreases near saturation due to charge spill effects.

The fraction of charge left behind also depends on the clocking frequency. At high frequencies, efficiency is limited by the electron mobility rate. That is, it takes time for the electrons to move from one storage site to the next. Thus, there is a tradeoff between frame rate (dictated by clock frequency) and image quality (the MTF is affected by CTE). Using smaller pixels partially overcomes the limitation imposed by the electron mobility rate. However, a smaller pixel has a smaller charge well capacity.

CTE effects in large linear arrays can be minimized by using alternate readouts. In Figure 3-12, the alternate readouts reduce the number of transfers by a factor of two. Adjacent pixels experience the same number of transfers so that the signal is preserved at any particular location.

An MTF[16] that accounts for the incomplete transfer of electrons is

$$MTF_{CTE}(u_i) = \exp\left(-N_{TRANS}(1-\varepsilon)\left[1-\cos\left(\frac{\pi\, u_i}{u_{iN}}\right)\right]\right) \qquad (9\text{-}25)$$

where N_{TRANS} is the total number of charge transfers from a detector to the output amplifier and u_{iN} is the array Nyquist frequency [$u_{iN} = 1/(2d_{CCH})$].

Transfers may be either in the vertical direction (column readout) or in the horizontal direction (row readout). Horizontal readout affects $MTF(u_i)$ and vertical readout affects $MTF(v_i)$. MTF_{CTE} only applies to the transfer readout direction. Figure 9-20 illustrates MTF_{CTE} at Nyquist frequency for several values of transfer efficiency. For consumer devices, CTE > 0.9999 and N_{TRANS} > 1500. For scientific applications, CTE > 0.999999 and N_{TRANS} can exceed 5000.

The number of transfers across the full array is the number of pixels multiplied by the number of gates. For example, for an array that is 1000×1000, the charge packet farthest from the sense node must travel 2000 pixels. If four gates (a four-phase device) are used to minimize crosstalk, the farthest charge packet must pass through 8000 wells (neglecting isolation pixels). The net efficiency varies with the target location. If the target is in the center of the array, then it only experiences one-half of the maximum number of transfers on the array. If the target is at the leading edge, it is read out immediately with virtually no loss of information.

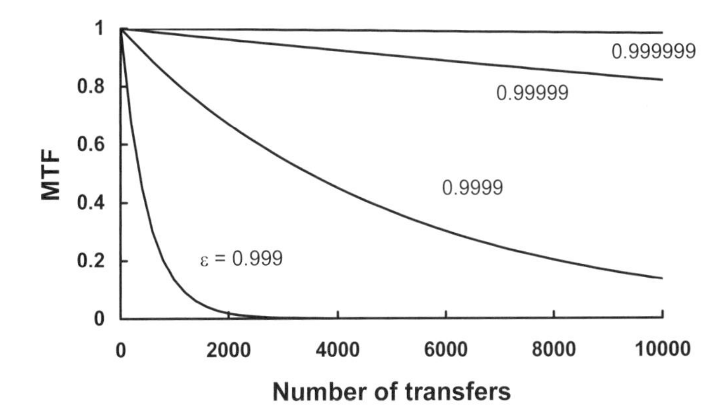

Figure 9-20. MTF_{CTE} at Nyquist frequency (u_{iN}) for several values of transfer efficiency. As the number of transfers increases, the efficiency must also increase to ensure that the MTF is not adversely affected.

Example 9-1
Number of Transfers

What is the maximum number of transfers that exist in an array that contains 1000×1000 detector elements?

For a charge packet that is furthest from the sense node, the charge must be transferred down one column (1000 pixels) and across 1000 pixels, for a total of 2000 pixels. The number of transfers is 4000, 6000, and 8000 for two-phase, three-phase, and four-phase devices, respectively. Generally, only ½ of the number is used in Equation 9-25 to obtain an "array average" MTF.

9.7. TDI

TDI operation was described in Section 3.3.6. *Time delay and integration.* It is essential that the clock rate match the image velocity so that the photoelectron charge packet is always in sync with the image. Any mismatch will blur the image. Image rotation with respect to the array axes affects both the horizontal (TDI direction) and vertical (readout direction) MTFs simultaneously. For simplicity they are treated as approximately separable: $MTF(u_i,v_i) \approx MTF(u_i)\, MTF(v_i)$.

If the TDI direction is horizontal, velocity errors degrade[17] the detector MTF by

$$MTF_{TDI}(u_i) = \frac{\sin(\pi\, N_{TDI} d_{ERROR} u_i)}{N_{TDI} \sin(\pi\, d_{ERROR} u_i)} \qquad (9\text{-}26)$$

The value d_{ERROR} is the difference between the expected image location and the actual location at the first pixel (Figure 9-21). After N_{TDI} elements, the image is displaced $N_{TDI} d_{ERROR}$ from the desired location. This is a one-dimensional MTF that apples in the TDI direction only. In image space

$$d_{ERROR} = |\Delta V| t_{INT} \quad \text{mm} \qquad (9\text{-}27)$$

The variable ΔV is the relative velocity error between the image motion and the charge packet "motion" or "velocity." The charge packet "velocity" is d_{CCH}/t_{INT} where t_{INT} is the integration time for each pixel. MTF_{TDI} is simply the MTF of an averaging filter for N_{TDI} samples displaced d_{ERROR} one relative to the next. As an averaging filter, the effective sampling rate is $1/\, d_{ERROR}$. Imagery illustrating scan error is provided in Reference 18.

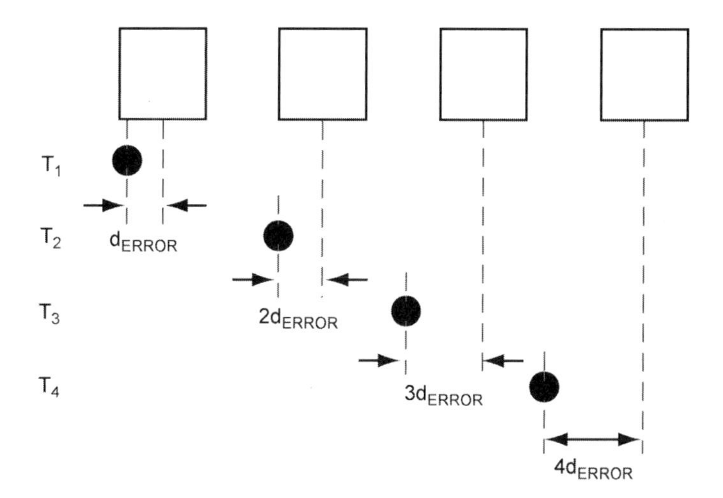

Figure 9-21. Definition of d_{ERROR}. With a velocity mismatch, the target falls further and further away from the desired location (on the center of the detector). After four TDI elements, the target is displaced from the detector center by $4d_{ERROR}$.

Figure 9-22 portrays the MTF for two different relative velocity errors for 32 and 96 TDI stages. As the number of TDI stages increases, d_{ERROR} must decrease to maintain the MTF. Equivalently, the accuracy of knowing the target velocity increases as N_{TDI} increases. Because the image is constantly moving, a slight image smear occurs[19] because charge packet is transferred at discrete times (i.e., at intervals of t_{CLOCK}). This linear motion MTF (discussed in Section 9.8.1. *Linear Motion*) is often small and therefore is usually neglected.

TDI is also sensitive to image rotation. If the image motion subtends an angle θ with respect to the TDI rows (TDI direction), the image moves vertically $N_{TDI}d_{CCH}\tan(\theta)$ across the array. The MTF degradation in the vertical direction is

$$MTF_{TDI}(v_i) = \frac{\sin(\pi N_{TDI} d_{CCH} \tan(\theta) v_i)}{N_{TDI} \sin(\pi d_{CCH} \tan(\theta) v_i)} \tag{9-28}$$

where $MTF_{TDI}(v_i)$ is simply the MTF of an averaging filter for N_{TDI} samples displaced $d_{CCH}\tan(\theta)$ one relative to the next. Figure 9-23 illustrates the vertical MTF degradation due to image rotation for 32 and 96 stages. The spatial frequency is normalized to $d_{CCH}v_i$ where d_{CCH} is the *horizontal* detector pitch.

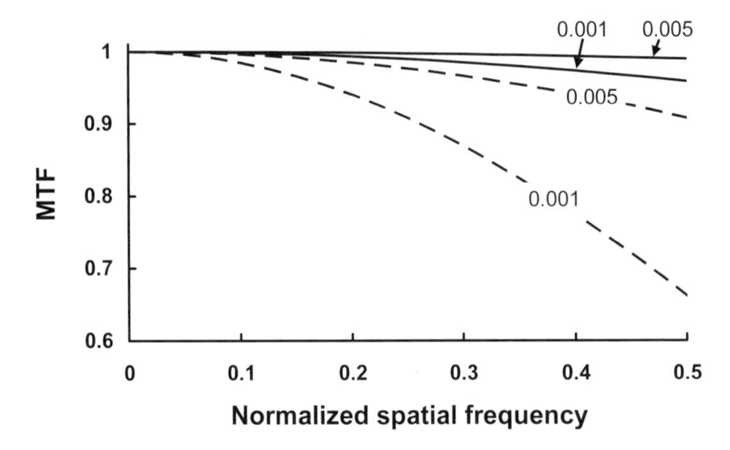

Figure 9-22. MTF degradation due to mismatch between the image velocity and the charge packet "velocity." The value d_{ERROR}/d_{CCH} equals 0.005 and 0.01. The spatial frequency axis is normalized to $d_{CCH}u_i$. The solid lines represent $N_{TDI} = 32$ and the dashed lines are for $N_{TDI} = 96$. The Nyquist frequency occurs at $d_{CCH}u_i = 0.5$. The MTF scale has been expanded.

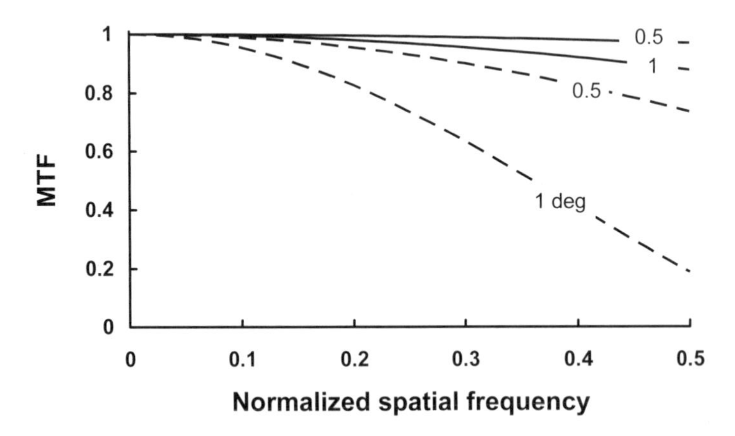

Figure 9-23. Vertical MTF degradation in the readout direction due to image rotation of 0.5 and 1 deg with respect to the array axis. The spatial frequency axis is normalized to $d_{CCH}v_i$. The solid lines represent $N_{TDI} = 32$ and the dashed lines are for $N_{TDI} = 96$.

9.8. MOTION

Motion blurs imagery. When motion is fast compared to the detector integration time, details become blurred in a reproduced image. In real-time imagery, edges may appear fuzzy to the observer even though any one frame may provide a sharp image. The eye integration time blends many frames. The eye integration time is exploited with a motion picture. Individual (static) frames are presented at 24 Hz and yet, motion is perceived. The effects of motion during the entire integration and human interpretation process must be considered.

Linear motion includes both target movement (relative to the imaging system) and motion of the imaging system (stationary target). High-frequency random motion is simply called jitter. Typically, an imaging system is subjected to linear and random motion simultaneously

$$MTF_{MOTION} = MTF_{LINEAR} MTF_{RANDOM} \qquad (9\text{-}29)$$

In a laboratory, the system and test targets may be mounted on a vibration-isolated stabilized table. Here, no motion is expected so that $MTF_{MOTION} = 1$.

9.8.1. LINEAR MOTION

The OTF degradation due to horizontal linear motion is

$$OTF_{LINEAR}(u_i) = \frac{\sin(\pi\, d_L u_i)}{\pi\, d_L u_i} \qquad (9\text{-}30)$$

where d_L is the distance a target edge moved across the image plane. It is equal to $v_R \Delta t$, where v_R is the relative image velocity between the sensor and the target. For scientific and machine vision applications, the exposure time is equal to the detector integration time. But, an observer's eye blends many frames of data and the eye integration time should be used. Although the exact eye integration time is debatable, the values most often cited are between 0.1 and 0.2s.

Figure 9-24 illustrates the OTF due to linear motion as a fraction of the detector size. Linear motion only affects the OTF in the direction of the motion. As a rule of thumb, when the linear motion causes an image shift less than about 20% of the detector size, it has minimal effect on system performance (Figure 9-25).

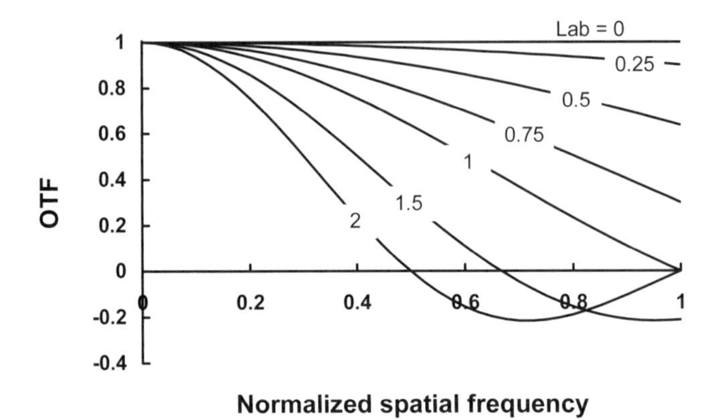

Figure 9-24. $\boldsymbol{OTF_{LINEAR}}$ as a function of normalized spatial frequency $\boldsymbol{du_i}$. The curves represent values of $\boldsymbol{d_L/d_H}$. In the laboratory $\boldsymbol{d_L}$ is usually zero. The negative OTF represents phase reversal. The MTF is always positive.

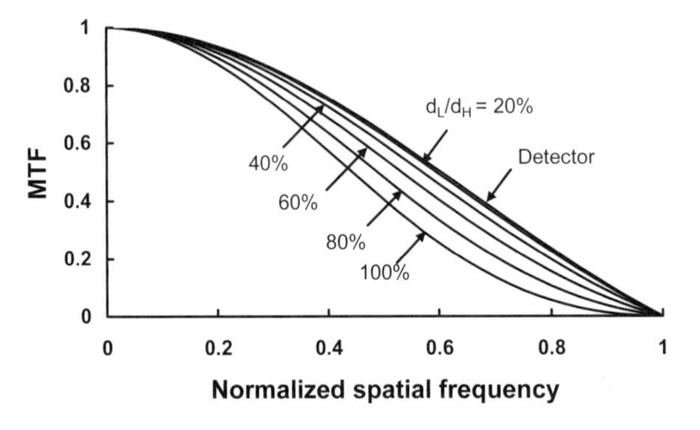

Figure 9-25. $\mathrm{MTF_{LINEAR}MTF_{DETECTOR}}$. Nyquist frequency is not shown. An MTF cannot be considered in isolation (Figure 9-24). The effect a subsystem has on the overall system can only be determined by plotting $\mathrm{MTF_{SYS}}$ and all the subsystem MTFs.

As the motion increases, a phase reversal occurs which is represented by a negative OTF. Here, periodic black bars appear white and periodic white bars appear black. Figure 9-26 illustrates a test pattern smeared by linear motion. At the first zero ($\boldsymbol{u_i} = 1/\boldsymbol{d_L}$), the bar pattern cannot be resolved. Phase reversal occurs above the first zero. Note that phase reversal will only be seen when viewing periodic targets (test patterns, picket fences, railroad ties, Venetian blinds, etc). Complex imagery will just appear smeared.

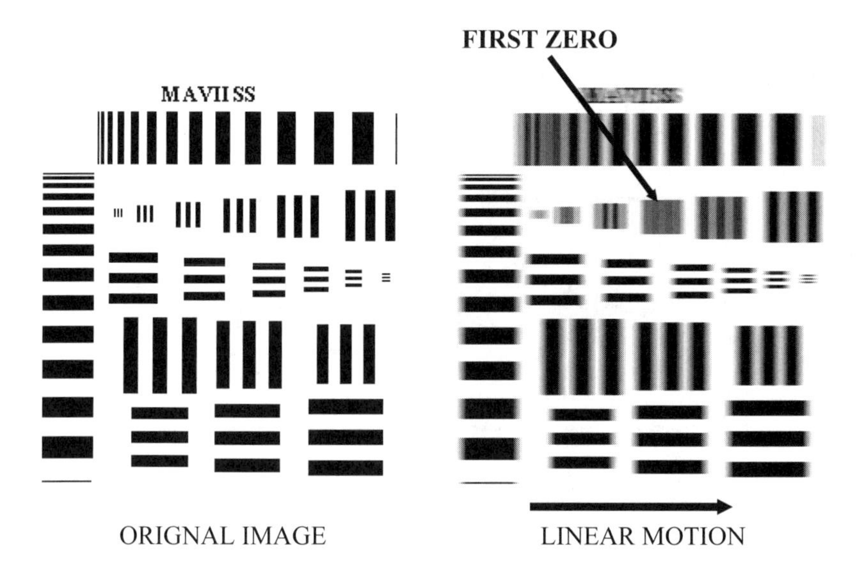

Figure 9-26. A test pattern image when only linear motion is present ($MTF_{OPTICS}MTF_{DETECTOR}$ = 1 over frequencies of interest). Since the motion is horizontal, the vertical resolution is maintained. When the bar target frequency is equal to the first zero of the OTF_{LINEAR}, the bars blend into a blob. Higher spatial frequency bars illustrate phase reversal: There appears to be two black bars on a gray background. This is the image that appears at the detector plane. The detector response further degrades the image. Imagery created[1] by MAVIISS. See ***NOTE*** in Figure 9-4 caption.

9.8.2. RANDOM MOTION (JITTER)

With high-frequency motion, it is assumed that the image has moved often during the integration time (Figure 9-27) so that the central limit theorem is valid. The central limit theorem says that many random movements can be described by a Gaussian distribution

$$MTF_{RANDOM}(u) = \exp\left(-2\pi^2\sigma_R^2 u^2\right) \tag{9-31}$$

where σ_R is the rms random displacement in millimeters. Figure 9-28 illustrates the random motion MTF. As a rule of thumb, when the rms value of the random motion is less than about 10% of the detector size, system performance is not significantly affected (Figure 9-29). Jitter is considered equal in all directions and is included in both the horizontal and vertical system MTFs. High-frequency motion blurs imagery. Figure 9-30 illustrates a bar target that has been degraded by the detector MTF and random motion. Sampling has not been included. If included, the image would look blocky with obvious sampling artifacts (moiré patterns).

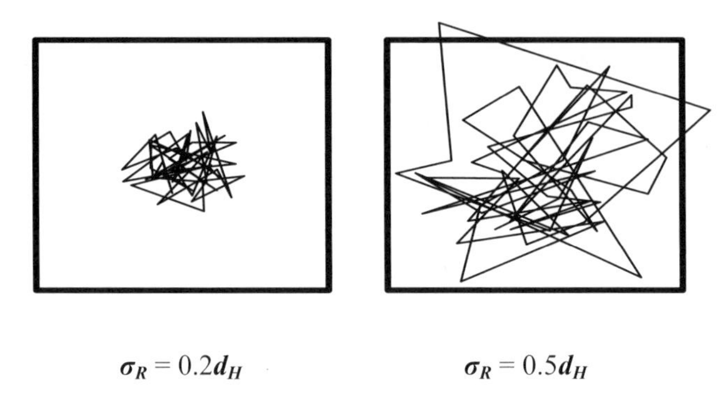

Figure 9-27. Instantaneous line-of-sight location when random motion is present. The heavy line outlines the detectors. Approximately 99.98% of the line-of-sight locations fall between ±3.5σ.

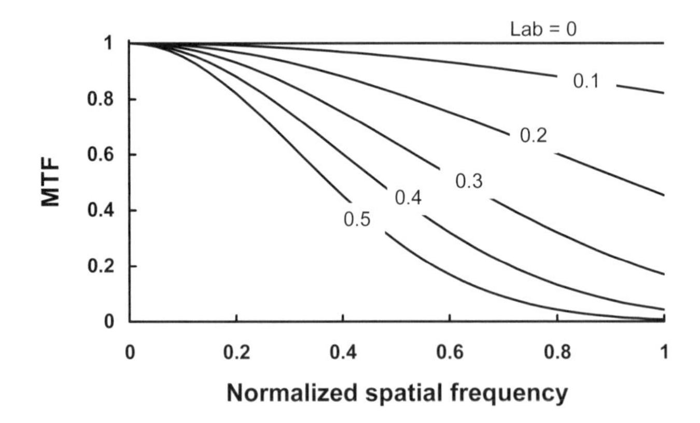

Figure 9-28. MTF degradation due to high-frequency random (Gaussian) movement as a function of $\sigma_R u_i$. In the laboratory, σ_R is usually zero.

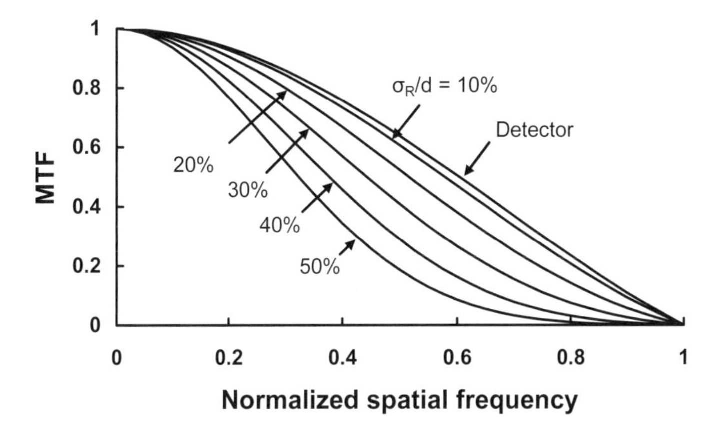

Figure 9-29. $MTF_{RANDOM}MTF_{DETECTOR}$. Nyquist frequency is not shown.

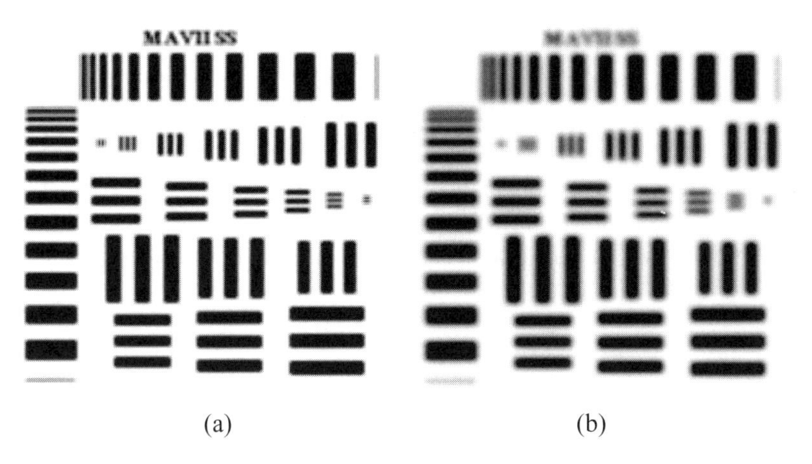

(a) (b)

Figure 9-30. Random motion blurs imagery. The original is shown in Figure 9-26. (a) Image degraded $MTF_{DETECTOR}$ and (b) $MTF_{DETECTOR}$ MTF_{RANDOM} degradation when $\boldsymbol{\sigma_R/d}$ = 0.5. Sampling artifacts will make these images appear blocky. Imagery created[1] by MAVIISS. See ***NOTE*** in Figure 9-4 caption.

9.9. ELECTRONIC FILTERS

Electronic filter response is denoted by $H(f_e)$ or $H(f_v)$. When the electrical frequency is expressed in object space or image space units, the filter impulse response is labeled as an MTF. Boost filters may be used to enhance the signal at selected spatial frequencies. Depending upon the camera design, both analog and digital filters may be used.

9.9.1. ANTI-ALIAS ANALOG FILTER

Analog systems with an internal analog-to-digital converter (ADC) will benefit from an anti-alias filter placed before the ADC. As described in Section 10.1. *Sampling theorem*, the ideal anti-alias filter approaches unity MTF up to the array Nyquist frequency and then drops to zero. This filter maximizes the SNR by attenuating out-of-band amplifier noise and passing the signal without attenuation. The ideal filter is unrealizable but can be approximated by an Nth-order Butterworth filter

$$H_{LOWPASS}(f_e) = \frac{1}{\sqrt{1+\left(\frac{f_e}{f_{e3dB}}\right)^{2N}}} \quad (9\text{-}32)$$

where f_{e3dB} is the frequency at which the power is one-half or the amplitude is 0.707. As N → ∞, $H_{LOWPASS}(f_e)$ approaches the ideal filter response with the cutoff frequency of f_{e3dB} (Figure 9-31).

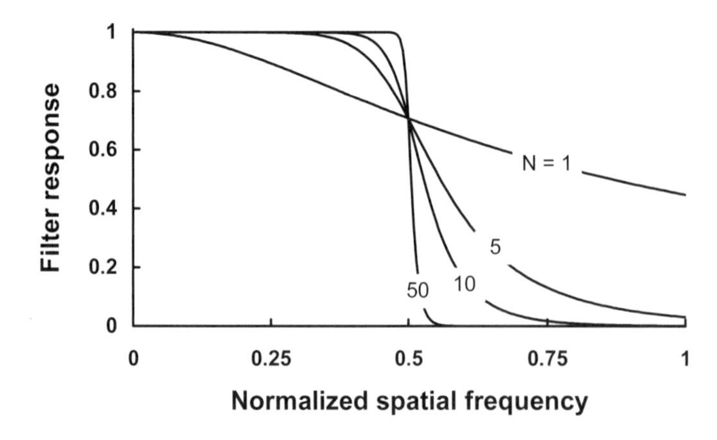

Figure 9-31. Butterworth filters for N = 1, 5, 10, and 50 when f_{e3dB} = 0.5.

9.9.2. DIGITAL FILTERS

There are many available image processing algorithms. Only a few can be described mathematically in closed form, and only these can be included in an end-to-end system performance model. These are included in this section. The performance of the remaining algorithms can be inferred only by viewing the system output for a few representative inputs.

The sampling process replicates the frequency spectrum about multiples of the sampling frequency (discussed in Section 10.1. *Sampling theorem*). As a result, digital filter response is symmetrical about the Nyquist frequency. The highest frequency of interest is the Nyquist frequency.

Digital filters process data that reside in a memory. The units assigned to the filter match the units assigned to the data arrays. With one-to-one mapping of pixels to datels, the filter sampling frequency is the same as the array sampling frequency. With this mapping, each filter coefficient processes one pixel value. Digital filters can be two-dimensional.

There are two general classes of digital filters[20]: infinite impulse response (IIR) and finite impulse response (FIR). Both have advantages and disadvantages. The FIR has a linear phase shift, whereas the IIR does not. IIR filters tend to have excellent amplitude response whereas FIR filters tend to have more ripples. FIR filters are typically symmetrical in that the weightings are symmetrical about the center sample. They are also the easiest to implement in hardware or software.

Figure 9-32 illustrates two one-dimensional FIR filters. The digital filter design software provides the coefficients, A_i. The central data point is replaced by the digital filter coefficients as they operate on the neighboring data points. The filter is then moved one data point and the process is repeated until the entire data set has been operated on. Edge effects exist with any digital filter. The filter illustrated in Figure 9-32a requires seven inputs before a valid output can be achieved. At the very beginning of the data set, there are insufficient data points to have a valid output at data point 1, 2, or 3. The user must be aware of edge effects at both the beginning and the end of his data record. In effect, this states that edges cannot be filtered.

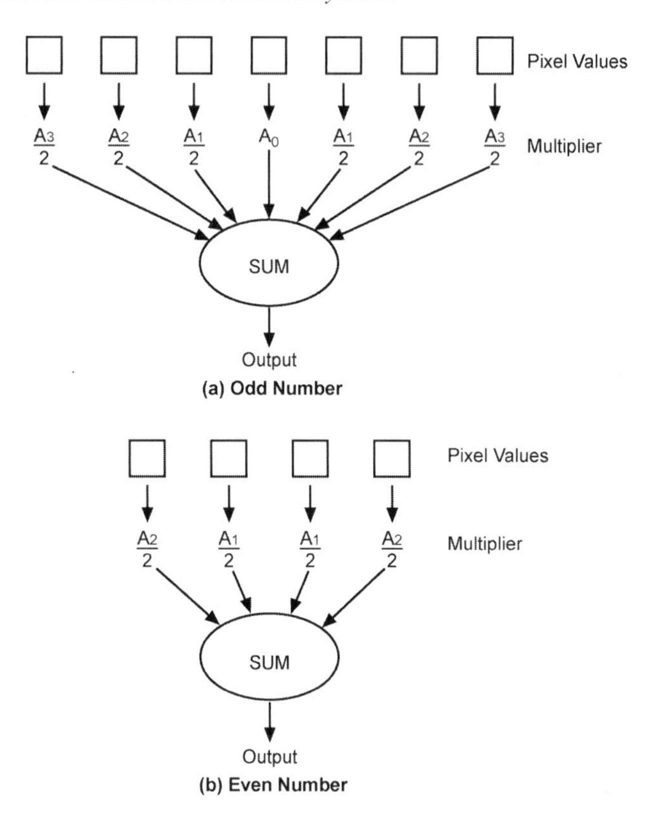

Figure 9-32. Symmetrical digital filters. (a) 7-tap (odd number) filter and (b) 4-tap (even number) filter.

Digital filters operate on a data array and the effective sampling rate is related to the timing of the readout clock frequency

$$f_{eS} = \frac{1}{T_{CLOCK}} \quad Hz \tag{9-33}$$

The following MTF equations are only valid where there are no edge effects. For FIR filters where the multiplicative factors (weightings) are symmetrical about the center, the filter is mathematically represented by a cosine series (sometimes called a cosine filter). With its frequency response transposed to electrical frequency, an FIR filter with an odd number of samples (also called taps) provides

$$H_{DFILTER}(f_e) = \left| \sum_{k=0}^{\frac{N-1}{2}} A_k \cos\left(\frac{2\pi k f_e}{f_{eS}}\right) \right| \tag{9-34}$$

For an even number of samples

$$H_{DFILTER}(f_e) = \left| \sum_{k=1}^{\frac{N}{2}} A_k \cos\left(\frac{2\pi (k-1) f_e}{f_{eS}}\right) \right| \tag{9-35}$$

These are one-dimensional filters. Two-dimensional filters cannot be evaluated in closed forms unless they are separable; $H(u,v) = H(u)\,H(v)$. The sum of all the coefficients should equal unity so that $H_{DFILTER}(f_e = 0)$ is one

$$\sum A_k = 1 \tag{9-36}$$

Although the above equations provide the filter response in closed form, the response of a real filter is limited by the ability to implement the coefficients. Figure 9-33 illustrates the response of a 7-tap filter that provides boost (peaking).

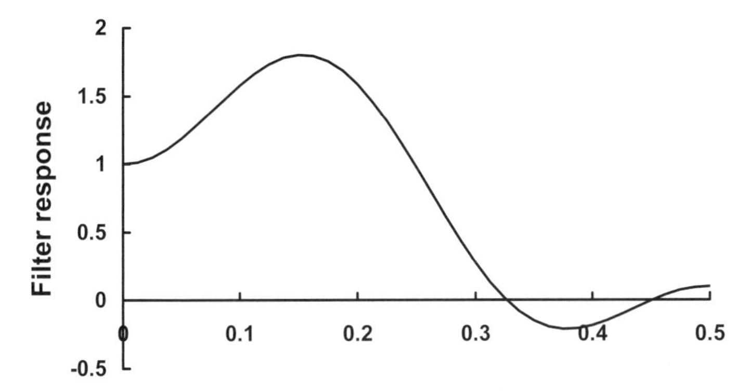

Figure 9-33 A symmetric (cosine) 7-tap digital filter normalized to f_e/f_{se}. $A_0 = 0.7609$, $A_1 = 0.9115$, $A_2 = -0.2100$, and $A_3 = -0.4624$. The "MTF" is illustrated only up to f_N. A response greater than unity is possible.

The simplest boost filter is the unsharp mask. For a 3-tap filter the coefficients are

$$A_1 = -\frac{2\alpha}{3(1-\alpha)} \qquad A_o = \frac{3-\alpha}{3(1-\alpha)} \tag{9-37}$$

or

$$H_{DFILTER}(f_e) = \frac{3-\alpha\left[1+2\cos\left(\frac{2\pi f_e}{f_S}\right)\right]}{3(1-\alpha)} \tag{9-38}$$

As the unsharp parameter, α, approaches one the amount of boost increases dramatically (Figure 9-34).

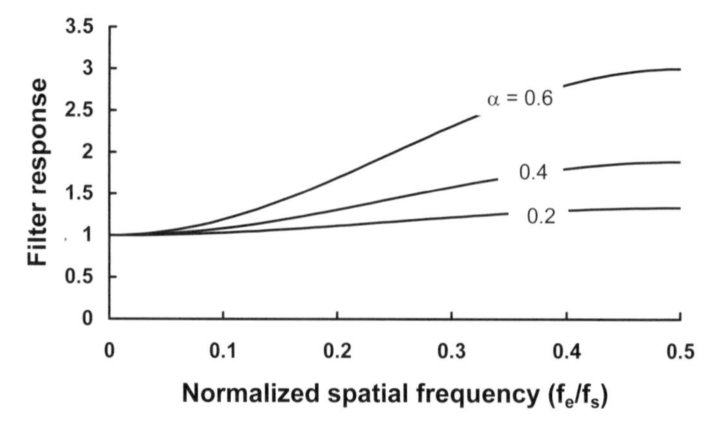

Figure 9-34. One-dimensional unsharp mask response.

With an averaging filter,[17] all the multipliers shown in Figure 9-32b are equal. When N_{AVE} datels are averaged, the equivalent MTF is

$$H_{AVE}(f_e) = \left|\frac{\sin\left(N_{AVE}\pi\frac{f_e}{f_{eS}}\right)}{N_{AVE}\sin\left(\pi\frac{f_e}{f_{eS}}\right)}\right| \tag{9-39}$$

The first zero of this function occurs at f_{eS}/N_{AVE}. Averaging samples together is called binning or super pixeling. It provides the same MTF as if the detector elements were proportionally larger. Cameras that operate in either the pseudo-interlace or the progressive scan modes will have a different vertical MTF for each mode. In the pseudo-interlace mode, two pixels are averaged together. If $d_H = d_{CCH}$, then

$$MTF_{DETECTOR} MTF_{DFILTER} = \mathrm{sinc}(d_H f_e) \frac{\mathrm{sinc}\left(N_{AVE} \dfrac{f_e}{f_{eS}} \right)}{\mathrm{sinc}\left(\dfrac{f_e}{f_{eS}} \right)} = \mathrm{sinc}(N_{AVE} d_H f_e) \tag{9-40}$$

That is, averaging N_{AVE} samples together provides the same MTF as a single detector element that is N_{AVE} as large.

9.9.3. SAMPLE-AND-HOLD

The digital-to-analog converter (DAC) performs two functions: 1) the quantized signal amplitude is converted into an analog voltage, and 2) extends the discrete pulses into a time-continuous signal. It is present in all cameras that offer a video output. The second process is called image reconstruction. The ideal reconstruction filter has unity MTF up to the Nyquist frequency and then drops to zero. The ideal filter is unrealizable. The simplest filter is the zero-order sample-and-hold filter

$$MTF_{S\&H} = \mathrm{sinc}\left(\frac{u_i}{u_{iS}} \right) \tag{9-41}$$

where $u_{iS} = 1/d_{CCH}$. Figure 9-35 illustrates the sample-and-hold MTF as a function of normalized spatial frequency. Although it is a poor approximation to the ideal reconstruction filter, it is widely used.

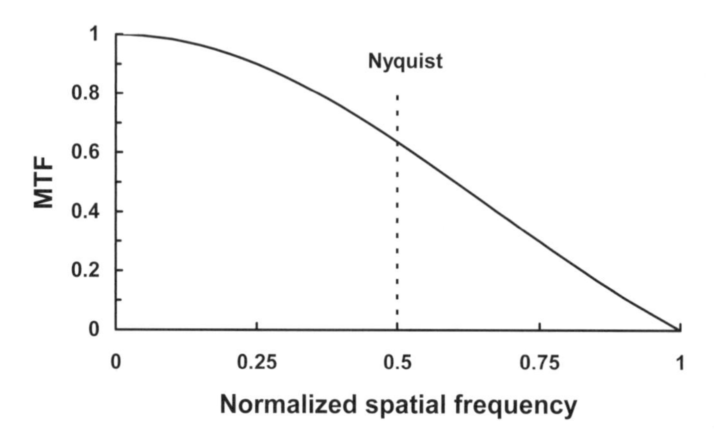

Figure 9-35. Zero-order sample-and-hold MTF normalized to $d_{CCH}u_i$. The sample-and-hold acts as a low-pass filter.

9.9.4. POST-RECONSTRUCTION FILTER

After the digital-to-analog conversion, the image appears blocky when using the zero-order sample-and-hold filter. The ideal post-reconstruction filter removes all the higher-order frequencies such that only the original smooth signal remains. The output is delayed, but this is not noticeable on most imagery (Figure 9-36). This additional filter is used in all video cameras to match the bandwidth of the video standard.

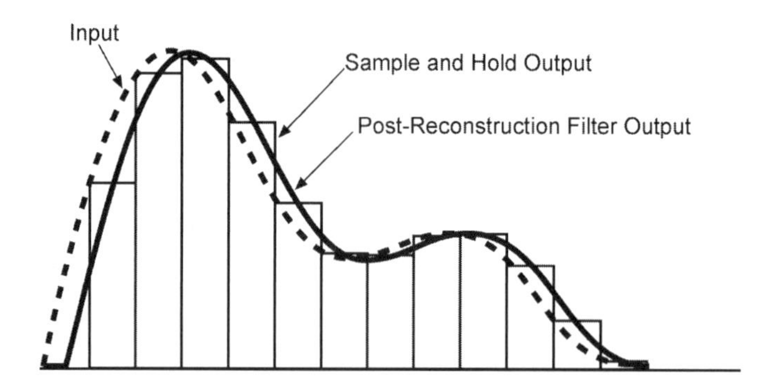

Figure 9-36. The post-reconstruction filter removes the blocky (stair-step) effect created by the sample-and-hold circuitry within the DAC.

The ideal post-reconstruction filter has an MTF of unity up to the array Nyquist frequency and then drops to zero. Real filters will have some roll-off and f_{v3dB} should be sufficiently large so that it does not affect the in-band MTF. The post-reconstruction filter can be of the same functional form as the anti-alias filter (see Section 9.9.1.) but serves a different purpose. It removes the spurious frequencies created by the sampling process. The ideal filter is unrealizable but can be approximated by Nth-order Butterworth filters where its frequency response is specified by the video timing

$$H_{LOWPASS}(f_v) = \frac{1}{\sqrt{1+\left(\frac{f_v}{f_{v3dB}}\right)^{2N}}} \tag{9-42}$$

For analog video output, a low-pass filter is used as a post-reconstruction filter. This filter affects only the horizontal signal. While the video bandwidth has been standardized (NTSC, PAL, and SECAM are 4.2, 5.5, and 6 MHz, respectively), the filter design has not. Thus, different cameras may have different post-reconstruction filters and therefore different horizontal MTFs even though they are built to the same video standard. No reconstruction filter is used in the vertical direction. Here, the display and human visual response provide the reconstruction.

9.9.5. BOOST

MTF degradation caused by the various subsystems can be partially compensated with electronic boost filters. Boost filters also amplify noise so that the signal-to-noise ratio may degrade. For systems that are contrast-limited, boost may improve image quality. However, for noisy images, the advantages of boost are less obvious. As a result, these filters are used only in high-contrast situations (typical of consumer applications) and are not used in scientific applications where low signal-to-noise situations are often encountered.

The boost amplifier can either be an analog or a digital circuit with peaking that compensates for any specified MTF roll-off. The MTF of a boost filter, by definition, exceeds one over a limited spatial frequency range. When used with all the other subsystem MTFs, the resultant MTF_{SYS} is typically less than one for all spatial frequencies. Excessive boost can cause ringing at sharp edges.

A boost filter can be placed in any part of the circuit where a serial stream of data exists. By appropriate selection of coefficients, digital filters can also provide boost (Figures 9-33 and 9-34). MTF_{BOOST} can also approximate the inverse of MTF_{SYS} (without boost). The combination of MTF_{BOOST} and MTF_{SYS} can provide unity MTF over spatial frequencies of interest

$$H_{BOOST}(f) = \frac{1}{MTF_{MOTION} MTF_{OPTICS} MTF_{DETECTOR}} \tag{9-43}$$

This suggests that the reproduced image will precisely match the scene in every spatial detail. When used to compensate for MTF losses, the boost filter provides "aperture correction". Motion compensation is more difficult. The amount of motion must be known precisely. Otherwise the "correction" may make the image worse.

9.10. CRT DISPLAY

This section considers monochrome displays. With minor modification, it also applies to color displays. The CRT display spot size is a complex function of design parameters, phosphor choice, and operating conditions. It is reasonable to assume that the spot intensity profile is Gaussian distributed and is radially symmetric with radius r

$$L(r) = L_o \exp\left(-\frac{1}{2}\left(\frac{r}{\sigma_{SPOT}} \right)^2 \right) \tag{9-44}$$

As illustrated in Figure 9-37, the spot diameter is defined as the full width at half-maximum (FWHM) and is

$$S = FWHM = \sqrt{8\ln(2)}\ \sigma_{SPOT} = 2.35\sigma_{SPOT} \tag{9-45}$$

Then

$$L(r) = L_O \exp\left[-4\ln(2)\left(\frac{r}{S} \right)^2 \right] \tag{9-46}$$

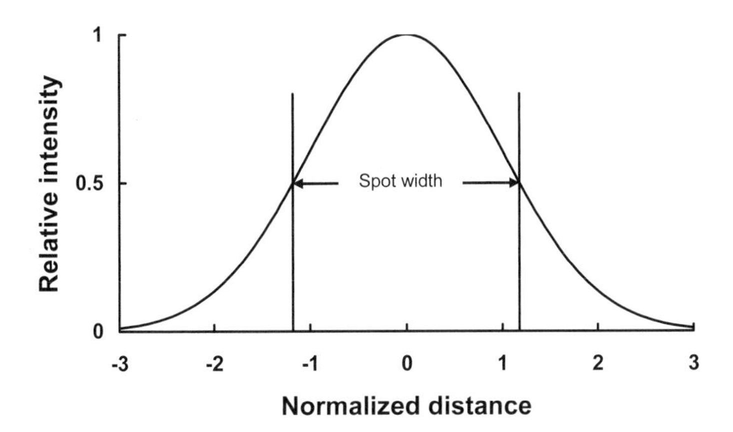

Figure 9-37. The spot diameter is the full width at one-half maximum intensity. The profile has been normalized to σ_{SPOT}.

While the shrinking raster method is listed as a resolution test method, its primary value is in determining the FWHM spot size. With the shrinking raster method, every line is activated (Figure 9-38). If the lines are far apart, the valleys between the lines are visible and the raster pattern is obvious. The lines are slowly brought together until a flat field is encountered. That is, the entire screen appears to have a uniform intensity. The resolution is the number of displayed lines divided by the picture height of the shrunk raster

$$R_{FF} = \frac{X}{Y} \quad \frac{\text{lines}}{\text{cm}} \tag{9-47}$$

As the lines coalesce, the MTF drops. Under nominal viewing conditions, experienced observers can no longer perceive the raster when the luminance variation (ripple) is less than 5%. The line spacing has been standardized[21] to $2\sigma_{SPOT}$ (Figure 9-39). Given a shrunk raster resolution of R_{FF} lines/cm, the line spacing in the shrunk raster is $1/R_{FF}$ cm. Then $\sigma_{SPOT} = 1/2R_{FF}$ cm and the spot size is

$$S = FWHM = \frac{2.35}{2R_{FF}} \tag{9-48}$$

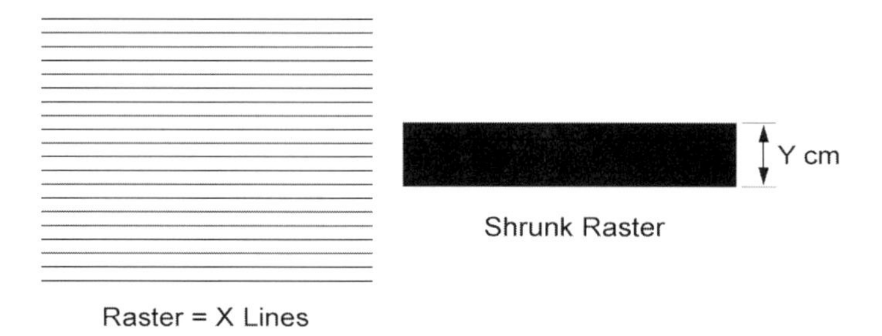

Figure 9-38. Shrinking raster resolution test method. The raster is shrunk until the discrete lines can no longer be discerned.

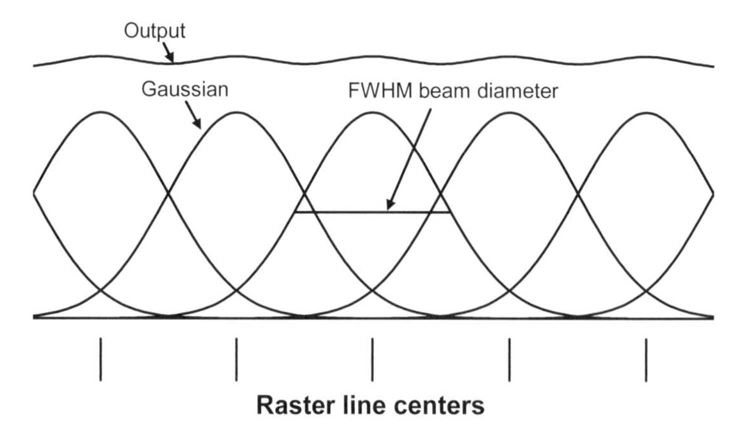

Figure 9-39. When the line spacing is $2\sigma_{SPOT}$, the spots appear to merge and this produces a flat field. The peak-to-peak ripple is 3% and the MTF associated with the residual raster pattern is 0.014. The output is the sum of all the discrete line intensities.

While a flat-field condition is desirable, the line spacing may be larger. The shrinking raster methodology is applicable for CRTs only and cannot be used for flat-panel displays because the "raster" is fixed in those displays. Sometimes the spot profile tends to be flat-topped. A Gaussian approximation is adequate if the standard deviation, σ_{SPOT}, is calculated from the full width at the 5% intensity level[22]

$$\sigma_{SPOT} = \frac{FW\ @\ 5\%\ level}{\sqrt{8\ln(20)}} \approx \frac{FW\ @\ 5\%\ level}{4.9} \tag{9-49}$$

The beam cannot be truly Gaussian because the Gaussian distribution is continuous from $-\infty$ to $+\infty$. Amplifiers may reasonably reproduce the Gaussian beam profile within the $-3\sigma_{SPOT}$ to $+3\sigma_{SPOT}$ limits or a smaller portion. No simple relationship exists between electronic bandwidth and spot size. That is, monitor performance cannot be determined simply by the electronic bandwidth. The bandwidth number quoted depends on the manufacturer's philosophy on rise time issues. Monitors with the same electronic bandwidth may have different resolutions.

9.10.1. ADDRESSABILITY

Addressability is a characteristic of the display controller and represents the ability to select and activate a unique area on the screen. It defines how precisely an electron beam can be positioned on the CRT screen. Addressability is given as the number of discrete lines per picture height or disels per unit length. With this definition, the inverse of addressability is the center-to-center spacing between adjacent disels. Addressability is the image format listed in Table 11-7.

Addressability and resolution are independent of each other. If the resolution is low, successive lines will overwrite preceding lines. If addressability is low, adjacent raster lines will not merge and they will appear as stripes. This was illustrated in Figure 9-39, where the line spacing is the inverse of the addressability and resolution is the inverse of σ_{SPOT}. Figure 9-39 illustrates good addressability (because lines can be placed close together) but poor resolution (because lines cannot detected when the modulation is low). CRT resolution is discussed in Section 11.6.1. *Disels*.

There are two opposing design requirements. The first design requirement is that the raster pattern be imperceptible to the observer. This is called the adjacent disel requirement. If the display meets this requirement, the picture will appear uniform and solid. It provides the flat field condition. Alphanumeric characters will appear well constructed and highly legible. The second design requirement is the alternating disel requirement. Here, individual lines (one disel on and one disel off) should be highly visible.

Increased addressability favors the adjacent disel requirement. Placing large spots close together will eliminate the visibility of the raster. But, this also reduces the visibility of alternating disels. Similarly, an increase in resolution favors the alternating requirement but may make the raster pattern visible.

The resolution/addressability ratio[23] (RAR) is the ratio of FWHM spot size to disel spacing, P

$$RAR = \frac{S}{P} = \frac{2.36\,\sigma_{SPOT}}{P} \tag{9-50}$$

The variable P is the disel pitch. For example, if a display is 27.5 cm high and it displays (addresses) 1024 lines, then P is 27.5 cm/1024 lines = 0.27 mm/line. Assuming a 0.28-mm wide spot then RAR = 0.28/0.27 = 1.04.

As shown in Figure 9-39, the summation of periodically spaced spots creates a modulated output. Figure 9-40 illustrates the adjacent disel (on-on-on-on) and alternating disel (on-off-on-off) modulation as a function of the RAR. A RAR of 1.0 is considered desirable. In Figure 9-39, the RAR is 1.18. Figure 9-41 illustrates imagery with different RARs.

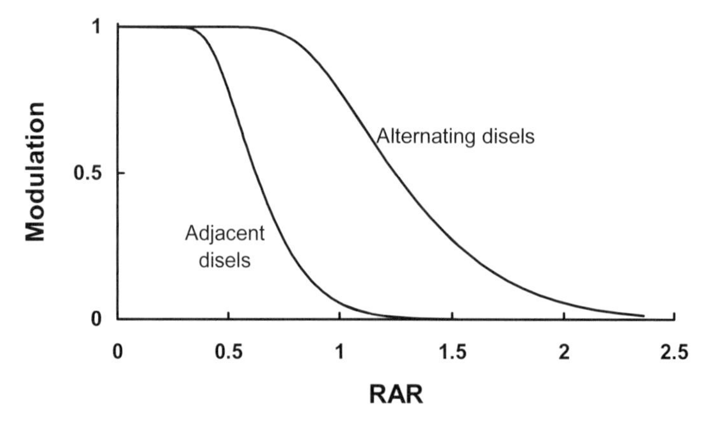

Figure 9-40. Modulation based on Gaussian beam profiles for the adjacent disel and alternating disels. A RAR of one is considered desirable. The modulation of the alternating disel is associated with the display's Nyquist frequency.

Note that the modulation transfer function is traditionally defined for sinusoidal inputs. The residual ripple is not truly sinusoidal and therefore the modulation is not exactly what would be expected if the input was sinusoidal. That is, the modulation shown in Figure 9-40 is not a true MTF.

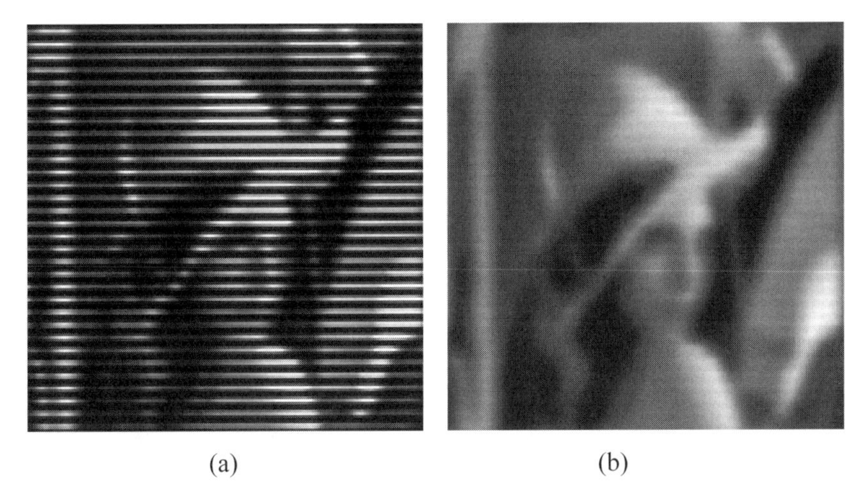

(a) (b)

Figure 9-41. (a) RAR = 0.5 and (b) RAR = 1.18. As illustrated in Figure 9-39, a slight vertical modulation exists (visible raster lines). Imagery created[2] by MAVIISS. See ***NOTE*** in Figure 9-4 caption.

While any disel format can be specified, the RAR determines if the disels are visible. For the configuration shown in Figure 7-14, if the shadow mask pitch is 0.28 mm, then the disels will be separated by 0.336 mm when the RAR is one. If using a 13-inch diagonal display, the width is 264 mm and 786 horizontal disels are viewable when the ***RAR*** = 1. More disels can be displayed, but with reduced modulation (Table 9-2 and Figure 9-42). Thus the disel format, by itself, does not fully characterize the display capability. For optimized displays, it is reasonable to assume that the RAR is about one.

Table 9-2
ALTERNATING DISEL MODULATION
13-inch diagonal display and 0.28-mm shadow-mask pitch
5% intensity level illuminating 2.5 shadow-mask holes

DISELS	RAR	MODULATION
800	1.02	0.75
900	1.15	0.61
1000	1.27	0.47
1010	1.40	0.35
1200	1.53	0.25

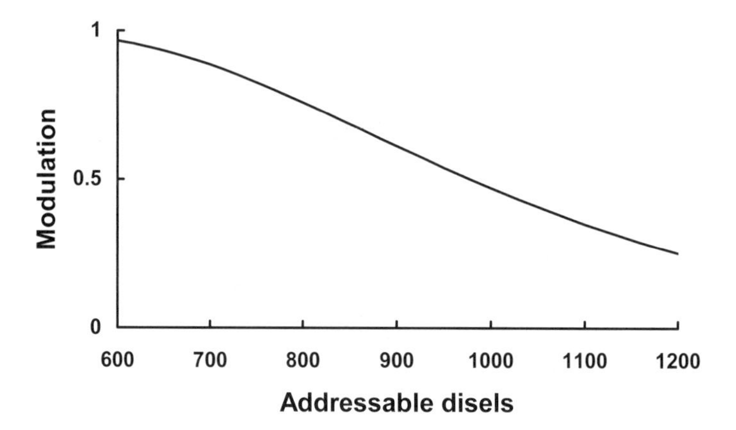

Figure 9-42. 13-inch diagonal display alternating disel (Gaussian beam) modulation. The shadow-mask pitch is 0.28 mm. The 5% intensity level illuminates 2.5 shadow-mask holes.

ANSI/HFS 100-1988 requires that the adjacent disel modulation be less than 20%. This deviates significantly from the flat field condition (adjacent disel modulation is 0.014). ISO 9241 recommends that the alternating disel modulation be greater than 0.4 for monochrome and greater than 0.7 for color systems. For monochrome systems, the RAR will vary from 0.8 to 1.31. For color systems, the RAR varies from 0.8 to 1.04. These standards were developed to assure legibility of alphanumeric characters.

If the RAR is too high, then the output of two adjacent "on" lines is much greater than the output of two separated lines due to the summing effect (Figure 9-43). For the flat field condition, **RAR** = 1.18 and the summed lines are 27% brighter than an individual line. If the RAR is very large, the intensity of several adjacent lines "on" can be significantly brighter than a single line "on". Thus, there is also a tradeoff between intensity and RAR.

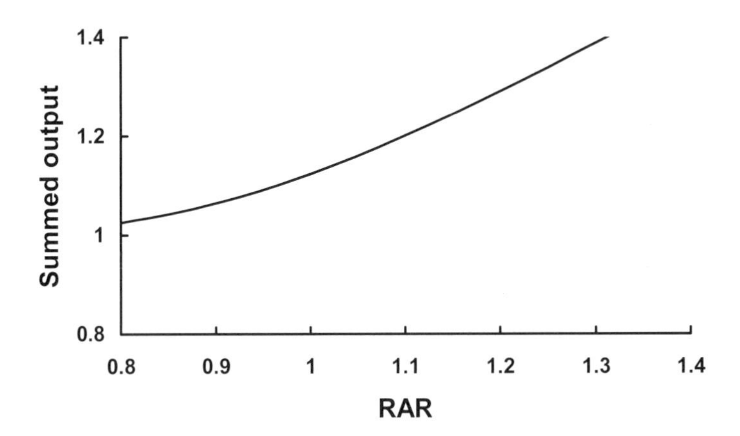

Figure 9-43. Summed output of contiguous lines "on". Ideally, the intensity should remain constant whether one or multiple lines are "on."

Multi-sync monitors automatically adjust the internal line rate to the video line rate. These monitors can display any format from 525 up to 1225 lines. Because the displayed image size remains constant, the vertical line spacing changes. At the 1225 line format, the raster lines are close together and form a uniform (flat) field. At the lower line rates, the individual raster lines are separated and may be perceived. Because the same video amplifiers are used for all line rates, the horizontal resolution is independent of the line rate. That is, the RAR in the horizontal direction stays constant whereas the RAR in the vertical direction changes with the number of lines. If the vertical RAR is one at 1225 lines, then it must be 525/125 = 0.43 at 525 lines.

9.10.2. CHARACTER RECOGNITION

Computer monitors are designed for maximum legibility of alphanumeric characters and enhanced graphical capability. Readable characters can be formed in any block of disels from 5×7 to 9×16 and still be aesthetically pleasing. More complex symbols such as Japanese characters may require 16×16 disels.

The ability to see detail is related to the RAR. Figure 9-44 illustrates three letters and the resultant intensity traces. The RAR must be near one so that a reasonable contrast ratio exists between "on" and "off" disels. With reasonable contrast, the inner detail of the character is seen and the character is legible.

Similarly, with a reasonable contrast ratio, adjacent letters will appear as separate letters. With characters, the alternating disel pattern is called one stroke separated by a space.

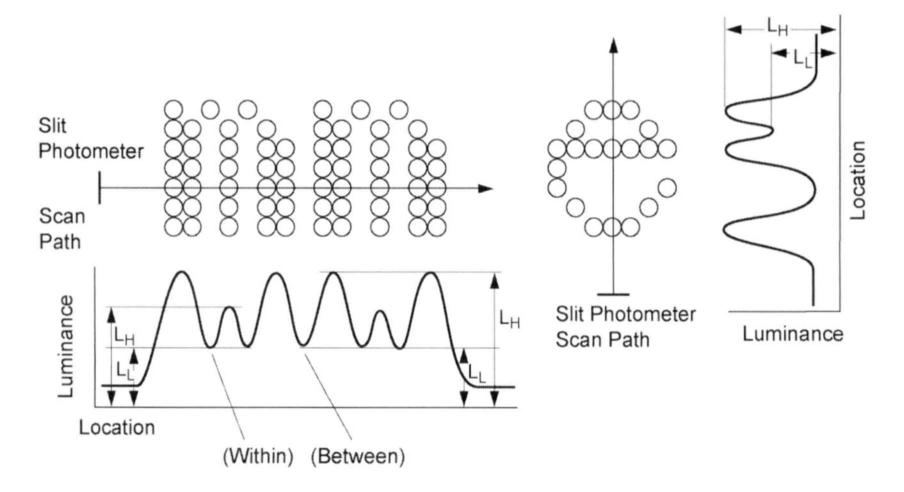

Figure 9-44. The RAR must be near one so that the inner details of characters are visible and the double width appears solid. Similarly, with RAR near one, two adjacent characters appear separate. Each dot represents a disel. Lower RARs will make the characters appear very sharp. However, the raster pattern will become visible and the characters will not appear solid.

Figure 9-45 illustrates how reduced modulation affects character legibility. For easy interpretation, the system contains a shadow mask but only white and black can be seen. The image is magnified so that the individual phosphor dots can be seen. The sampled nature of the Gaussian electron beam is evident. As the beam width increases, the modulation decreases and the RAR increases. The characters become blurry as the RAR increases.

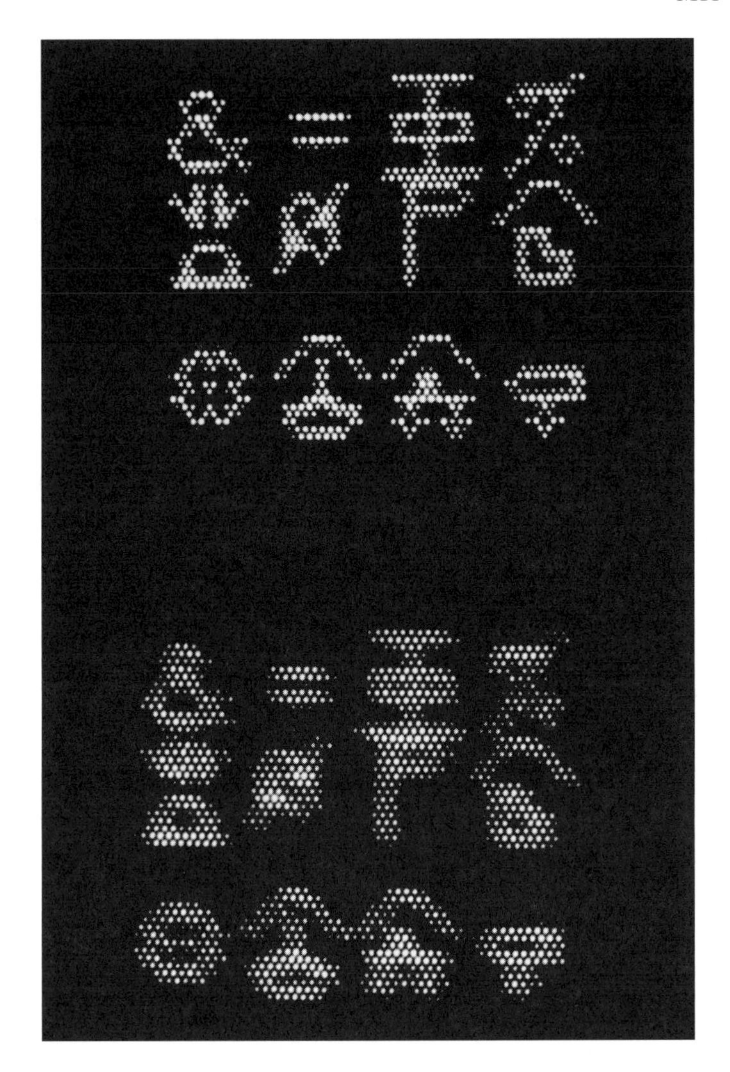

Figure 9-45. Complex symbols are more legible when the display modulation is high. Each dot represents a hole in the shadow mask. Top: modulation = 0.5 at the Nyquist frequency. Bottom: modulation = 0.1 at the Nyquist frequency. When viewed at 10 feet, the individual dots are no longer perceptible. At this distance, the top characters appear sharp (RAR ≈ 1.25) and the bottom ones appear fuzzy (RAR ≈ 1.8). Courtesy of the Mitre Corporation.

9.10.3. DISPLAY MTF

The display MTF is a composite that includes both the internal amplifier and the CRT responses. Implicit in the MTF is the conversion from input voltage to output display brightness. Although not explicitly stated, the equation implies radial symmetry and the MTFs are the same in both the vertical and horizontal directions. That is, the MTF is considered separable and $MTF_{CRT}(u_d,v_d) = MTF_{CRT}(u_d)MTF_{CRT}(v_d)$.

The amplifier bandwidth is not a number that is often useful to the user. We must assume that the manufacturer selected an adequate bandwidth to provide the display resolution. For example, for a constant frame rate, a 1600×1200-disel display must have a wider bandwidth than a 1280×1024-disel display by a factor of approximately 1600×1200/1280×1024 = 1.47. MTF_{CRT} includes both the electronics MTF and spot diameter. For an infinitesimal input, the spot diameter is affected by the video bandwidth. That is, the minimum spot diameter is the equivalent point spread function.

With a Gaussian spot

$$MTF_{CRT}(u_d) \approx \exp\left(-2\pi^2\sigma_{SPOT}^2 u_d^2\right) = \exp\left(-2\pi^2\left(\frac{S}{2.35}u_d\right)^2\right) \qquad (9\text{-}51)$$

If the spot size is not specified, then σ_{SPOT} can be estimated from the TV limiting resolution. With the TV limiting resolution test, an observer views a wedge pattern and selects that spatial frequency at which the converging bar pattern can no longer be seen. The display industry has standardized the bar spacing to $2.35\sigma_{SPOT}$.

If N_{TV} lines are displayed on a monitor, then $H_{MONITOR} \approx 2.35\,\sigma_{SPOT}N_{TV}$. When transposed to image space

$$MTF_{CRT}(u_i) \approx \exp\left(-2\pi^2\left(\frac{VFOV}{2.35}\frac{fl}{N_{TV}}u_i\right)^2\right) \qquad (9\text{-}52)$$

For staring arrays

$$MTF_{CRT}(u_i) \approx \exp\left(-2\pi^2\left(\frac{(N_V-1)d_{CCV}+d_V}{2.35N_{TV}}u_i\right)^2\right) \qquad (9\text{-}53)$$

It is convenient to match the number of vertical detectors to the number of active video lines ($\mathbf{N_{TV}} = \mathbf{N_V}$). Thus, the video standard uniquely defines the CRT MTF. For 100% fill factor arrays

$$\boldsymbol{MTF_{CRT}(u_i) \approx \exp\left(-2\pi^2\left(\frac{d_{CCV}}{2.35}u_i\right)^2\right)} \tag{9-54}$$

At Nyquist frequency, $\boldsymbol{MTF_{CRT}(u_{iN})} = 0.409$ (Figure 9-46). That is, the "standard" design effectively reduces sampling effects by attenuating the amplitudes above the Nyquist frequency. Reducing the fill factor increases the detector MTF, but does not change the Nyquist frequency. MTF_{CRT} should not be confused with the alternating disel RAR modulation. With the RAR, disel outputs are summed resulting in an apparent modulation. The MTF describes the displayed amplitude of sinusoidal inputs.

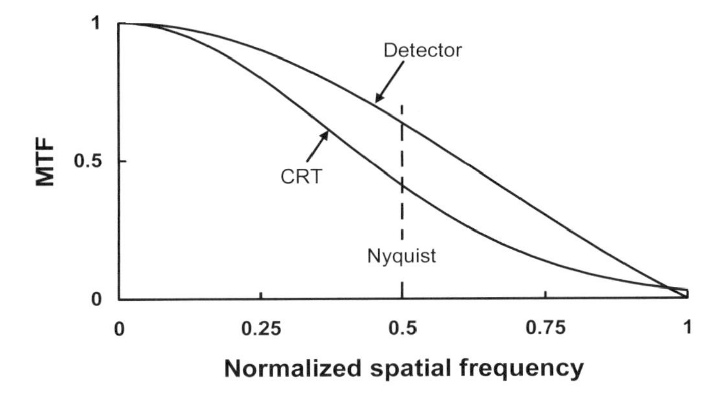

Figure 9-46. $MTF_{DISPLAY}$ as a function of $\boldsymbol{d_{CCV}v_i}$. The display MTF limits most systems. The Nyquist frequency occurs at $\boldsymbol{d_{CCV}v_i} = 0.5$.

9.11. FLAT PANEL DISPLAYS

Both flat-panel displays and LEDs are assumed to have rectangular elements. If the width of an individual element is $\boldsymbol{d_{FPW}}$

$$\boldsymbol{MTF_{FLAT\ PANEL}(u_d) = \frac{\sin(\pi\, d_{FPW}\, u_d)}{\pi\, d_{FPW}\, u_d}} \tag{9-55}$$

If the display width is D_W and the horizontal field of view is $HFOV$, then the image-space horizontal MTF is

$$MTF_{FLAT\ PANEL}(u_i) = \frac{\sin\left(\pi\, d_{FPW} \dfrac{HFOV}{D_W} fl\, u_i\right)}{\pi\, d_{FPW} \dfrac{HFOV}{D_W} fl\, u_i} \qquad (9\text{-}56)$$

The vertical MTF is obtained by substituting d_{FPV}, $VFOV$, D_V, and v into the equation. Assuming the number of elements matches the number of detectors

$$MTF_{FLAT\ PANEL}(u_i) \approx \frac{\sin(\pi\, d_{CCH}\, u_i)}{\pi\, d_{CCH}\, u_i} \qquad (9\text{-}57)$$

For 100% fill factor arrays, $d_{CCH} = d_H$ and $MTF_{FLAT\ PANEL}$ is identical to $MTF_{DETECTOR}$.

9.12. PRINTER MTF

A printer acts as a reconstruction filter by placing spots on a piece of paper. The printer's capability to space the spots closely is a measure of its MTF. Inkjet printers use[24] a dispersed-dot half-toning algorithm. A series of vertical sinusoidal targets [$L(x) = \sin(x/x_P)$] were printed and then scanned at 1200 dpi for computer analysis (Figure 9-47).

The columns were averaged and the minimum and maximum values provided the MTF. Although data was only reported up to u_P = 150 lines/inch, the average MTF approximately follows

$$MTF_{INKJET\ PRINTER}(u_P) \approx \exp\left(-\frac{u_P}{187}\right) \qquad (9\text{-}58)$$

Whether this affects the printed quality depends upon the image size and the spatial frequencies associates with the scene. At normal viewing distances, the sinusoids in Figure 9-47 are much finer than that which can be discerned. That is, the observer is the reconstruction filter. Neither the inkjet manufacturer nor model number was specified. Presumably an HP printer ,since one author works for HP. Hasegawa et. al. measured[26] the MTF of three color laser printers using a variety of techniques (response to square waves, checkerboard, and stripes). Their results approximately follow Equation 9-58.

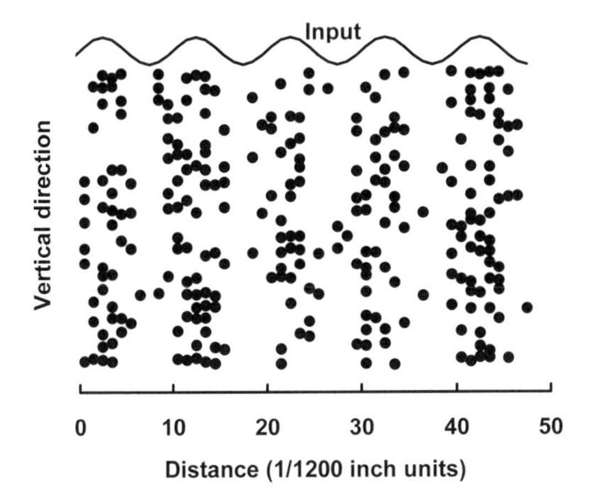

Figure 9-47. Dispersed-dot half-toning for a vertical sinusoid: $L(x,y) = \sin(x/x_P)$. The vertical direction represents 10 rows. This figure is approximately 54 times larger than that created by the printer. Equivalently, it would have to be viewed at a distance 54 times greater than normal reading distance to have the same visual angle. If the normal reading distance is 15 inches, this figure should be viewed at 67.5 feet. (After reference 24).

9.13. THE OBSERVER

The observer should probably be the starting point for system design. Display design is based on both perceptual and physical attributes (Table 9-3). For most applications, the design is driven by perceptual parameters. These parameters are partially related to the physical parameters. For example, sharpness is related to MTF, but the precise relationship has not been quantified. While the physical parameters are important to all applications, they tend to be quantified only for scientific and military displays. Table 9-3 provides different ways of specifying the same requirement: the need for a high quality display.

Maximum image detail is only perceived when the observer is at an optimum viewing distance. If the observer is too far away, the detail is beyond the eye's limiting resolution. It the observer moves closer to the display than the optimum, the image does not become clearer because there is no further detail to see. Display size and resolution requirements are based on an assumed viewing distance.

Table 9-3
PERCEPTIBLE AND PHYSICAL PARAMETERS

PERCEPTUAL PARAMETERS	PHYSICAL PARAMETERS
Brightness Contrast Sharpness Color rendition Flicker	Luminance Resolution Uniformity Addressability Gamma Color saturation Color accuracy Color convergence (CRT) Distortion (CRT) Refresh rate MTF

The contrast sensitivity function (CTF_{EYE}) is the HVS' threshold detection of sinusoids. Barten's model[26] is

$$CTF_{EYE}(f_{EYE}) = \frac{a}{b f_{eye}\, e^{-c f_{EYE}} \sqrt{1 + 0.06\, e^{c f_{EYE}}}} \tag{9-59}$$

where

$$a = 1 + \frac{12}{w\left(1 + \frac{f_{EYE}}{3}\right)^2} \qquad b = \frac{540}{\left(1 + \frac{0.7}{L}\right)^{0.2}} \qquad c = 0.3\left(1 + \frac{100}{L}\right)^{0.15} \tag{9-60}$$

The variable $\boldsymbol{w}$ is the square root of the picture area in degrees and $\boldsymbol{L}$ is the display luminance in cd/m^2 (Table 9-4 and Figure 9-48). Some authors approximate the HVS MTF as $MTF_{EYE} \approx 1/CTF_{EYE}$ normalized to one (Figure 9-49). This approximation does not include the eye's optical MTF[27].

Table 9-4
TYPICAL DISPLAY BRIGHTNESS (for comfortable viewing)
(1 foot-lambert = 3.43 cd/m^2)

AMBIENT LIGHTING	TYPICAL DISPLAY BRIGHTNESS
Dark night	0.3 to 1 cd/m2 ($\approx$ 0.1 to 0.3 ftL)
Dim light	3 to 30 cd/m2 ($\approx$ 1 to 10 ftL)
Normal room light	100 cd/m2 ($\approx$ 30 ftL)

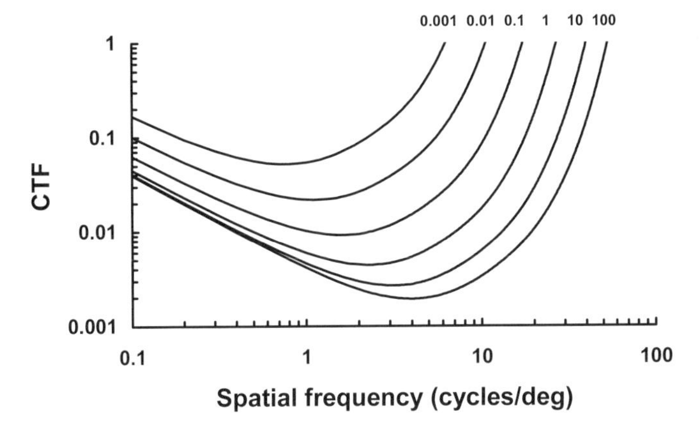

Figure 9-48. Eye CTF where w = 10 deg and the luminance ranges from 0.001 to 100 cd/m^2. Changing the picture size, w, has a small effect at low spatial frequencies.

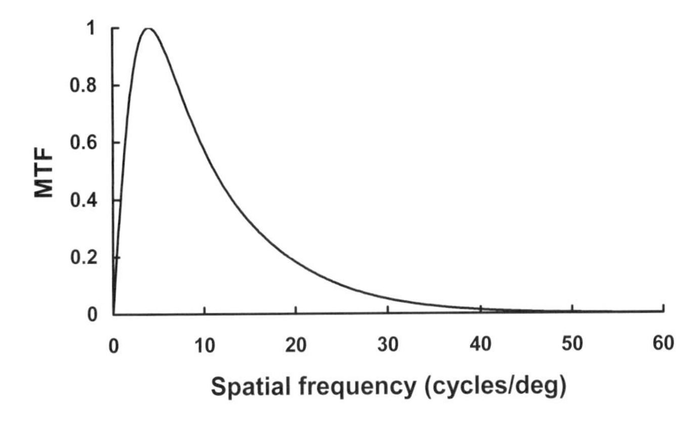

Figure 9-49. A rough approximation for the HVS "MTF". It is 1/CTF for a luminance of 100 cd/m^2.

9.14. INTENSIFIED CCD

The intensified CCD (ICCD) is more difficult to analyze. The MTF of an image intensifier tube (IIT) varies with tube type,[28] manufacturing precision,[29] and materials used. The newer tubes (i.e., OMNI IV) can be approximated by a simple Gaussian expression

$$MTF_{INTENSIFIER}(u) \approx e^{-2\pi^2 \sigma_{IIT}^2 u^2} \qquad \text{(9-61)}$$

Measured MTFs are provided in Figure 9-50. The fiberoptic taper MTF is approximated by a Gaussian MTF

$$MTF_{FO}(u) \approx e^{-2\pi^2 \sigma_{FO}^2 u^2} \tag{9-62}$$

where σ_{FO} = 0.00358 mm for 5 μm fibers and σ_{FO} = 0.00716 mm for 10 μm fibers (Figure 9-51). In this range, $\sigma_{FO} \approx$ 0.000716 mm per μm fiber diameter. Generally, the fiber diameter is ½ the size of the detector. MTF theory will provide an optimistic result of fiber optic coupled ICCDs. It does not include the CCD mismatch or the chicken wire effects. The MTF associated with lens coupling is probably more accurate. The MTF of an ICCD is $MTF_{INTENSIFIER}MTF_{FO}$ multiplied by the appropriate MTFs presented in this chapter.

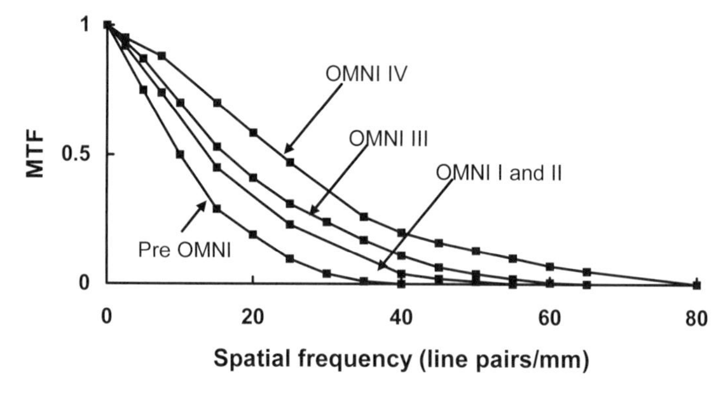

Figure 9-50. Image intensifier MTFs (From Reference 30).

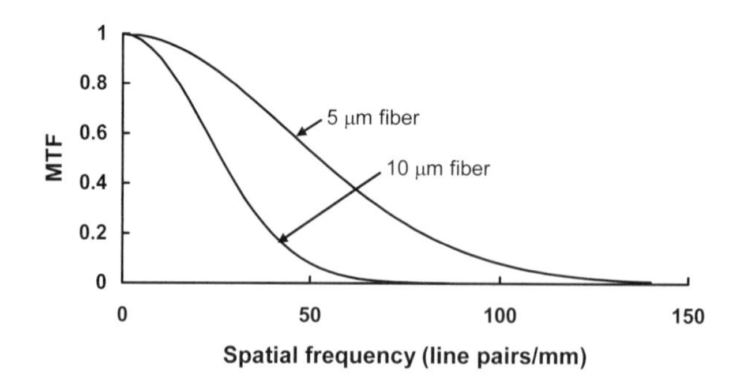

Figure 9-51. Fiber optic MTF for two different sized fibers. (From Reference 31).

9.15. REFERENCES

1. MAVIISS (MTF based Visual and Infrared System Simulator) is an interactive software program available from JCD Publishing at www.JCDPublishing.com.
2. G. C. Holst, *Electro-optical Imaging System Performance*, 4th edition pp. 92-97, JCD Publishing, Winter Park, Fl (2006).
3. H. H. Hosack, "Aperture Response and Optical Performance of Patterned-Electrode Virtual-Phase Imagers," *IEEE Transactions on Electron Devices*, Vol. ED-28(1) pp. 53-63 (1981).
4. N. Schuster, "The Geometrical noise Bandwidth – a New Tool to Characterize the Resolving Power of Analogue and Digital Imaging Devices," in *Sensors and Camera Systems for Scientific, Industrial, and Digital Applications III*, M. M. Blouke, J. Canosa, and N. Sampat, eds., SPIE Proceedings Vol. 4669, pp. 336-346 (2002).
5. O. Yadid-Pecht, "Geometrical Modulation Transfer Function for Different Pixel Active Area Shapes" *Optical Engineering*, Vol. 39, pp. 859-865 (2000).
6. I. Shcherback and O. Yadid-Pecht, "CMOS APS MTF Modeling," *IEEE Trans on Electron Devices*, Vol. 48, pp 2710-2715 (2001).
7. M Estribeau, P Magnan, "CMOS Pixels Crosstalk Mapping and its Influence on Measurements Accuracy for Space Applications" in *Sensors, Systems, and Next-Generation Satellites IX,* R. Meynart, S. P. Neeck, and H. Shimoda, eds., SPIE Proceeding Vol. 5978, paper 597813 (2005).
8. M Estribeau, P Magnan, "Pixel Crosstalk and Correlation with Modulation Transfer Function of CMOS Image Sensor," in *Sensors and Camera Systems for Scientific and Industrial Applications VI,* M. M. Blouke, ed., SPIE Proceedings Vol. 5677, pp, 98-108 (2005).
9. J. Li, J. Liu, and Z Hao, "Geometrical Modulation Transfer Function of Different Active Pixel of CMOS APS" in *2nd International Symposium on Advanced Optical Manufacturing and Testing Technologies: Optical Test and Measurement Technology and Equipment, X*, Hou, J. Yuan, J. C. Wyant, H. Wang, and S. Han, eds., SPIE Proceedings Vol. 6150 paper 61501Y (2006).
10. D. H. Sieb, "Carrier Diffusion Degradation of Modulation Transfer Function in Charge Coupled Imagers," *IEEE Transactions on Electron Devices*, Vol. ED-21(5), pp. 210-217 (1974).
11. L. W. Schumann and T. S. Lomheim, "Modulation Transfer Function and Quantum Efficiency Correlation at Long Wavelengths (Greater Than 800 nm) in Linear Charge Coupled Imagers," *Applied Optics*, Vol. 28(9), pp. 1701-1709 (1989).
12. T. E. Dutton, J. Kang, T. S. Lomheim, R. Boucher, R. M. Shima, M. D. Nelson, C. Wrigley, X. Zheng, and B. Pain, "*Measurement and Analysis of Pixel Geometric and Diffusion Modulation Transfer Function (MTF) Components in Photodiode Active Pixel Sensors*", 2003 IEEE Workshop on CCDs and Advanced Image Sensors, Elmau, Germany, May 16, 2003.
13. M. M. Blouke and D. A. Robinson, "A Method for Improving the Spatial Resolution of Frontside-Illuminated CCD's", *IEEE Transactions On Electron Devices,* Vol. ED-28, pp. 251 -256 (1981).
14. E.G. Stevens and J.P. Lavine, "An Analytical, Aperture, and Two-Layer Carrier Diffusion MTF and Quantum Efficiency Model for Solid State Image Sensors', *IEEE Transactions on Electron Devices*, Vol. 41, pp. 1753 – 1760 (1994).
15. C. S. Lin, B. P. Mathur, and M. F. Chang, "Analytical Charge Collection and MTF Model for Photodiode-Based CMOS Imagers," *IEEE Transactions on Electron Devices*, Vol. 49, pp. 754 – 761, (2002).
16. E. L. Dereniak and D. G. Crowe, *Optical Radiation Detectors*, pp. 199-203, John Wiley & Sons, New York (1984).
17. H. V. Kennedy, "Miscellaneous Modulation Transfer Function (MTF) Effects Relating to Sampling Summing," in *Infrared Imaging Systems: Design, Analysis, Modeling, and Testing*, G. C. Holst, ed., SPIE Proceedings Vol. 1488, pp. 165-176 (1991).

18. T. S. Lomheim, J. D. Kwok, T. E. Dutton, R. M. Shima, J. F. Johnson, R. H. Boucher, and C. Wrigley, "Imaging Artifacts due to Pixel Spatial Sampling Smear and Amplitude Quantization in Two-dimensional Visible Imaging Arrays," in *Infrared Imaging Systems: Design, Analysis, Modeling, and Testing* X, G. C. Holst ed., SPIE Vol. 3701, pp 36-60 (1999).
19. H.-S. Wong. Y. L. Yao, and E. S. Schlig, "TDI Charge-coupled Devices: Design and Applications," *IBM Journal of Research and Development*, Vol. 36(1), pp. 83-105 (1992).
20. A variety of texts on digital filter design are available. See, for example, *Digital Signal Processing*, A. V. Oppenheim and R. W. Schafer, Prentice-Hall, New Jersey (1975).
21. L. M. Biberman, "Image Quality," in *Perception of Displayed Information*, L. M. Biberman, ed., pp. 13-18, Plenum Press, New York (1973).
22. P. G. J. Barten, "Spot Size and Current Density Distributions of CRTs," in Proceedings of the SID, Vol. 25(3), pp. 155-159 (1984).
23. G. M. Murch and R. J. Beaton, "Matching Display Resolution and Addressability to Human Visual Capacity," *Displays*, Vol. 9, pp. 23-26 (Jan 1988).
24. W. Jang and J. P. Allebach, "Characterization of printer MTF", in *Image Quality and System Performance* III; L. C. Cui and Y. Miyake, eds., SPIE Proc. Vol. 6059, paper 60590D (2006).
25. J. Hasegawa, T.-Y. Hwang, H.-C. Kim, D.-W. Kim, and M.-H. Choi, "Measurement-based Objective Metric for Printer Resolution," in *Image Quality and System Performance* IV, L. C. Cui, Y. Miyake, eds., SPIE Proceedings Vol. 6494, paper 64940D (2007).
26. P. Barten, "The SQRI as a Measure for VDU Image Quality," *Society of Information Display 92 Digest*, pp. 867-870 (1992).
27. G. C. Holst, *Electro-optical Imaging System Performance*, 4th edition pp. 122-123, JCD Publishing, Winter Park, Fl (2006).
28. G. M. Williams, Jr., "A High Performance LLLTV CCD Camera for Nighttime Pilotage," in *Electron Tubes and Image Intensifiers*, C. B. Johnson and B. N. Laprade, eds., SPIE Proceedings Vol. 1655, pp. 14-32 (1992).
29. I. P. Csorba, *Image Tubes*, pp. 79-103, Howard W. Sams, Indianapolis, IN (1985).
30. MTFs are provided in IICamIP documentation. This software was developed by the U.S. Army Night Vision and Electronic Sensor Directorate (Ft. Belvoir, VA). It can be obtained from SENSIAC (www.sensiac.gatech.edu/external/index.jsf).

10
SAMPLING

Sampling is an inherent feature of all electronic imaging systems. The scene is spatially sampled in both directions due to the discrete locations of the detector elements. This sampling creates ambiguity in target edges and produces moiré patterns when viewing periodic targets. Aliasing becomes obvious when image features approach the detector size. Spatial aliasing is rarely seen in photographs or motion pictures because the grains are randomly dispersed in the photographic emulsion.

If the detector size and spacing are different in the horizontal and vertical directions, sampling effects will be different in the two directions. This leads to moiré patterns that are a function of the target orientation with respect to the array axis. Although sampling is two-dimensional, for simplicity sampling effects are presented in one dimension.

The highest frequency that can be faithfully reconstructed is one-half the sampling rate. Any input signal above the Nyquist frequency, f_N (which is defined as one-half the sampling frequency, f_S), will be aliased down to a lower frequency. That is, an undersampled signal will appear as a lower frequency after reconstruction (Figure 10-1).

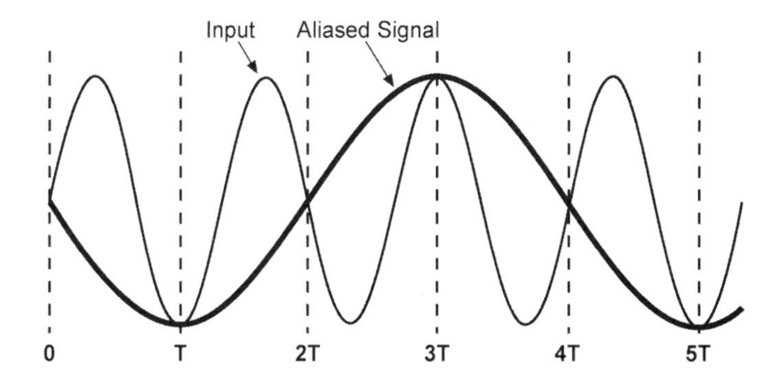

Figure 10-1. An undersampled sinusoid will appear as a lower frequency after reconstruction. The sampling frequency is $f_S = 1/\boldsymbol{T}$.

Signals can be undersampled or oversampled. Undersampling is a term used to denote that an input frequency is greater than the Nyquist frequency. It does not imply that the sampling rate is inadequate for any specific application. Similarly, oversampling does not imply that there is excessive sampling. It simply means that there are more samples available than required by Shannon's sampling theorem.

After aliasing, the original signal can never be recovered. Undersampling creates moiré patterns (Figure 10-2). Diagonal lines appear to have jagged edges or "jaggies." Periodic structures are rare in nature and aliasing is seldom reported when viewing natural scenery, although aliasing is always present. It may become apparent when viewing periodic targets such as test patterns, picket fences, plowed fields, railroad tracks, and Venetian blinds. Distortion effects can only be analyzed on a case-by-case basis. Since aliasing occurs at the detector, the signal must be band-limited by the optical system to prevent it.

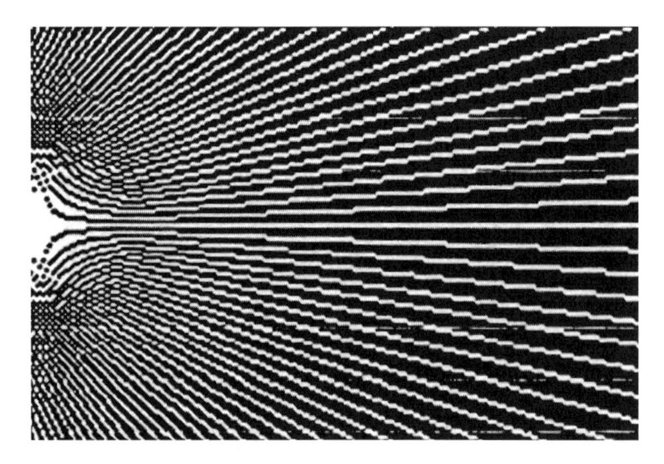

Figure 10-2. A raster scan system creates moiré patterns when viewing wedges or starbursts.

We have become accustomed to the aliasing in commercial televisions. Periodic horizontal lines are distorted due to the raster. Clothing patterns such as herringbones and stripes produce moiré patterns. Cross-color effects occur in color imagery. Many videotape recordings are undersampled to keep the price modest and yet the imagery is considered acceptable when observed at normal viewing distances.

Aliasing is less bothersome in monochrome imagery. It is extremely objectionable in color imagery and must be reduced to an acceptable level. For specific applications it may be considered unacceptable. In medical imagery, an aliased signal may be misinterpreted as a medical abnormality that requires medication, hospitalization, or surgery. It becomes bothersome when the scene geometric properties must be maintained as with mapping. It affects the performance of most image processing algorithms.

Sampled data systems are nonlinear and do not have a unique MTFs.[1-6] The "MTF" depends on the phase relationships of the scene with the sampling lattice. Superposition does not hold and, in principle, any MTF derived for the sampler cannot be used to predict results for general scenery. As the sampling rate increases, the MTF becomes better defined. As the sampling rate approaches infinity, the system becomes an analog system and the MTF is well defined. Equivalently, as the detector size and detector pitch decrease, signal fidelity increases. Reference 7 provides an in-depth discussion on sampling effects.

Sampling theory states that the *frequency* can be unambiguously recovered for all input frequencies below Nyquist frequency. It was developed for band-limited electrical circuits. The extension to imaging systems is straightforward. After aliasing, the original signal can never be recovered. The mathematics suggest that aliasing is an extremely serious problem. However, the extent of the problem depends upon the final interpreter of the data. The appearance depends on the display medium.

The eye is primarily sensitive to intensity variations and less so to frequencies. Therefore, sampling artifacts in imagery are usually tolerable. In contrast, the ear is a frequency detector and any distortion is immediately obvious. The sampling theorem must be strictly followed for auditory-based processes.

10.1. SAMPLING THEOREM

The sampling theorem as introduced by Shannon[8] was applied to information theory. He stated that if a time-varying function, $v(t)$, contains no frequencies higher than f_{MAX} (Hz), it is completely determined by giving its ordinates at a series of points spaced $1/2\, f_{MAX}$ sec apart. The original function can be reconstructed by an ideal low-pass filter. Shannon's work is an extension of others,[9] and the sampling theorem is often called the Shannon-Whittaker theorem.

Three conditions must be met to satisfy the sampling theorem. The signal must be band-limited, the signal must be sampled at an adequate rate, and a low-pass reconstruction filter must be present. When any of these conditions are not present, the reconstructed analog data will not appear exactly as the original.

In a sampled-data system, the sampling frequency interacts with the signal to create sum and difference frequencies. Any input frequency, f_o, will appear as $nf_S \pm f_o$ after sampling (n = $-\infty$ to $+\infty$). Figure 10-3 illustrates a band-limited system with frequency components replicated by the sampling process. The base band ($-f_H$ to f_H) is replicated at nf_S. To avoid distortion, the lowest possible sampling frequency is that value where the base band adjoins the first side band (Figure 10-3c). This leads to the sampling theorem that a band-limited system must be sampled at twice the highest frequency ($f_S \geq 2 f_H$) to avoid distortion in the reconstructed image.

After digitization, the data reside in data arrays (e.g., computer memory location) with nonspecific units. The user assigns units to the arrays during image reconstruction. That is, the data are read out of the memory in a manner consistent with display requirements. The data are transformed into an analog signal by the display medium. If the original signal was oversampled and if the reconstruction filter limits frequencies to f_N, then the reconstructed image can be identical to the original image (Figure 10-4).

Example 10-1
FREQUENCY REPRODUCTION

A staring array consists of detectors that are 10 μm square. The detector pitch is 15 μm. The focal length is 15 cm. The aperture diameter is 3 cm and the average wavelength is 0.5 μm. What is the highest spatial frequency that will be faithfully reproduced?

The highest frequency is the smaller of the optical cutoff, detector cutoff, or Nyquist frequency. In image space, the optical cutoff is $D_o/(\lambda fl)$ = 400 cycles/mm. The detector cutoff is $1/d_H$ = 100 cycles/mm. The detector pitch provides sampling every 15 μm for an effective sampling rate of 66.7 cycles/mm. Because the Nyquist frequency is one-half the sampling frequency, the highest frequency is 33.3 cycles/mm.

Since object space is related to image space by the lens focal length, the optical cutoff is 30 cycles/mrad, detector cutoff is 15 cycles/mrad, sampling is 10 cycles/mrad, and the Nyquist frequency is 5 cycles/mrad.

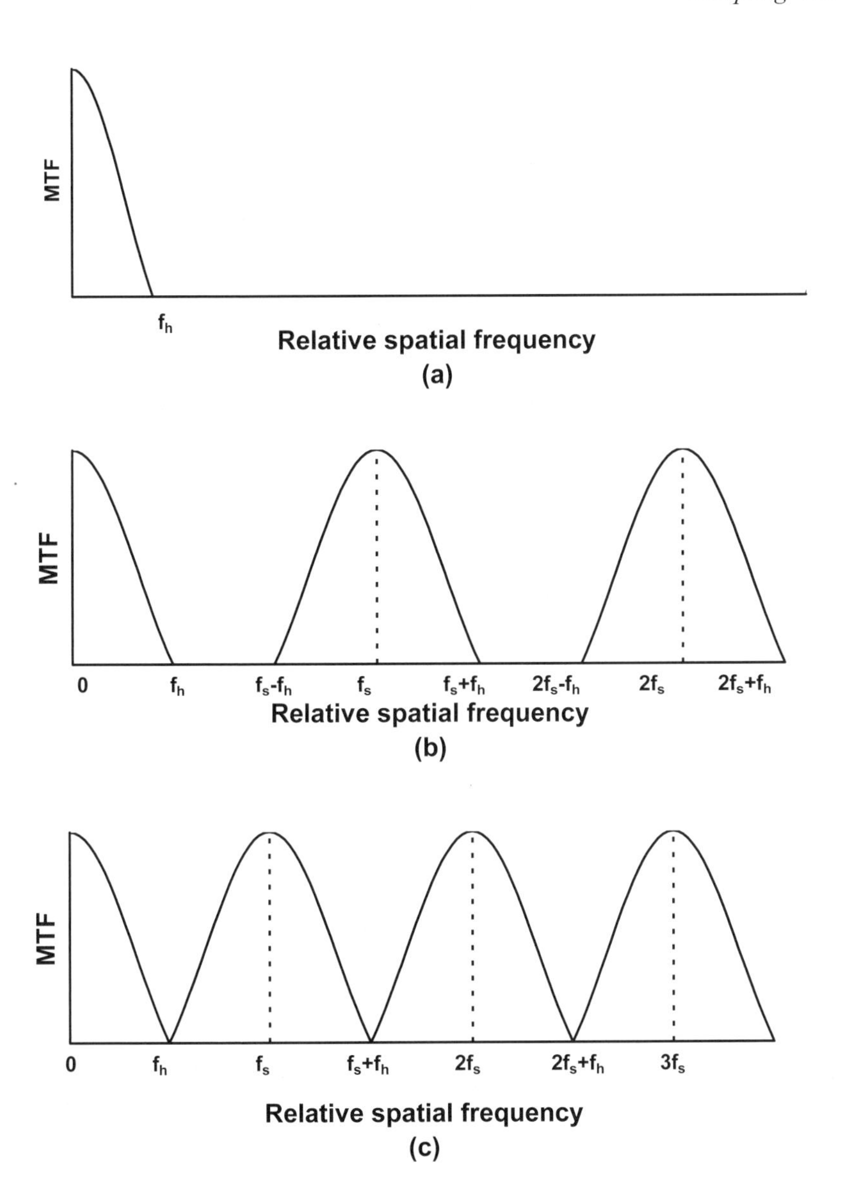

Figure 10-3. Sampling replicates frequencies at $nf_S \pm \underline{\mathbf{f}}$. (a) Original band-limited signal, (b) frequency spectrum after sampling, and (c) when $f_S = 2f_H$, the bands just adjoin.

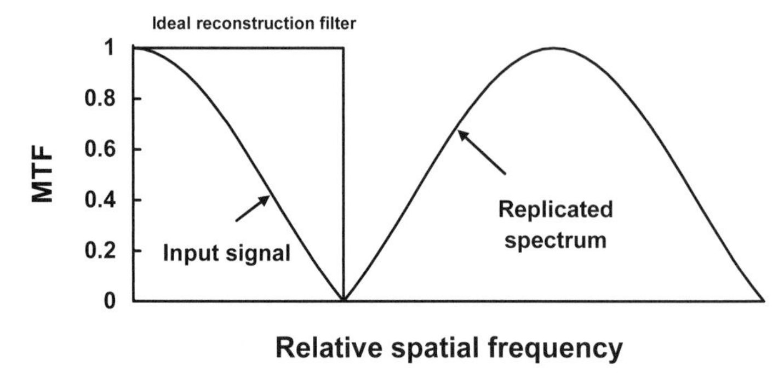

Figure 10-4. Signal reconstruction. The ideal reconstruction filter passes all the signals below f_N and no signal above f_N.

10.2. ALIASING

As the sampling frequency decreases, the first side band starts to overlap the base band. It is the summation of these power spectra that create the distorted image (Figure 10-5).

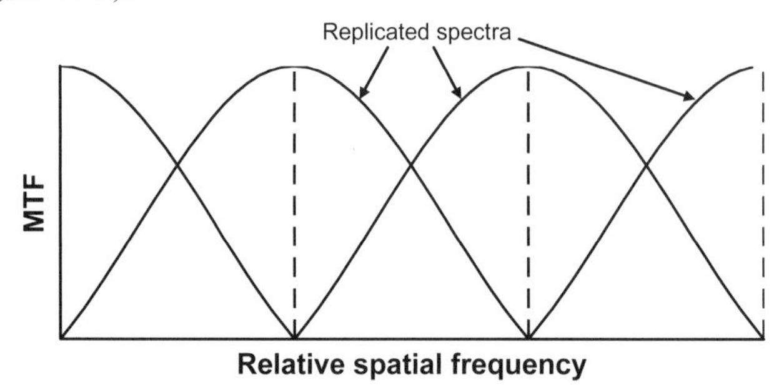

Figure 10-5. Overlapping spectra. The amount of aliasing (and hence image quality) is related to the amount of overlap.

Within an overlapping band, there is an ambiguity in frequency. It is impossible to tell whether the reconstructed frequency resulted from an input frequency of f or $nf_S \pm f$. This is aliasing. Once aliasing has occurred, it cannot be removed. All frequency components above f_N are folded back into the region bounded by $[0, f_N]$. This in-band region contains

$$I_{IN-BAND}(f) = \sum_{n=-\infty}^{\infty} I(nf_S \pm f) \qquad where\ nf_S \pm f \le f_N \qquad (10\text{-}1)$$

Although no universal method exists for quantifying aliasing, it is reasonable to assume that it is proportional to the MTF that exists above f_N. In Figure 10-6 a significant MTF exists above the Nyquist frequency. Whether this aliasing is objectionable, depends on the application. Figure 10-7 illustrates the spectra after an ideal low-pass filter. No aliasing can occur. The "information" in the higher frequencies is lost even though these frequencies would have been aliased to lower frequencies. That is, aliased signal still contains information. But it is still unknown how to interpret this "information".

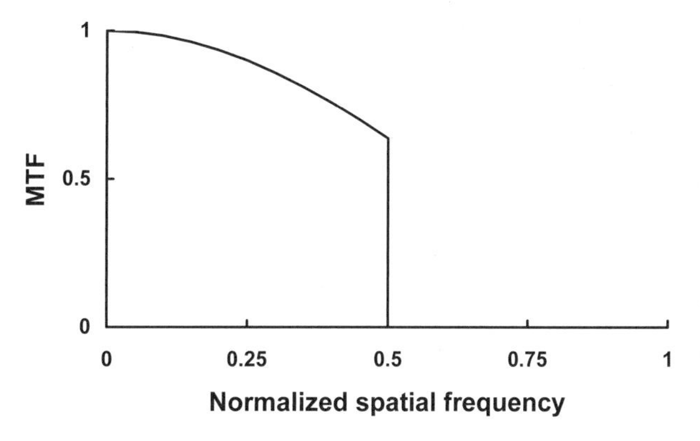

Figure 10-6. Signals above Nyquist frequency are aliased down to the base band. The area bounded by the MTF above f_N may be considered an "aliasing" metric.

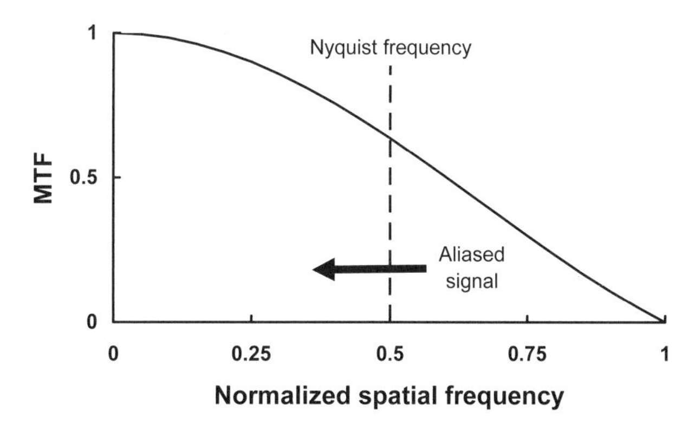

Figure 10-7. Detector MTF as a function of f/f_S when an ideal low-pass optical filter is used.

When the input is a sinusoid, an input frequency f_o above Nyquist frequency will appear at $f_S - f_o$. A bar pattern consists of an infinite number of frequencies. While the fundamental may be less than the Nyquist frequency, higher-order terms will not. These higher-order terms are aliased and distort the signal. In Figure 10-8 the input bar pattern fundamental is $1.68f_N$ and the aliased fundamental is $0.32f_N$. Since higher-order frequencies are present, the reconstructed bars appear more triangular than sinusoidal.

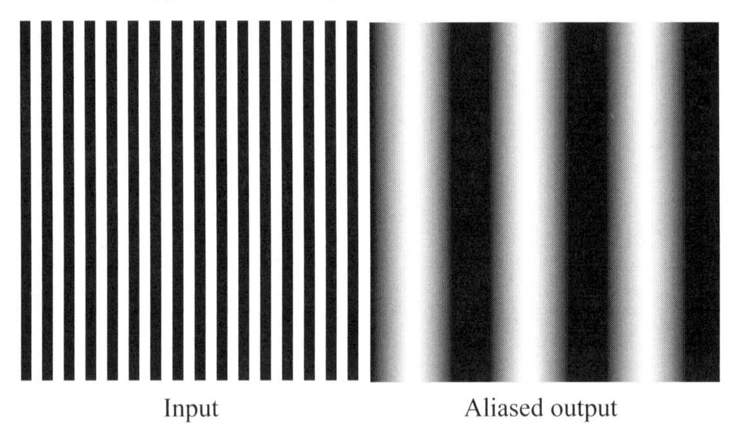

Figure 10-8. Aliasing. An ideal reconstruction filter was used. If viewed on a flat-panel display, the imagery would be blocky. Imagery created[10] by MAVIISS.

10.3. ARRAY NYQUIST FREQUENCY

A staring array acts a sampler whose sampling rate is $1/d_{CC}$ (cycles/mm), where d_{CC} is the detector center-to-center spacing (detector pitch). The pitch in the horizontal and vertical directions, d_{CCH} and d_{CCV}, respectively, may be different. Microlenses increase the effective detector size but do not affect the Nyquist frequency.

10.3.1. MONOCHROME

Figure 10-9 illustrates the Nyquist frequency for two arrays with different detector pitch. Aliasing in the vertical direction can be reduced by summing alternate detector outputs (field integration) in a two-field camera system (see Figure 3-19). Here, the effective detector height is $2d_V$ and the cutoff is $1/2d_V$ (Figure 10-10). This eliminates aliasing but significantly reduces the vertical MTF and, therefore, reduces image sharpness in the vertical direction. Horizontal sampling still produces aliasing.

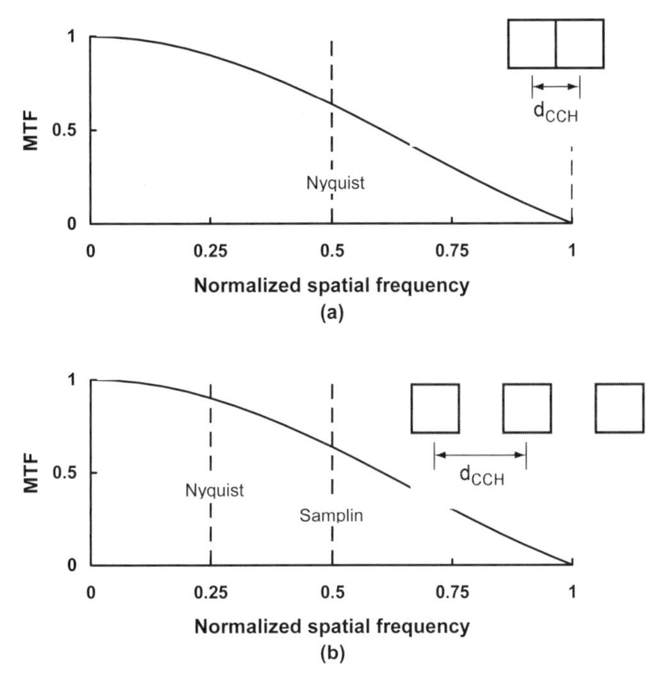

Figure 10-9. Two arrays with different horizontal center-to-center spacings. The detector size is the same for both. (a) $d_H/d_{CCH} = 1$. This typifies frame transfer devices. (b) $d_H/d_{CCH} = 0.5$. This is representative of interline transfer CCD arrays. The detector MTF is plotted as a function of u_i/u_{iD}.

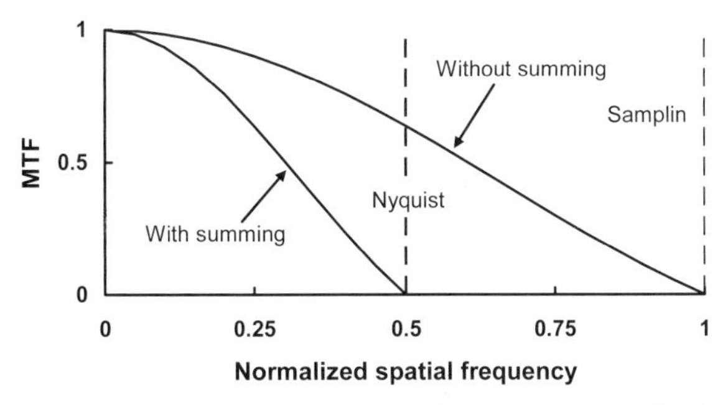

Figure 10-10. Alternate detector summing can reduce aliasing in the vertical direction. Vertical MTF with and without summing normalized to v_i/v_{iD}.

10.3.2. CFA

Most single-chip color filter arrays have an unequal number of red, green, and blue detectors. A typical array designed for NTSC operation will have 768 detectors in the horizontal direction with 384 detectors sensitive to green, 192 sensitive to red, and 192 sensitive to the blue region of the spectrum. Suppose the arrangement is G-B-G-R-G-B-G-R (see Figure 5.9). The spacing of the "blue" and "red" detectors is twice the "green" detector spacing. This produces a "blue" and "red" array Nyquist frequency that is one-half of the "green" array Nyquist frequency (Figure 10-11). Other detector layouts will create different array Nyquist frequencies. The "color" Nyquist frequencies can be different in the horizontal and vertical directions. These unequal array Nyquist frequencies create color aliasing in single-chip cameras that is wavelength specific. Black-and-white scenes can appear as green, red, or blue imagery.[11]

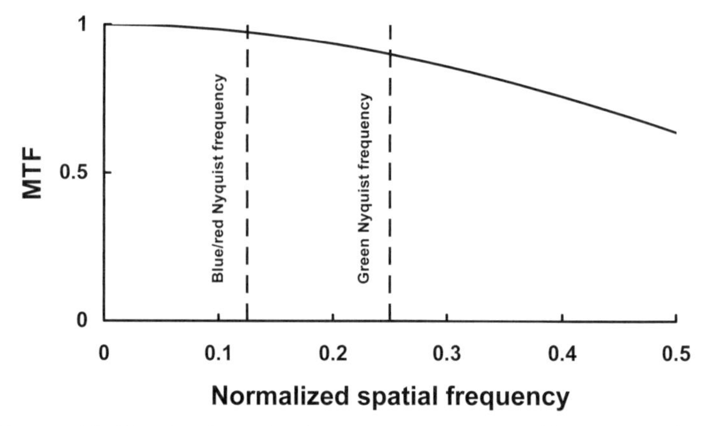

Figure 10-11. Horizontal detector MTF normalized to u_i/u_{iD}. Unequally spaced red, green, and blue sensitive detectors create different array Nyquist frequencies. This creates different amounts of aliasing and black-and-white scenes may break into color.

10.3.3. SUPER CCD

A hexagonal lattice has a higher Nyquist frequency than an equivalent rectangular lattice[12-15]. The Super CCD has octagonal detectors that are placed at 45 degrees with respect to the horizontal (Figure 10-12) and approximate hexagonal sampling. Assuming that the detector pitch, d_{CCP} is the same for a rectangular array, the Super CCD horizontal and vertical equivalent pitches are

$$d_{CCH} = d_{CCV} = \frac{\sqrt{2}}{2} d_{CCP} = 0.707 d_{CCP} \tag{10-2}$$

The Nyquist frequencies are

$$u_{iNH} = u_{iNV} = \frac{2}{\sqrt{2}} u_{iNP} = 1.414 u_{iNP} \tag{10-3}$$

If the pixel pitch is the same for a rectangular array as the Super CCD, then the Super CCD has a higher Nyquist frequency. The Super CCD has a higher MTF than the square detector (Figure 10-13). However, the interpolation required to create a rectangular data array reduces the MTF somewhat so that MTFs are essentially equivalent. The Super CCD has a higher Nyquist frequency and therefore should provide better imagery (less aliasing).

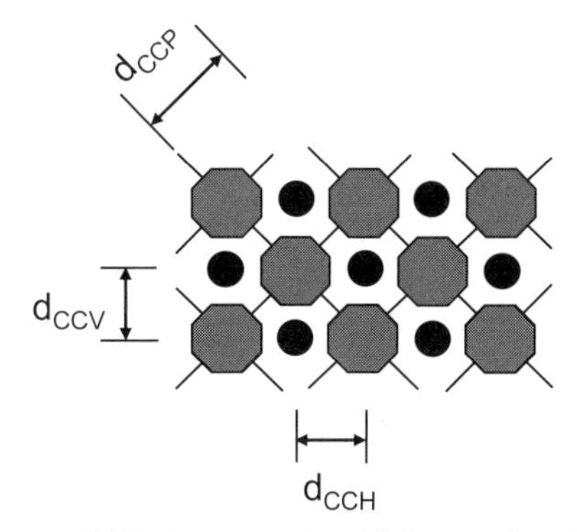

Figure 10-12. A super CCD (octagons) with interpolated values (circles).

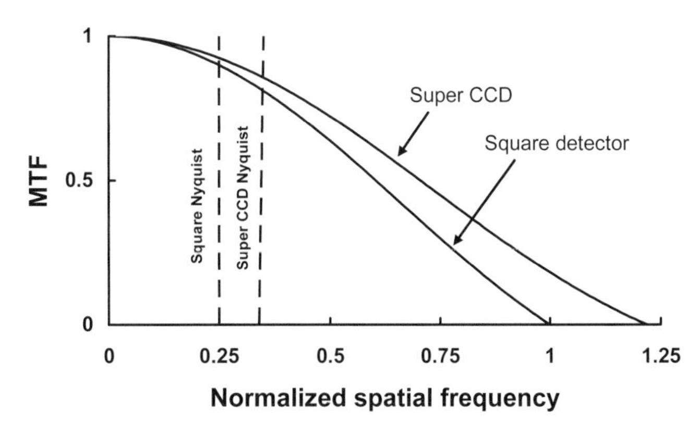

Figure 10-13. Comparison of Super CCD to a square detector. The detector sizes are the same and the fill factor is 50%. The spatial frequency is normalized to the square detector cutoff.

10.4. OPTICAL LOW PASS FILTER (OLPF)

If a white light point source impinged on a "blue" detector, the image would be blue. Likewise, if it landed on a "red" detector, the image would be red. To avoid this CFA color crosstalk problem, the blur diameter must be substantially increased to cover many detectors (Figure 10-14a). Now the color correction algorithm can interpolate the "color" values to create a white spot. Since the blur diameter has increased, the spatial resolution has decreased. Increasing the blur diameter is equivalent to reducing the optical MTF. While color aliasing is generally not a major problem, color crosstalk is considered unacceptable.

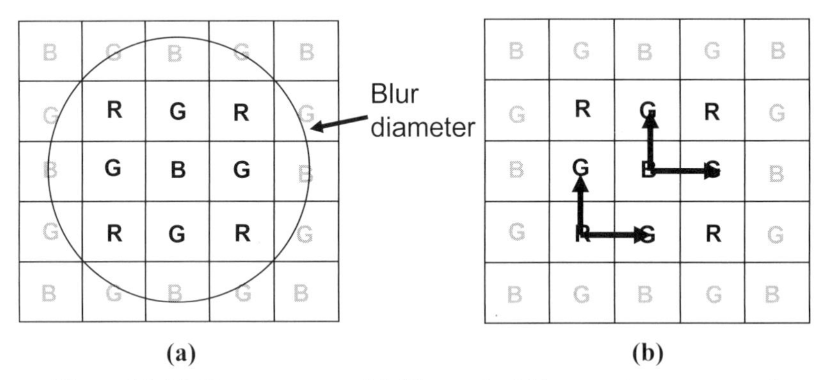

Figure 10-14. Bayer pattern, (a) The optical blur must cover several detectors in a CFA. (b) Two birefringent crystals create 3 spots.

Blurring can be achieved by using small-diameter optics, defocusing, or by inserting a birefringent crystal between the lens and array. Birefringent crystals break a beam into two components: the ordinary and the extraordinary. The Bayer pattern requires two crystals oriented at 0 and 90 degrees. Each incident beam is broken up into three beams and places the flux on the adjoining detectors as indicated by the arrows in Figure 10-14b. Two separate incident beams are shown. In reality, the array is flooded with contiguous beams.

As illustrated in Figure 10-15, the detector collects light from a larger beam. Rays that would have missed the detector are refracted onto the detector. This makes the detector appear[11] optically larger. While this changes the system MTF, the system sensitivity does not change. The light that is lost to adjoining areas is replaced by light refracted onto the detector.

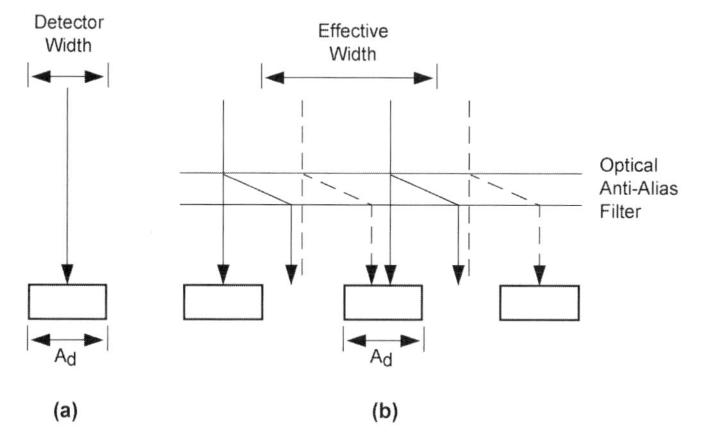

Figure 10-15. The birefringent filter increases the effective optical area of the detector. (a) No filter and (b) filter. Rays that would fall between detectors are refracted onto the detector. The separation between the beams is a function of the crystal refractive indices and thickness. This figure illustrates parallel light falling on the OLPF. However, the light is actually converging due to the lens system. The error[18] is less than 25% of the pixel pitch.

The filter design depends upon the CFA design. If detectors sensitive to like colors are next to each other, the filter may not be required. Since the filter creates two spots[16,17], it is equivalent to an even digital filter with 2 taps. It acts like an optical low-pass filter whose MTF is

$$MTF_{OLPF} = \cos(\pi\, d_{OLPF}\, u_i) \tag{10-4}$$

The variable d_{OLPF} is the separation of the 2 spots and is equal to the like-colored detector pitch. Figure 10-16 illustrates how the detector MTF is modified by the birefringent crystal when the detector pitch is twice as large as the element size. While the OLPF is designed to increase the blur diameter, it also reduces aliasing. As such, it is often called an optical anti-alias filter. Unfortunately, it also reduces the overall system MTF. It is the optics/OLPF combination that determines the amount of aliased signal. It can be reduced by increasing the focal ratio (reduce the optical cutoff).

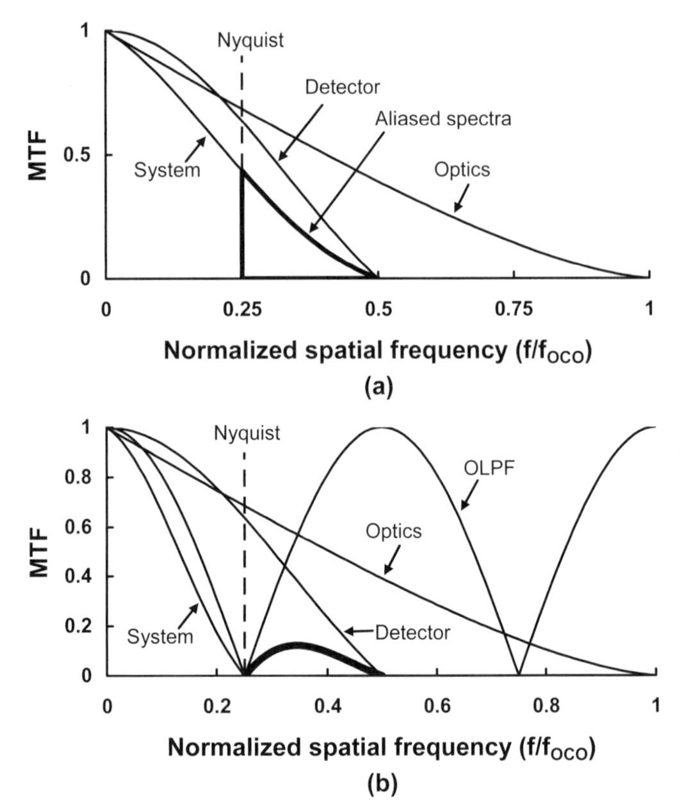

Figure 10-16. (a) No OLPF. The heavy line outlines the signal that will be aliased. (b) Reduction of aliased signal using an OLPF for a typical design. The heavy line outlines the signal that will be aliased with the OLPF.

Multiple crystals are used to change the effective detector size both vertically and horizontally. Because the spacing depends on the CFA design, there is no unique low-pass filter design. Clearly, if the detectors are unequally spaced, the Nyquist frequency for each color is different. Similarly, with different effective sizes, the MTFs are also different for each color. Complex algorithms are required to avoid color aliasing (see Example 7-1) and equalize the MTFs. The OLPF is essential for CFAs to avoid color crosstalk. It can be used in any camera to reduce aliasing. Comparisons between the Bayer pattern and Foveon's direct image sensors can be found in References 18 and 19.

The OLPF thickness may be too great for miniaturized cameras (like those in mobile phones). An alternate approach is to use a grating optical low-pass filter (GOLF), which is a thin film. GOLFs multiple spots[20] produce a poorer MTF when compared to an OLPF. The advantage is the thickness.

The low-cost camera (e.g., cell phone) market typically uses 3- or 4-element optical systems. The optical MTF is reduced at the Nyquist frequency and thereby acts as a low-pass filter. This avoids the additional birefringent crystal. Figure 10-17 illustrates diffraction limited optics and detector MTF for the STMicroelectronics VS6724 camera with a Bayer pattern CFA. The optics specification is indicated.

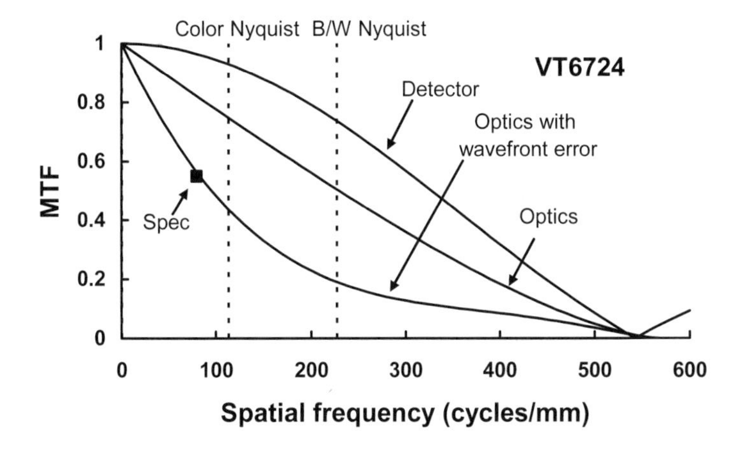

Figure 10-17. VT6724 CMOS 2.2 μm pixel predicted MTF. The focal ratio is 3.2. Fill factor assumed to be 70%. Manufacturers typically provide only one or two MTF values. A wavefront error of 0.52 waves (peak-to-peak) has been added (Equation 9-15) to match the optics specification. While this curve is hypothetical, it indicates the significantly reduced MTF at Nyquist frequency which is typical for low cost, single chip cameras.

10.5. RECONSTRUCTION

Digital data cannot be seen because it resides in a computer memory. Any attempt to represent a digital image requires a reconstruction filter[21]. Imaging systems may use an electronic low-pass filter, but most rely on the display medium and human visual system to produce a *perceived* continuous image. Display media include laser printers, half-toning, fax machines, cathode ray tubes (CRTs), and flat-panel displays. The display medium creates an image by painting a series of light spots onto a screen or ink spots onto paper. Each "spot" has an MTF associated with it. Because of its finite size, the spot acts as a low-pass filter. Different low-pass filters (e.g., different display mediums) will create different images. The "quality" of the imagery is scene dependent.

The zero-order sample-and-hold creates a blocky image (Figures 10-18a and 10-19a). It also simulates a flat panel display. The zero-order filter contributes significant out-of-band (above f_N) information and this makes the image blocky or pixelated. A CRT will remove the higher frequencies (above f_N) but also attenuates the in-band frequencies to create a somewhat blurry image (Figures 10-18b and 10-19b). The ideal reconstruction filter abruptly drops to zero at f_N. As illustrated in Figure 10-20, a sharp drop in one-domain produces ringing in the other (Gibbs phenomenon). In these figures, MTF_{OPTICS} and $MTF_{DETECTOR}$ are unity over the spatial frequencies on interest. This emphasizes sampling artifacts and the influence of the type of reconstruction filter on image quality.

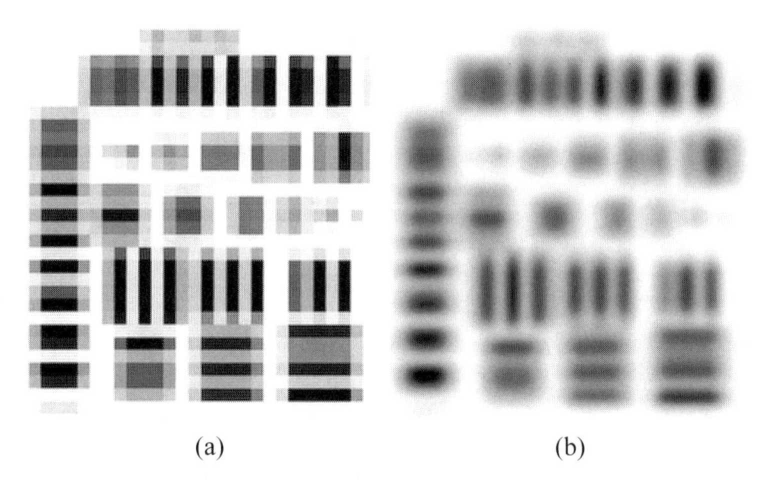

Figure 10-18 Reconstruction with (a) flat panel display and (b) CRT. There are 32×32 detectors across the image. Imagery created[10] by MAVIISS. NOTE: The imagery is enlarged so that your eye MTF and the printing do not significantly affect the image quality. This allows you to see the distortion created by sampling and the system MTF degradation. With 20/20 vision, you can just perceive 1 arc sec (0.291 mrad). The images represent 32×32 pixels and are 50 mm in size (varies). Each pixel is 50/32 =1.56 mm. At normal reading distance, (about 15 inches = 380 mm), each pixel subtends 1.56/380= 4.1 mrad. You must move 4.1/0.291 = 14 times away (about 17 feet) to appreciate what is seen on a typical display. The image will start to become uniform at ½ that distance.

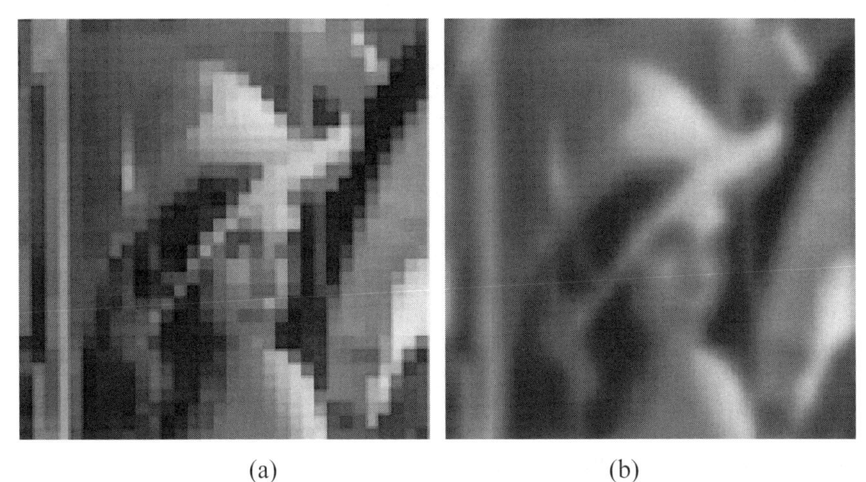

(a) (b)

Figure 10-19. Reconstruction with (a) flat panel display and (b) CRT. Imagery created[10] by MAVIISS. See NOTE in Figure 10-18 caption.

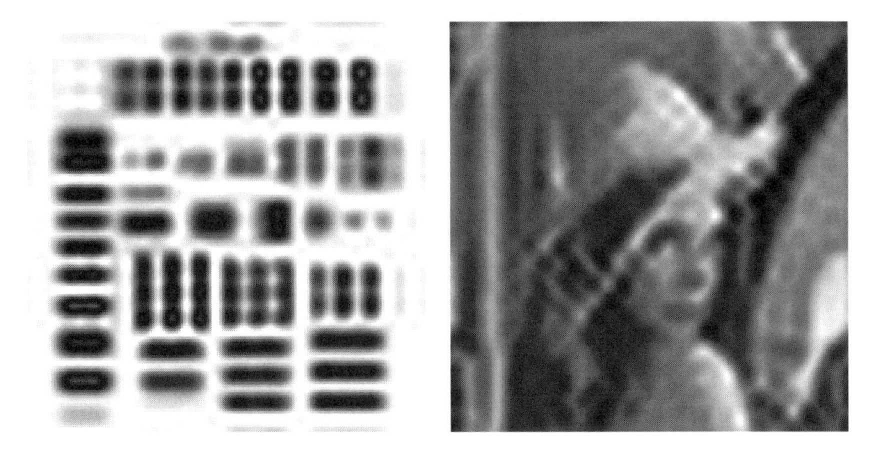

Figure 10-20. Reconstruction with an ideal reconstruction filter. Imagery created[10] by MAVIISS. See ***NOTE*** in Figure 10-18 caption.

Since the ideal filter creates ringing, many authors[22] have attempted to design the "optimum" filter. The design is based upon the amount of aliased signal as implied by the MTF at Nyquist frequency. The resultant imagery was viewed and if the deemed "good", then the filter design was "optimum."

10.6. IMAGE DISTORTION

The square wave is the most popular test target and it is characterized by its fundamental frequency f_o. When expanded into a Fourier series, the square wave consists of an infinite number of frequencies (See Section 8.4 *Superposition applied to optical systems*). Although the square wave fundamental may be oversampled, the higher harmonics will not. The appearance of the square wave after reconstruction depends on the relative values of the optical, detector, and reconstruction MTFs. There will be intensity variations from bar-to-bar and the bar width will not remain constant.

Sampling replicates frequencies. When two frequencies are close together, they create a beat frequency that is equal to the difference. The fundamental and its replication create a beat frequency of $f_{BEAT} = (f_S - f_o) - f_o$. The beat frequency period lasts for N_{CYCLES} input frequency cycles

$$N_{CYCLES} = \frac{f_o}{2(f_N - f_o)} \qquad (10\text{-}5)$$

Figure 10-21 illustrates N as a function of f/f_N. As f/f_N approaches one, the beat frequency becomes obvious. Lomheim et. al.[23,24] created these beat frequencies when viewing multiple bar patterns with a CCD camera. An ideal reconstruction filter will eliminate the beat frequency. Thus the appearance depends upon the type of reconstruction (zero-order sample-and-hold, CRT, or flat-panel display).

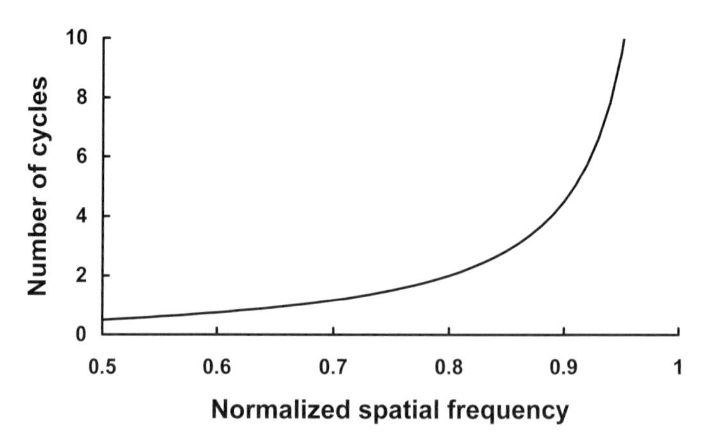

Figure 10-21. Number of input frequency cycles required to see one complete beat frequency cycle as a function of f/f_N.

Figure 10-22 illustrates an ideal staring array output when the optical MTF is unity for all spatial frequencies and therefore represents a worst-case scenario. That is, all higher harmonics that would have been attenuated by the optical system are available for aliasing by the detector. If f/f_N = 0.952, the beat frequency is equal to 9.9 cycles of the input frequency. The target must contain at least 10 cycles to see the entire beat pattern.

Standard characterization targets, however, consist of several bars. Therefore the beat pattern may not be seen when viewing a 3-bar or 4-bar target. The image of the bar pattern would have to be moved ±½***d*** (detector extent) to change the output from a maximum value (in-phase) to a minimum value (out-of-phase). This can be proven by selecting just four adjoining bars in Figure 10-22.

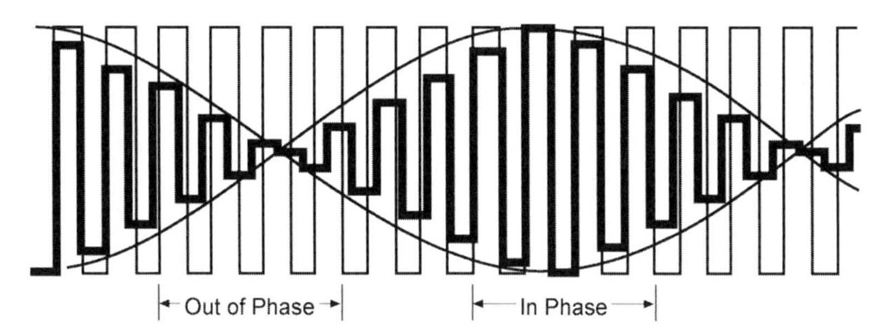

Figure 10-22. Beat frequency produced by an ideal staring system when f/f_N = 0.952. The value N = 9.9 and d_H/d_{CCH} = 1. The input frequency f is fundamental of the square wave. The light line is the input and the heavy line is the detector output. The beat frequency envelope is shown. It requires 10 cycles to see this beat frequency. References 23 and 24 provide numerous line traces and imagery for a variety of f/f_N values.

When f/f_N is approximately between 0.6 and 0.9, a 3-bar or 4-bar target never looks "right." At least one bar is different in either amplitude or width compared to the others. This impacts[25] testing and test results.

10.7. SAMPLING "MTF"

For a sampled-data system, a wide range of MTF values is possible[1,4,6,26] for any given spatial frequency. To account for this variation, a sample-scene MTF may be included. The detector spatial frequency response is represented by $\boldsymbol{MTF_{DETECTOR}MTF_{PHASE}}$. This MTF provides an average performance response that may be used to calculate the system response general imagery.

In general, the "MTF" is[27]

$$MTF_{PHASE}(u_i) \approx \cos\left(\frac{u_i}{u_{iN}}\theta\right) \tag{10-6}$$

where θ is the phase angle between the target and the sampling lattice. At Nyquist, the MTF is a maximum when $\theta = 0$ (in-phase) and a zero when $\theta = \pi/2$ (out-of-phase).

To approximate a median value for phasing, θ is set to $\pi/4$. Here, approximately one-half of the time the MTF will be higher and one-half of the time the MTF will be lower. The median sampling MTF is

$$MTF_{MEDIAN}(u_i) \approx \cos\left(\frac{\pi}{2}\frac{u_i}{u_{iS}}\right) \tag{10-7}$$

At Nyquist frequency, $\boldsymbol{MTF_{PHASE}}$ is 0.707. This is the Kell factor so often reported when specifying the resolution of monitors. An MTF averaged over all phases is represented by

$$MTF_{AVERAGE}(u_i) \approx \operatorname{sinc}\left(\frac{\pi}{2}\frac{u_i}{u_{iS}}\right) \tag{10-8}$$

Figure 10-23 illustrates the difference between the two equations. Because these are approximations, they may be considered roughly equal over the range of interest (zero to the Nyquist frequency).

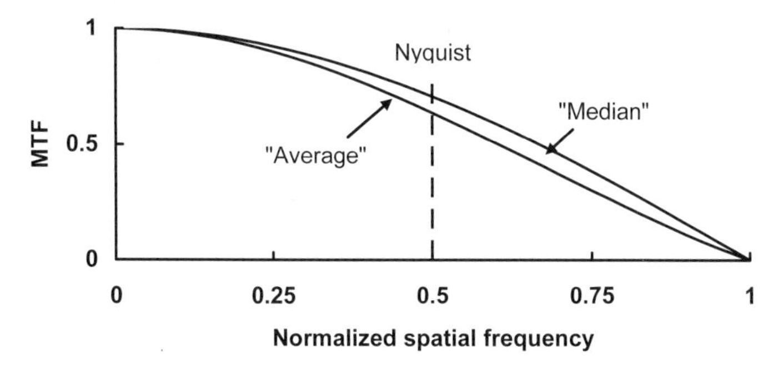

Figure 10-23. Average and median scene-sample phase MTFs normalized to u_i/u_{iD}. The MTF is defined only up to the Nyquist frequency.

Sampling effects are phase-dependent and, as a result, have no unique MTF. Because average MTF is assigned, only an "average" system performance can be predicted. Performance can vary dramatically depending upon the target phase with respect to the sampling lattice. For laboratory measurements, it is common practice to "peak-up" the target (in-phase relationship). This is done to obtain repeatable results.

10.8. SPURIOUS RESPONSE

The previous sections and Equation 10-1 described the in-band response only $[0, f_N]$. In one dimension, the reconstructed image is

$$I(f) = MTF_{POST}(f)\sum_{n=0}^{\infty} MTF_{PRE}(nf_S \pm f)O(nf_S \pm f) \qquad (10\text{-}9)$$

where $O(f)$ is the Fourier transform of the object and $I(f)$ is the Fourier transform of the resultant displayed image. MTF_{PRE} contains all the MTFs up to the sampler (the detector) and MTF_{POST} represents all the filters after the sampler (Table 10-1). Equation 10-9 can be written as

$$\begin{aligned} I(f) &= MTF_{POST}(f)MTF_{PRE}(f)O(f) \\ &+ MTF_{POST}(f)\sum_{n=1}^{\infty} MTF_{PRE}(nf_S \pm f)O(nf_S \pm f) \end{aligned} \qquad (10\text{-}10)$$

The first term is the spectrum of the image when no sampling is present and is the direct response (also called the base band). Sampling created the remaining terms and they represent aliasing. The amount of aliasing is scene specific scene and may or may not be bothersome.

Table 10-1
STARING ARRAYS PRE- and POST MTFs

MTF_{PRE}	MTF_{OPTICS}, MTF_{MOTION}, $MTF_{DIFFUSION}$, MTF_{CTE}, MTF_{TDI}
MTF_{POST} (reconstruction filter)	$MTF_{LOWPASS}$, $MTF_{DFILTER}$, $MTF_{S\&H}$, $MTF_{MONITOR}$

Shade[28] defined spurious response for the first fold back frequency (***n*** =1) as

$$SR = \frac{\int_{-\infty}^{\infty} \left(\textit{Spurious response function}\right) df}{\int_{-\infty}^{\infty} \left(\textit{Base band}\right) df}, \tag{10-11}$$

Assuming that the object contains all frequencies over the region of interest,

$$\textit{Spurious response} = \frac{\int_0^{\infty} MTF_{POST}(f) MTF_{PRE}(f_S - f) O(f_S - f) df}{\int_0^{\infty} MTF_{POST}(f) MTF_{PRE}(f) O(f) df} \tag{10-12}$$

The spurious response is a measure of artifacts introduced into the image by the sampling process and may be considered as an aliasing metric. Figure 10-24 illustrates the spurious response for a system that has an ideal reconstruction filter (***MTF****POST*) and Figure 10-25 illustrates the spurious response when a practical post-reconstruction filter is used.

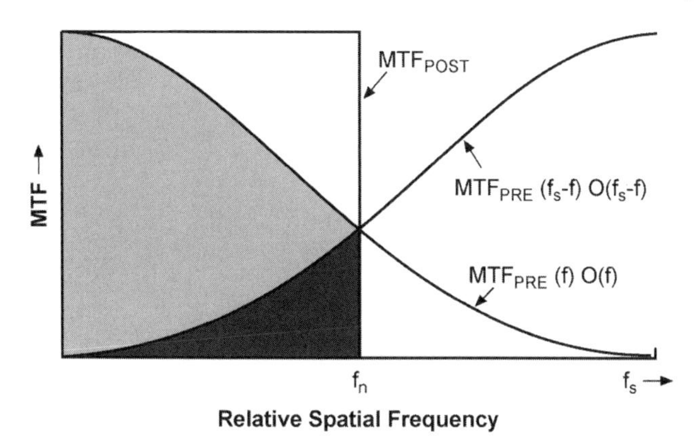

Figure 10-24. Ideal reconstruction filter. The ratio of the shaded areas is the spurious response.

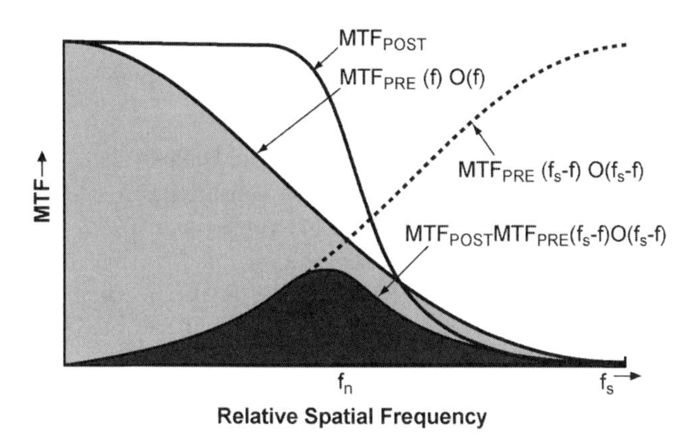

Figure 10-25. Practical reconstruction filter. The ratio of the shaded areas is the spurious response.

The in-band spurious response is limited to the Nyquist frequency (Figure 10-26)

$$SR_{IN-BAND} = \frac{\int_{-f_n}^{f_n} \left(\textit{Spurious response function} \right) df}{\int_{-\infty}^{\infty} \left(\textit{Base band} \right) df} \qquad (10\text{-}13)$$

and the out-of-band spurious response is simply the remainder

$$SR_{OUT-OF-BAND} = SR - SR_{IN-BAND} \quad (10\text{-}14)$$

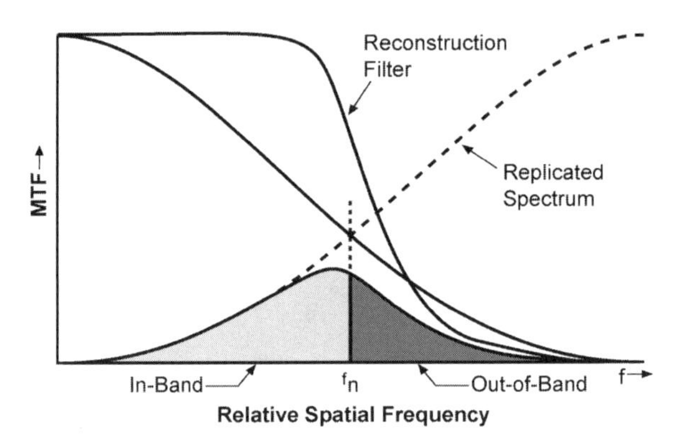

Figure 10-26. Definition of in-band and out-of band spurious response.

A comprehensive end-to-end analysis should include both aliased signal and aliased noise. Aliased signal[29-31] can be considered as part of the noise spectrum because it interferes with the ability to perceive targets. MTF boost is used to enhance that spatial region where the MTF is low. This tends to be the same region that contains aliased signal and aliased noise. These aliased components limit the extent to which a sampled image can be sharpened. Excessive peaking can cause ringing at sharp edges.

10.9. MULTIPLE SAMPLERS

Nearly all analyses focus on the spatial sampling created by the detector array. There can be additional samplers in the overall system that add additional sampling artifacts. If the camera output is analog, then to process the imagery in a computer requires a frame capture board. The board can digitize the analog signal at a rate that is different than the detector sampling lattice (see Section 7.8. *Computer interface*). This can only be ascertained with careful measurements.

There are four spatial samplers in the intensified CCD camera: 1) sampling by the discrete channels in the MCP, 2) the fiberoptic output window within the intensifier, 3) the input to the fiberoptic minifier, and 4) the detector array. At these interfaces, the image is re-sampled. Depending on the location (alignment) of the fiberoptic minifier, moiré patterns can develop and the MTF can degrade. In addition, manufacturing defects in the fiberoptic material can appear[33] as a "chicken wire" effect.

10.10. REFERENCES

1. S. K. Park and R. A. Schowengerdt, "Image Sampling, Reconstruction and the Effect of Sample-scene Phasing," *Applied Optics*, Vol. 21(17), pp. 3142-3151 (1982).
2. S. E. Reichenbach, S. K. Park, and R. Narayanswamy, "Characterizing Digital Image Acquisition Devices," *Optical Engineering*, Vol. 30(2), pp. 170-177 (1991).
3. W. Wittenstein, J. C. Fontanella, A. R. Newberry, and J. Baars, "The Definition of the OTF and the Measurement of Aliasing for Sampled Imaging Systems," *Optica Acta*, Vol. 29(1), pp. 41-50 (1982).
4. J. C. Felz, "Development of the Modulation Transfer Function and Contrast Transfer Function for Discrete Systems, Particularly Charge Coupled Devices," *Optical Engineering*, Vol. 29(8), pp. 893-904 (1990).
5. S. K. Park, R. A. Schowengerdt, and M. Kaczynski, "Modulation Transfer Function Analysis for Sampled Image Systems," *Applied Optics*, Vol. 23(15), pp. 2572-2582 (1984).
6. L. deLuca and G. Cardone, "Modulation Transfer Function Cascade Model for a Sampled IR Imaging System," *Applied Optics*, Vol. 30(13), pp. 1659-1664 (1991).
7. G. C. Holst, *Sampling, Aliasing, and Data Fidelity*, JCD Publishing, Winter Park, FL (1998).
8. C. E. Shannon, "Communication in the Presence of Noise," *Proceedings of the IRE*, Vol. 37, pp. 10-21 (January 1949).
9. A. B. Jerri, "The Shannon Sampling Theorem - Its Various Extensions and Applications: A Review," *Proceedings of the IEEE*, Vol. 85(11), pp. 1565-1595 (1977).
10. MAVIISS (MTF based Visual and Infrared System Simulator) is an interactive software program available from JCD Publishing at www.JCDPublishing.com.
11. J. E. Greivenkamp, "Color Dependent Optical Prefilter for the Suppression of Aliasing Artifacts," *Applied Optics*, Vol. 29(5), pp 676-684 (1990).
12. O. Hadar, A. Dogariu, and G. D. Boreman, "Angular Dependence of Sampling Modulation Transfer Fucntion," *Applied Optics*, Vol. 36, pp. 7210-7216 (1997).
13. O. Hadar and G. D. Boreman, "Oversampling Requirements for Pixelated-Imager Systems," *Optical Engineering* Vol. 38, pp. 782-785 (1999).
14. R. Vitulli, U. Del Bello, P. Armbruster, S. Baroni, and L. Santurri, "Aliasing Effects Mitigated by Optimized Sampling Grids and Impact on Image Acquisition Chains," in *Geoscience and Remote Sensing Symposium, 2002. IGARSS* '02, pp. 979-981 (2002).
15 S. Vitek and J. Hozman, "Image Quality Influenced by Selected Image-Sensor Parameters," in *Photonics, Devices, and Systems* II, M. Hrabovsky, D. Senderakova, and P. Tomanek, eds., SPIE Proceedings Vol. 5036, pp. 14-19 (2003).
16. X. Qi, B. Lin, J. Liang, B. Chen, and X. Cao, "Fabrication and Characterization of Two-dimensional Optical Low-pass Filter, in *Passive Components and Fiber-based Devices* III; S. B. Lee, Y. Sun, K. Qiu, S.C. Fleming, and I. H. White, eds., SPIE Proceedings Vol, 6251, paper 63510W (2006).
17. T. Zhao, R.Wang, Y. Liu, and F. Yu, "Characteristic-analysis of Optical Low-pass Filter used in Digital Camera," in ICO20: *Optical Design and Fabrication*, J. Breckinridge, and Y. Wang, eds., SPIE Proceedings Vol. 6034, paper 60340N (2006).

18. P. M. Hubel, J. Liu, and R. J. Guttosch, "Spatial frequency response of color image sensors: Bayer color filters and Foveon X3," in *Sensors and Camera Systems for Scientific, Industrial, and Digital Photography Applications* V; M. M. Blouke, N. Sampat, and R. J. Motta, eds., SPIE Proceedings Vol. 5301, pp. 402-407 (2004).
19. R. J. Guttosch, "Investigation of Color Aliasing of High Spatial Frequencies and Edges for Bayer-Pattern Sensors and Foveon X3 Direct Image Sensors," Available at www.Foveon.com
20. C.-C. Lee, and S.-H. Chen, "A New Design of Thin-film Grating Optical Low-Pass Filter and its Fabrication, in *Advances in Thin-Film Coatings for Optical Applications* III; M. J. Ellison, ed., SPIE Proceedings Vol. 6286, paper 628609 (2006).
21. G. C. Holst. "Are Reconstruction Filters Necessary?" in *Infrared Imaging Systems: Design, Analysis, Modeling, and Testing* XVII, G. C. Holst, ed., SPIE Proceedings Vol. 6207, paper 62070K (2006).
22. Reconstruction filter requirements are discussed by many authors in *Perception of Displayed Information*, L. C. Biberman, ed., Plenum Press, New York (1973).
23. T. S. Lomheim, L. W. Schumann, R. M. Shima, J. S. Thompson, and W. F. Woodward, "Electro-Optical Hardware Considerations in Measuring the Imaging Capability of Scanned Time-delay-and-integrate Charge-coupled Imagers," *Optical Engineering*, Vol. 29(8), pp. 911-927 (1990).
24. T. S. Lomheim, J. D. Kwok, T. E. Dutton, R. M. Shima, J. F. Johnson, R. H. Boucher, and C. Wrigley, "Imaging Artifacts due to Pixel Spatial Sampling Smear and Amplitude Quantization in Two-dimensional Visible Imaging Arrays," in *Infrared Imaging Systems: Design, Analysis, Modeling, and Testing* X, G. C. Holst ed., SPIE Vol. 3701, pp 36-60 (1999).
25. G. C. Holst, *Testing and Evaluation of Infrared Imaging Systems*, 2nd edition, pp 46-48, JCD Publishing, Winter Park, FL (1998).
26. J. C. Feltz and M. A. Karim, "Modulation Transfer Function of Charge-coupled Devices," *Applied Optics*, Vol. 29(5), pp. 717-722 (1990).
27. F. A. Rosell, "Effects of Image Sampling," in *The Fundamentals of Thermal Imaging Systems*, F. Rosell and G. Harvey, eds., p. 217, NRL Report 8311, Naval Research Laboratory, Washington D.C. (1979).
28. O. H. Shade, Sr., "Image Reproduction by a Line Raster Process," in *Perception of Displayed Information*, L. C. Biberman, ed., pp. 233-278, Plenum Press, New York (1973).
29. S. K. Park and R. Hazra, "Aliasing as Noise: A Quantitative and Qualitative Assessment," in *Infrared Imaging Systems: Design, Analysis, Modeling and Testing IV*, G. C. Holst, ed., SPIE Proceedings Vol. 1969, pp. 54-65 (1993).
30. S. K. Park, "Image Gathering, Interpolation and Restoration: A Fidelity Analysis," in *Visual Information Processing*, F. O. Huck and R. D. Juday, eds., SPIE Proceedings Vol. 1705, pp. 134-144 (1992).
31. S. K. Park and R. Hazra, "Image Restoration Versus Aliased Noise Enhancement," in *Visual Information Processing III*, F. O. Huck and R. D. Juday, eds., SPIE Proceedings Vol. 2239, pp. 52-62 (1994).
32. S. K. Park and Z. Rahman, "Fidelity Analysis of Sampled Imaging Systems," Optical Engineering, Vol. 38(5), pp. 786-800 (1999).
33. G. M. Williams, "A High-Performance LLLTV CCD Camera for Nighttime Pilotage," in *Electron Tubes and Image Intensifiers*, C. B. Johnson and B. N. Laprade, eds., SPIE Proceedings Vol. 1655, pp. 14-32 (1992).

11

IMAGE QUALITY

Our perception of good image quality is based upon the real-world experiences of seeing all colors, all intensities, and textures. An imaging system has limited field of view, limited temporal and spatial resolutions, and presents a two-dimensional view of a three-dimensional world. In the real world our eyes scan the entire scene. Not all of the available information is captured by an imaging system. Furthermore, an imaging system introduces noise; the loss of image quality due to noise can only be estimated. Cameras sensitive to wavelengths less than 0.4 μm and greater than 0.7 μm provide imagery that we cannot directly perceive. The quality of the ultraviolet or infrared imagery can only be estimated because we do not know how it really appears.

Image quality is a subjective impression ranging from poor to excellent. It is a somewhat learned ability. It is a perceptual one, accomplished by the brain, affected by and incorporating inputs from other sensory systems, emotions, learning, and memory. The relationships are many and not well understood. Perceptual quality of the same scene varies between individuals and temporally for the same individual. Large variations exist in an observer's judgment as to the correct rank ordering of image quality from poor to best, and therefore image quality cannot be placed on an absolute scale. Visual psychophysical investigations have not measured all the properties relevant to imaging systems.

Many formulas exist for predicting image quality. Each is appropriate under a particular set of viewing conditions. These expressions are typically obtained from empirical data in which multiple observers view many images with a known amount of degradation. The observers rank the imagery from worst to best and then an equation is derived which relates the ranking scale to the amount of degradation.

If the only metric for image quality was resolution, then we would attempt to maximize resolution in our system design. Many tests have provided insight into image quality metrics. In general, images with higher MTFs and less noise are judged as having *better* image quality. There is no single *ideal* MTF shape that provides best image quality. For example, Kusaka[1] showed that the MTF that produced the most aesthetically pleasing images depended on the scene content.

The metrics suggested by Granger and Cupery, Schade, and Barten offer additional insight on how to optimize an imaging system. Granger and Cupery developed[2] the Subjective Quality Factor (SQF): an empirically derived relationship using individuals' responses when viewing many photographs. Schade used[3] photographs and included high-quality TV images. Barten's approach[4-6] is more comprehensive in that it includes a variety of display parameters. It now includes[7] contrast, luminance, viewing ratio, number of scan lines, and noise.

There are potentially two different system design requirements: 1) good image quality and 2) performing a specific task. Sometimes these are equivalent, other times they are not. All image quality metrics incorporate some form of the system MTF. The underlying assumption is that the image spectrum is limited by the system MTF. Equivalently, it is assumed that the scene contains all spatial frequencies and that the displayed image is limited by system MTF. This may not be a reasonable assumption. Computer monitors are usually designed for alphanumeric legibility.

Often, the display is the limiting factor in terms of image quality and resolution. No matter how good the electronic imaging system is, if the display resolution is poor then the overall system resolution is poor. A high-resolution display does not offer any "extra" system resolution. A high-resolution display just ensures that all the information available is displayed. Image quality depends upon the display medium (CRT, flat-panel display, or printer). Each modifies the image in a different way. System resolution may be limited by the human visual system (HVS). If the observer is too far from the screen, not all of the image detail can be discerned.

The HVS appears to operate as a tuned spatial-temporal filter where the tuning varies according to the task at hand. Because the HVS approximates an optimum filter, no system performance improvement is expected by *precisely* matching the image spectrum to the HVS preferred spectrum. Rather, if the displayed spectrum is within the limits of the eye spectrum, the HVS will automatically tune to the image. Clearly, the overall system magnification should be set such that the frequency of the maximum interest coincides with the peak frequency of the HVS's contrast sensitivity. Although this implies a specific frequency, the range of optimization is broad.

A large number of metrics are related to image quality. Most are based on monochrome imagery such as resolution, MTF, and minimum resolvable contrast. Color reproduction and tonal transfer issues, which are very important to color cameras, are not covered here. The metrics in this chapter are MTF-based and do not include sensitivity nor noise. The results here must be consolidated with SNR considerations (Sections 6.8 through 6.11) before designing a camera.

11.1. RESOLUTION METRICS

An overwhelming majority of image quality discussions center on resolution. Resolution has been in use so long that it is thought to be fundamental and that it uniquely determines system performance. There are four different types of resolution: 1) temporal, which is the ability to separate events in time; 2) grayscale, which is determined by the analog-to-digital converter design, noise floor; or the monitor capability, 3) spectral; and 4) spatial. An imaging system operating at 30 Hz frame rate has a temporal resolution of 1/30 sec. Grayscale resolution is a measure of the dynamic range. The spectral resolution is simply the spectral band pass (e.g., UV, visible, or near infrared) of the system. This section covers spatial resolution.

Resolution provides valuable information regarding the finest spatial detail that can be discerned. A large variety of resolution measures exist,[8] and the various definitions may not be interchangeable. An electronic imaging system is composed of many subsystems and each has its own metric for resolution (Table 11-1). These can also be divided into analog (Table 11-2) and sampled-data resolution metrics (Table 11-3). A single measure of spatial resolution cannot be satisfactorily used to compare all sensor systems. Spatial resolution does not provide information about total imaging capability.

Table 11-1
SUBSYSTEM MEASURES of RESOLUTION

SUBSYSTEM	RESOLUTION METRIC
Optics	Rayleigh criterion Airy disk diameter Blur diameter
Detectors	Detector size Detector angular subtense Instantaneous field of view Detector pitch Pixel-angular-subtense
Electronics	Bandwidth
Displays	TV limiting resolution

Table 11-2
RESOLUTION MEASURES for ANALOG SYSTEMS

RESOLUTION	DESCRIPTION
Rayleigh criterion	Ability to distinguish two adjacent point sources
Airy disk	Diffraction-limited diameter produced by a point source
Blur diameter	Actual minimum diameter produced by a point source
Limiting resolution	Spatial frequency at which MTF= 0.02 to 0.10
TV limiting resolution	Number of resolved lines per picture height
Ground resolved distance	The smallest test target (1 cycle) that a photointerpreter can distinguish
Ground resolution	An estimate of the limiting feature size seen by a photointerpreter
Resel	Smallest region that contains unique information

Table 11-3
RESOLUTION MEASURES for SAMPLED DATA SYSTEMS

RESOLUTION	DESCRIPTION
Detector angular subtense	Angle subtended by one detector element
Instantaneous field of view	Angular region over which the detector senses radiation
Nyquist frequency	One-half of the sampling frequency
Detector pitch	Center-to-center spacing
Pixels, datels, and disels	Number of detector elements or number of digital data points

If the only metric for image quality were resolution, then we would attempt to maximize resolution in our system design. Given a system angular resolution element, $\boldsymbol{R_{SYS}}$, and requiring $\boldsymbol{N}$ resolutions elements across a target of size $\boldsymbol{W}$ to achieve a predetermined performance level, the range to the target is

$$\boldsymbol{Range} = \frac{\boldsymbol{W}}{\boldsymbol{N R_{SYS}}} \qquad (11\text{-}1)$$

Traditionally, the detector instantaneous field of view (IFOV) is used for $\boldsymbol{R_{SYS}}$. Resolution is independent of sensitivity. It does not consider radiometry and therefore the predicted range does not include the atmospheric transmittance.

Models[9] such as SSCamIP (solid state cameras) and IICamIP (image intensifier) combine radiometry with sensitivity to accurately predict target acquisition (discussed in Section 12.4. *Range predictions*).

Complex systems cannot be characterized by a single number (for example, an "-el" described in Section 1.5). Nevertheless, this chapter discusses various resolution metrics. Four observer-based metrics are compared: Schade's equivalent resolution, square root factor (SQF), MTFA, and the targeting task performance (TTP) metric. All metrics exhibit the same feature: there is an optics-limited and a detector-limited region of operation.

11.2. OPTICAL RESOLUTION

Table 11-4 provides various optical resolutions. As analog metrics, these are various definitions of a resel. Diffraction measures include the Rayleigh criterion and the Airy disk diameter. The Airy disk is the bright center of the diffraction pattern produced by an ideal optical system. In the focal plane of the lens, the Airy disk diameter is

$$d_{AIRY} = 2.44\frac{\lambda}{D_o}fl = 2.44\lambda F \quad (11\text{-}2)$$

While valid for paraxial rays (1st order approximation), when $F < 3$, the equation must be modified (see *Appendix).*

Table 11-4
OPTICAL RESOLUTION METRICS (in image space)

RESOLUTION	DESCRIPTION	DEFINITION (usual units)
Rayleigh criterion	Ability to distinguish two adjacent point sources	$u_{iRES} = 1.22\,\lambda\,F$ (Calculated: mm)
Airy disk	Diffraction-limited diameter produced by a point source	$u_{iRES} = 2.44\,\lambda\,F$ (Calculated: mm)
Blur diameter	Actual minimum diameter produced by a point source	Calculated from ray tracing (mm)

The Rayleigh criterion is a measure of the ability to distinguish two closely spaced objects when the objects are point sources. Optical aberrations and focus limitations increase the diffraction-limited spot diameter to the blur diameter. Optical designers use ray tracing programs to calculate the blur diameter. The blur diameter size is dependent on how it is specified (i.e., the fraction of encircled energy[10]). den Dekker and van den Bos provide[11] an in-depth review of optical resolution metrics.

Considerable literature[12] has been written on the image forming capability of lens systems. Image quality metrics include aberrations, Strehl ratio, and blur or spot diagrams. These metrics, in one form or another, compare the actual blur diagram to diffraction-limited spot size. As the blur diameter increases, image quality decreases and edges become fuzzy. The ability to see this degradation requires a high resolution sensor. Both the human eye and photographic film visually have this resolution. However, with electronic imaging systems, the detectors are often too large to see the degradation because the detectors are often larger than the blur diameter.

11.3. DETECTOR RESOLUTION

Detector arrays are often specified by the number of pixels and detector pitch. These are not meaningful until an optical system is placed in front of the array. Table 11-5 provides the most common resolution metrics expressed in object-space units. The detector angular subtense (DAS) is often used by the military to describe the resolution of systems when the detector is the limiting subsystem. If the detector's element horizontal and vertical dimensions are different, then the DAS in the two directions is different. Note that the PAS is different than the DAS Only with 100% fill factor arrays are they the same. Spatial sampling rates are determined by the pitch in image space or the PAS in object space. If resolution is defined by the DAS, then a vanishingly small detector is desired.

Table 11-5

DETECTOR ELEMENT RESOLUTION MEASURES (in object space)

RESOLUTION	DESCRIPTION	DEFINITION (usual units)
Detector angular subtense	Angle subtended by one detector element	d/fl (Calculated: mrad)
Instantaneous field of view	Angular region over which the detector senses radiation	Measured width at 50% amplitude (mrad)
Pixel angular subtense	Angle subtended by one pixel	d_{CC}/fl (Calculated: mrad)
Ground sampled distance	Projection of detector pitch onto the ground	$d_{CC}R/fl$ (calculated: inches)
Nyquist frequency	One-half of the angle subtended by detector pitch	$d_{CC}/2fl$ (calculated: mrad)

For single-chip color cameras, resolution is more difficult to define. The primaries or their complements spatially sample the scene unequally (Figure 10-11). Most chips contain more green detectors, and therefore the "green" resolution is more than just one-third of an equivalent monochrome system. Through data interpolation, any number of output data points can be created. This number (datels) should not be confused with the camera spatial resolution (pixels). The resolution of CFAs and the Super CCD is ill-defined since the datels are an interpolated version of the pixel number.

Example 11-1
DISELS and PIXELS

A staring array consists of 512×512 detectors. Using 4× electronic zoom it is presented on a digital monitor that displays 1024×1024 disels. What is the system resolution?

Each pixel is mapped onto four datels that are then mapped one-to-one onto disels. Here, the imaging system determines the system resolution, not the number of datels or disels. High quality monitors only ensure that the image quality is not degraded. Electronic zoom cannot increase resolution. But, decimation or minifying may reduce resolution. Non-integer magnification (e.g., 1.05, 0.95, etc) will distort imagery. Because image processing is performed on datels, the image analyst must be made aware of the system resolution.

11.4. ELECTRICAL RESOLUTION METRIC

For high data rate systems such as line scanners, the electronic bandwidth may limit system response. For electronic circuits, resolution is implied by its bandwidth. The minimum pulse width is approximately

$$\tau_{MINIMUM} = \frac{1}{2\,BW} \tag{11-3}$$

where $\tau_{MINIMUM}$ is measured in seconds and BW is measured in Hertz. The 3 dB point (half-power frequency) is often called the bandwidth. With this minimum value the output signal *looks* like the input. Other expressions (e.g., $\tau_{MINIMUM} = 1/BW$) simply imply there is a measurable output without any inference about signal fidelity.

11.5. MTF-BASED RESOLUTION

For many systems, the MTF is dominated by the optics, detector, and display MTFs. The human visual system (HVS) then interprets this information. Some resolution metrics include the HVS. Others do not. Both approaches provide insight into system performance. If the scene contains all spatial frequencies, then an MTF-based resolution may be appropriate. This may not be a reasonable assumption. Scene content varies and perceived image quality depends upon the scene. The following resolution metrics assume a 100% fill factor and is one-dimensional. Modifying the equations for finite fill factors is straight forward. The extension to two dimensions is less clear.

11.5.1. LIMITING RESOLUTION

The limiting resolution is the spatial frequency at which the MTF is, say, 10%. Figure 11-1 illustrates two systems that have the same limiting resolution. Which system is selected depends on the specific application. System A is better at high spatial frequencies, and system B is better at low spatial frequencies. Equivalently, if the scene contains mostly low frequencies (large objects), then system B should be used. If edge detection is important (high spatial frequencies), then system A should be considered. This clearly indicates the difficulty encountered when using resolution exclusively as a measure of system performance.

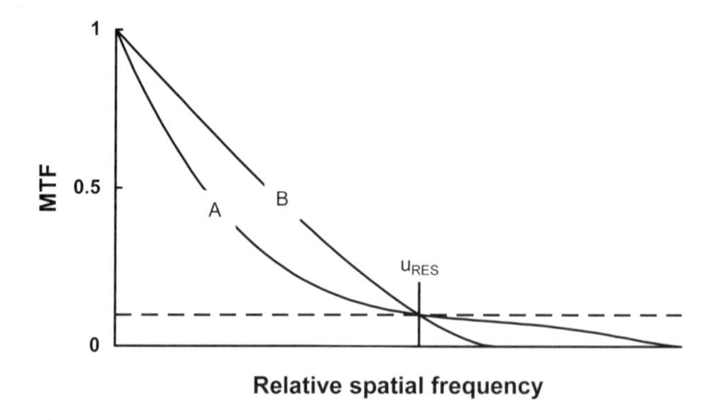

Figure 11-1. The MTFs of two different imaging systems with the same limiting resolution, $\boldsymbol{R}_{SYS} = 1/\boldsymbol{u}_{RES}$.

11.5.2. OPTICS-DETECTOR SUBSYSTEM

Image metrics may be described in the spatial domain where the optical blur diameter is compared to the detector size or in the frequency domain where the detector cutoff is compared to the optics cutoff. Either comparison provides an image quality metric that is a function of $F\lambda/d$. Table 11-6 summarizes the two limiting cases. Since the transition from one region to the other is gradual, it is difficult to precisely select an $F\lambda/d$ value that separates the two regions. It nominally selected at $F\lambda/d = 1$. From a sampling viewpoint, the important parameter[13] is $F\lambda/d_{CC}$ (also call $F\lambda/p$ and Q). For 100% fill factor arrays, $F\lambda/d = F\lambda/d_{CC}$. While valid for paraxial rays (1st order approximation), when $F < 3$, the equation must be modified (see *Appendix*).

Table 11-6.
OPTICS-LIMITED VERSUS DETECTOR-LIMITED PERFORMANCE

$F\lambda/d$	System performance	Spatial domain	Frequency domain
<1	Detector-limited	Airy disk much smaller than detector	Optical cutoff much greater than the detector cutoff
>1	Optics-limited	Airy disk much larger than detector	Optical cutoff much less than the detector cutoff

The MTF at Nyquist frequency is often used as a measure of performance (Figure 11-2). As the MTF increases, image quality should increase. Unavoidably, as the MTF increases, aliasing also increases and image quality suffers. This suggests there may be an optimum MTF, or, equivalently an optimum $F\lambda/d$. In reality, the range of "good" imagery is rather broad.

Figures 11-3 through 11-5 illustrate the MTF and imagery for three different F/d ratios when $\lambda = 0.55$ µm. All images were reconstructed with a flat-panel display. There are 32×32 pixels in the image. Image quality depends upon viewing distance. View the images at normal reading distance and then at several feet. The relationship between F and d is provided in Figure 11-6. For a 1 µm detector, the focal ratio is 3 to achieve imagery shown in Figure 11-5. Lower $F\lambda/d$ values provide a slightly better image (edges are slightly sharper). However, a 3 µm detector will produce an image like that shown in Figure 11-3 when the focal ratio is 3.

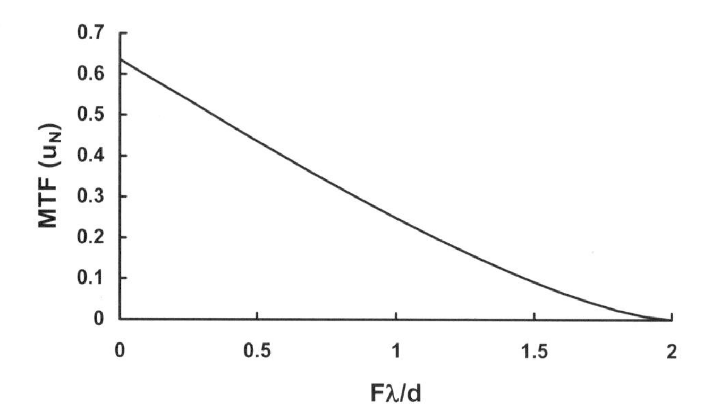

Figure 11-2. $MTF_{OPTICS}MTF_{DETECTOR}$ at Nyquist frequency as a function of $F\lambda/d$. The fill factor is 100%.

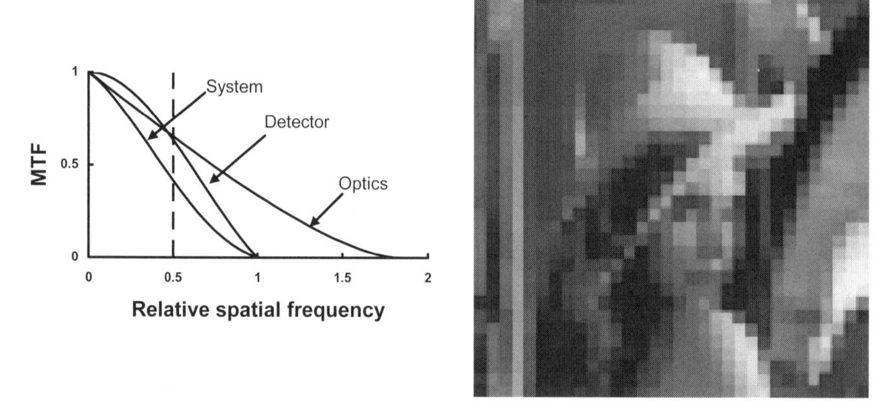

Figure 11-3. MTF and imagery for λ = 0.55 μm and F/d = 1 ($F\lambda/d$ = 0.55). The vertical dashed line is the Nyquist frequency (100% fill factor). The spatial frequency is normalized to u/u_{iD}. Imagery created[14] by MAVIISS. ***NOTE***: The imagery is enlarged so that your eye MTF and the printing do not significantly affect the image quality. This allows you to see the distortion created by sampling and the system MTF degradation. With 20/20 vision, you can just perceive 1 arc sec (0.291 mrad). The images represent 32×32 pixels and are about 50 mm across. Each pixel is about 1.56 mm. At normal reading distance, (about 380 mm), each pixel subtends 4.1 mrad. You must move 4.1/0.291 = 14 times away (about 17 feet) to appreciate what is seen on a typical display. The image will start to become uniform at ½ that distance.

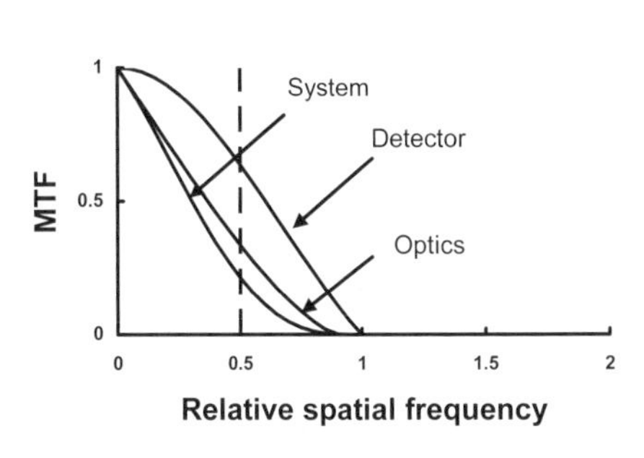

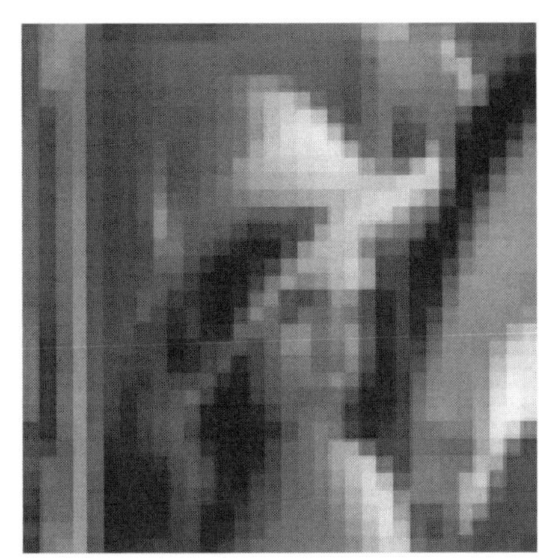

Figure 11-4. MTF and imagery for λ = 0.55 µm and F/d = 2 ($F\lambda/d$ = 1.1). The vertical dashed line is the Nyquist frequency (100% fill factor). The spatial frequency is normalized to u/u_{iD}. Imagery created[14] by MAVIISS. See ***NOTE*** in Figure 11-3 caption.

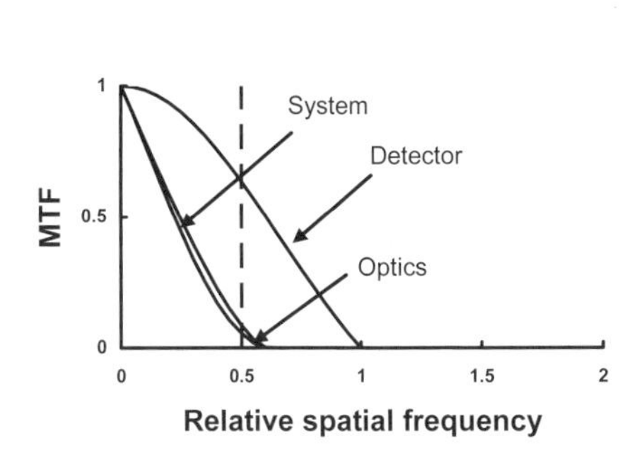

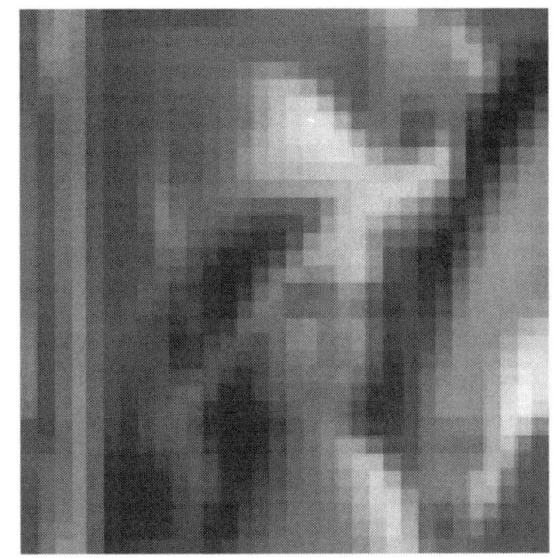

Figure 11-5. MTF and imagery for λ = 0.55 µm and F/d = 3 ($F\lambda/d$ = 1.65). The vertical dashed line is the Nyquist frequency (100% fill factor). The spatial frequency is normalized to u/u_{iD}. Imagery created[14] by MAVIISS. See ***NOTE*** in Figure 11-3 caption.

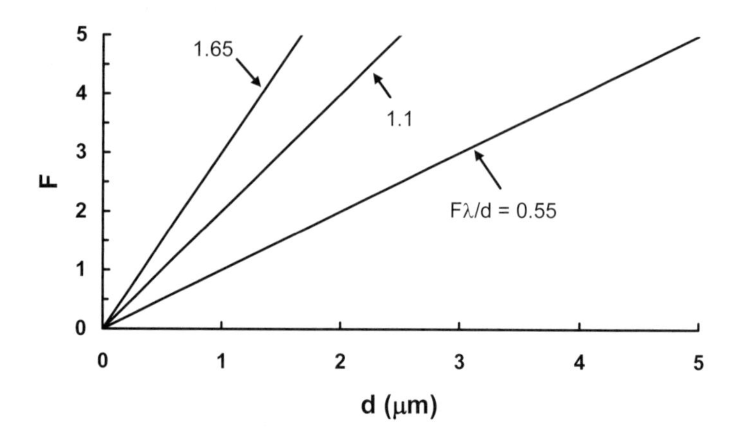

Figure 11-6. ***F*** as a function of **d** for three different ***Fλ/d*** ratios. λ = 0.55 μm.

There is no aliasing when $F\lambda/d \geq 2$. This limiting condition is known as "critical sampling" or "Nyquist sampling of the optical blur" in the astronomy community.[15] The latter term is not good terminology, since the blur diameter consists of all frequencies up to the optical cutoff (the blur is the Fourier transform of the optical MTF). The Nyquist frequency is only one frequency of many. Having $F\lambda/d = 2$ (optics-limited) may be appropriate for certain applications. The imagery can replicate the scene (assuming an ideal reconstruction filter). With no aliasing, what you see is exactly what is present in the scene. This may be an important consideration for medical imaging where a sampling artifact could be construed as a medical abnormality or space probes where it is impossible to obtain ground truth. Having $F\lambda/d \geq 2$ may be overly restrictive. In-band MTFs (frequencies slightly less then the Nyquist frequency) are reduced in amplitude. If the SNR is sufficiently high, a boost circuit can increase the MTF and sharpen the image.

With no reconstruction filter, the in-band band spurious response (Figure 11-7) is

$$SR_{NO\ RECON} = \frac{\int_{f_N}^{f_C} MTF_{OPTICS}(f) MTF_{DETECTOR}(f)\, df}{\int_{0}^{f_C} MTF_{OPTICS}(f) MTF_{DETECTOR}(f)\, df} \qquad (11\text{-}4)$$

The variable f_C is the optical cutoff. The spurious response is 1% when $F\lambda/d$ = 1.64 and the MTF at Nyquist frequency is 0.057 (Figure 11-2). From a practical viewpoint, this is probably acceptable. Recall that $F\lambda/d$ is the ratio of the detector cutoff to the optical cutoff. Then the ratio of the Nyquist frequency to the optical cutoff is one-half that value. When $F\lambda/d = 1.64$, $f_N = 0.82 f_C$.

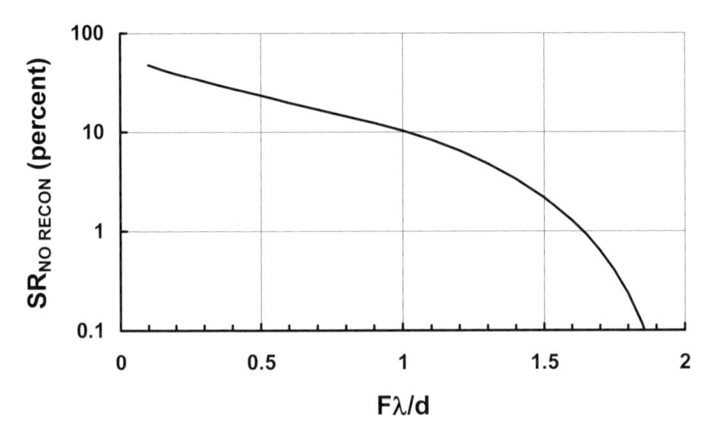

Figure 11-7. In-band spurious response.

While the detector size is usually fixed, the focal ratio can have a wide range of values leading to a wide range of $F\lambda/d$ values. For the visible region (λ = 0.55 μm), F/d ratios are provided in Figure 11-8. Some authors[16] have approached camera design by comparing the Airy disk diameter to the detector linear dimension (assumed square). Here $F\lambda/d$ = 0.41 (dashed line in Figure 11-8). While there is nothing inherently wrong with this approach, it does not directly apply to imagery. The scene consists of an infinite number of point sources and the image is the summation of the point source responses (Figure 8-2). The MTF approach assumes the scene contains all spatial frequencies.

It is clear that there will always be some aliasing in the visible region for practical focal ratios. While the desire to eliminate aliasing may be high, there is no overpowering need to do so for monochrome imagery (see Figures 11-3 through 11-5) except for special applications. Single-chip color arrays exhibit aliasing similar to that shown in Figure 11-8. Since the HVS is very sensitive to color variations, the single chip array requires a low-pass optical filter to avoid color cross-talk. The effectiveness to the OLPF depends upon the focal ratio or, equivalently, the expected spurious response. This analysis assumes a 100% fill factor array. For lower fill factors (e.g., CFA), the important parameter is $F\lambda/d_{CC}$.

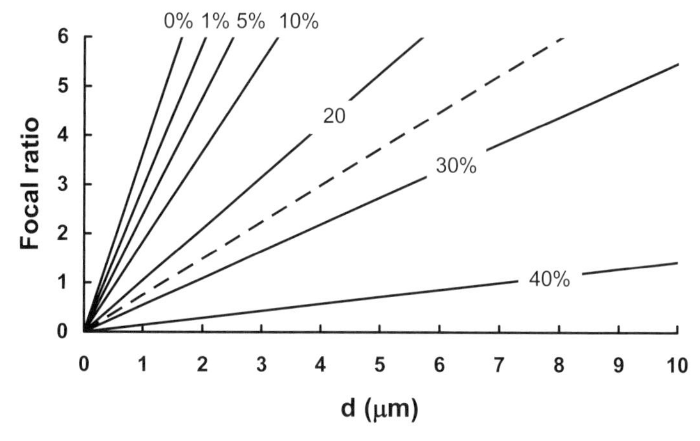

Figure 11-8. F versus d for a variety of in-band spurious responses when $\lambda = 0.55$ μm. Zero percent SR represents $F\lambda/d = 2$ (no aliasing). When the Airy disk diameter is equal to the detector linear dimension, $F\lambda/d = 0.41$ (dashed line). Figure 11-6 (linked to Figures 11-3 through 11-5) provides additional insight.

11.5.3. SCHADE'S EQUIVALENT RESOLUTION

Schade[3] equated apparent image sharpness of a television picture to an equivalent pass band. As reported by Lloyd,[17] Sendall modified Schade's equivalent resolution (inserted a factor of 2) such that

$$R_{EQ} = \frac{1}{2N_e} = \frac{1}{2\int_0^{\infty} \left|MTF_{SYS}(u)\right|^2 du} \tag{11-5}$$

The value R_{EQ} cannot be directly measured and is a mathematical construct that expresses overall performance. As R_{EQ} decreases, the resolution "improves" (smaller is better). As an approximation, the system resolution, $R_{EQ\text{-}SYS}$, may be estimated from the component equivalent resolutions, R_i, by

$$R_{EQ-SYS} = \sqrt{R_1^2 + \ldots + R_i^2 + \ldots + R_N^2} \tag{11-6}$$

Schade's approach using the square of the MTF emphasized those spatial frequencies at which the MTF is relatively high. The equivalent resolution approach assumes that the system is completely analog and it ignores sampling effects. Therefore, R_{EQ} becomes a resel. The metric was developed from observer responses and therefore implicitly includes the eye's response.

As a summary metric, $R_{EQ\text{-}SYS}$ provides a better indication of system performance than just a single metric such as the detector size. $R_{EQ\text{-}SYS}/fl$ probably should be used in Equation 11-1 to obtain a more realistic measure of range performance. The equivalent resolution for diffraction-limited optics and detector are

$$R_{OPTICS} = 1.845F\lambda \ mm \qquad R_{DETECTOR} = d \ \ mm$$

Note that Schade's approach provides an optical resolution value that is smaller than the Airy disk diameter ($d_{AIRY} = 2.44\lambda/D$). Recall that R_{EQ} is only a mathematical construct used to analyze system performance. The composite resolution provides

$$\begin{aligned} R_{EQ-SYS} &\approx \sqrt{R_{OPTICS}^2 + R_{DETECTOR}^2} \\ &= d\sqrt{\left(1.845\frac{F\lambda}{d}\right)^2 + 1} \quad mm \end{aligned} \qquad (11\text{-}7)$$

The equation is graphed in Figure 11-9. For small $F\lambda/d$ values, $R_{EQ\text{-}SYS}$ approaches d. For large values of $F\lambda/d$, the system becomes optics-limited and the equivalent resolution increases. Although any detector size, optical aperture, and focal length can be chosen to select the resolution limit, the same parameters affect the sensitivity (Equation 2-18). The focal ratio can approach zero. Figure 11-10 illustrates the equivalent resolution for small sensors. For $d >$ 3 μm, the sensor is operating in the detector-limited region for practical focal ratios.

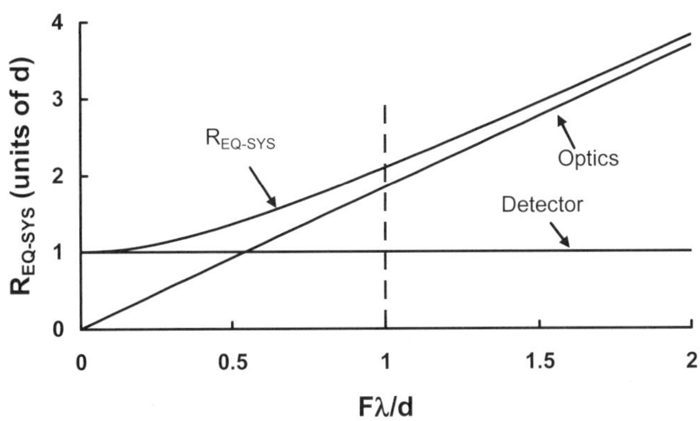

Figure 11-9. Equivalent resolution normalized to d as a function of $F\lambda/d$. The optics-limited region is to the right of the dashed line and detector-limited performance is on the left. The best resolution occurs when $F\lambda/d$ approaches zero ($F\rightarrow0$). As shown in the *Appendix*, F can approach zero.

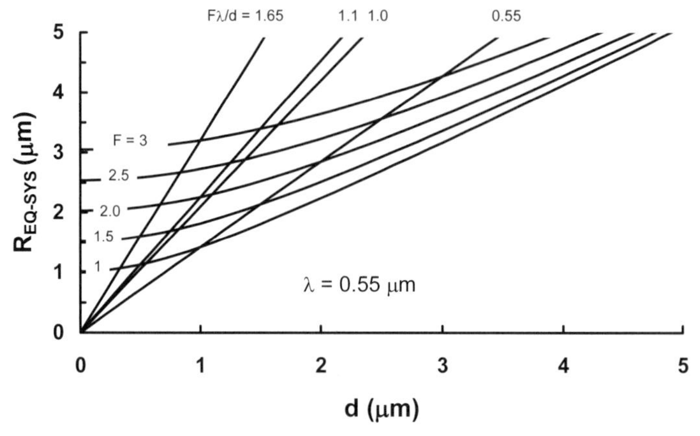

Figure 11-10. Schade's equivalent resolution for small detectors. The optics-limited region is to the left of the $F\lambda/d$ = 1 line and detector-limited performance is on the right. The lines $F\lambda/d$ = 0.55, 1.1, and 1.65 corresponds to Figures 11-3 through 11-5, respectively.

11.6. DISPLAY RESOLUTION

For most electronic imaging systems, the display medium acts as a reconstruction filter. Displays do not completely attenuate the replicated frequency amplitudes created by the sampling process. The HVS further reduces the sampling remnants so that the perceived image appears continuous with minimal sampling artifacts. Therefore, displays cannot be analyzed in isolation but must be considered as part of the display/observer system. Since a printer is an image forming device, it is a "display."

11.6.1. DISELS

The disel format is the arrangement of disels into horizontal and vertical rows. Table 11-7 lists the common formats found in computer color monitors. Disel density is usually specified in units of dots per inch. Table 11-8 provides the approximate relationship between "resolution" and disel density. Confusion exists because dots also refer to the individual phosphor sites. In this context, a "dot" is a triad (disel). The horizontal disel density is the number of horizontal disels (Table 11-7) divided by the horizontal display size.

Table 11-7
COMMON COMPUTER DISPLAY FORMATS

TYPE	FORMAT Triad number (disels)
CGA (color graphics card adapter)	340×200
EGA (extended graphics adapter)	640×350 and 640×400
VGA (video graphics array)	640×480
SVGA (super video graphics array)	800×600
XGA (extended graphics array) Typical of multi-sync displays	1024×768
High resolution	>1024×1024

Table 11-8
TYPICAL DISEL DENSITY

"RESOLUTION"	DISEL DENSITY (dpi)
Ultra-high	> 120
High	71 - 120
Medium	50 - 70
Low	< 50

The user is typically interested in performance and not in specific design specifications. For example, the dot pitch is typically superfluous information to the user. Disel size and the display modulation transfer function are usually more important. Monitor disels are not related to camera pixels. Each is generated by the respective designs. Cameras with a standard video output may not have the resolution suggested by the video standard. For example, a camera containing an array of 320×240 pixels may have its output formatted into EIA 170 timing. This standard suggests that it can support approximately 640×480 pixels. If the displayed image is greater than 320×240 disels the system resolution is limited to 320×240 pixels.

The disel size indicates how many individual pieces of information the system can support. It does not indicate the extent to which the detail will be resolved. The distance between adjoining disels is related to resolution. As the distance decreases, disels overlap and the resolution becomes poorer (see Section 9.10.1. *Addressability*). Disels can overlap in printers and CRTs but not flat panel displays.

The vertical disel size is defined by the raster pitch. It is simply the display height divided by the number of active scan lines. This is dictated by the video standard. With an optimum system, the electron beam diameter and video bandwidth are matched in the horizontal direction. For television receivers, the number of monitor disels is matched to the video standard. In this sense, a video standard has a "resolution."

The definition of a disel for color systems is more complicated. The representation shown in Figure 7-13 implies that the electron beam passes through one hole only. Due to alignment difficulties between the mask and phosphor dots and because the phosphor may have imperfections, CRTs have electron beams that encompass several mask holes. The beam width at the 5% intensity level typically illuminates about 2.5 holes. Encompassing a non-integral number of holes minimizes aliasing and color moiré patterns.[18] The shadow mask samples the continuous electron beam to create a sampled output (Figure 11-11). The phasing effects created by the shadow mask are obvious.

Figure 11-12 illustrates the relationship between the shadow-mask pitch and the beam profile. Assume the shadow-mask pitch is 0.28 mm. If the 5% diameter covers 2.5 holes, then, using Equation 9-49, $\sigma_{SPOT} = (0.28)(2.5)/4.9 = 0.143$ mm. The spot size (Equation 9-45) is $S = (2.35)(0.143) = 0.336$ mm. In Figure 11-10, the spots are contiguous and each is considered a disel.

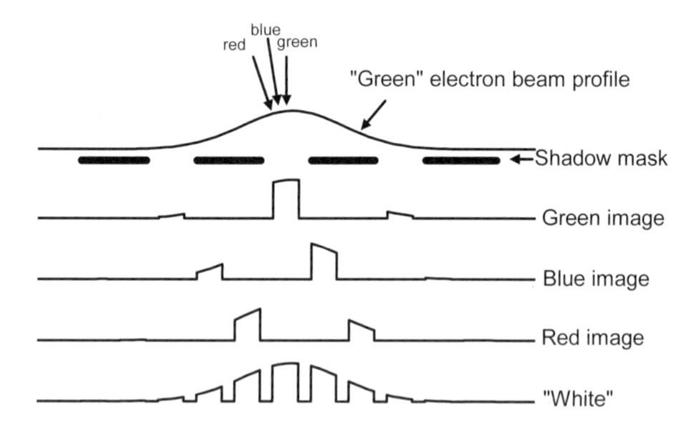

Figure 11-11. Representative image of a single spot on a color CRT. The "green" electron beam profile is shown. The red and blue beam profiles (not shown) arrive at slightly different angles (shown by arrows) and provide different intensity profiles. The "sum" appears white when the dot size is less than the eye's resolution limit.

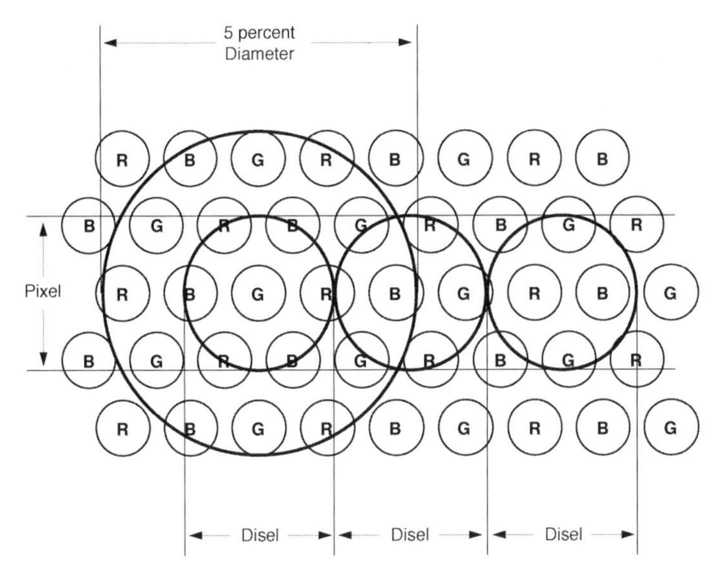

Figure 11-12. Five percent intensity level illuminating 2.5 shadow-mask holes. The spot size (FWHM) is equal to the disel size. If the shadow-mask pitch is $\boldsymbol{X}$ mm, then the beam 5% diameter is 2.5$\boldsymbol{X}$ mm, σ_{SPOT} = 0.51$\boldsymbol{X}$ mm, and the disel is 1.2$\boldsymbol{X}$ mm.

11.6.2. TV RESOLUTION

The vertical resolution is limited by the number of raster lines. For example, NTSC compatible monitors display 485 lines. However, to see precisely this number, the test pattern must be perfectly aligned with the raster pattern. Because the test pattern may be randomly placed, the pattern at Nyquist frequency can have a zero output if 180 degrees out-of-phase (discussed in Section 10.6. *Image distortion*). To account for random phases, the Kell factor[19] is applied to the vertical resolution to provide an average value. A value of 0.7 is widely used but not universally accepted[20] (see Equation 10.7).

$$\boldsymbol{R_{VERTICAL} = (active\ scan\ lines)(Kell\ \ factor)} \qquad (10\text{-}8)$$

Timing considerations force the vertical resolution to be proportional to the number of vertical lines in the video signal. Therefore, NTSC displays have an average vertical resolution of (0.7)(485) = 340 lines. PAL and SECAM displays offer an average of (0.7)(575) = 402 lines of resolution.

Whereas the flat field condition exists when two adjacent lines cannot be resolved, the TV limiting resolution is a measure of when alternate vertical bars are just visible (on-off-on-off). The standard resolution test target is a wedge pattern with spatial frequency increasing toward the apex of the wedge. It is equivalent to a variable square wave pattern.

The measurement is a perceptual one and the results vary across the observer population. The industry selected[21] the limiting resolution as a bar spacing of $1.18\sigma_{SPOT}$. This result is consistent with the flat field condition. The flat field condition was determined for two Gaussian beams, whereas square waves are used for the TV limiting resolution test. Suppose the Gaussian beams in Figure 9-39 were separated by an imaginary "black" beam. Then the beams would be separated by just σ_{SPOT} but the intensity distribution would be identical to Figure 9-39. The "black" beam does not contribute to the visible image. For TV limiting resolution, the "on" lines are separated by $(2)(1.18\sigma_{SPOT})$. Because bar targets are used, the resultant image is not a precise Gaussian beam and a larger line separation is required ($1.18\sigma_{SPOT}$ versus σ_{SPOT}) for the TV limiting resolution.

The flat field condition and high TV limiting resolution are conflicting requirements. For high TV limiting (horizontal) resolution, σ_{SPOT} must be small. But raster pattern visibility (more precisely, invisibility) suggests that σ_{SPOT} should be large.

11.6.3. FLAT PANEL DISPLAYS

Color flat-panel displays use rectangular emitting elements whose triad is square. Each emitting element roughly has a 3:1 aspect ratio. The number of disels (triads) is the same as that provided in Table 11-7. Resolution is inversely proportional to the triad pitch and has units of dots per inch (Table 11-8)

11.6.4. PRINTERS

A CRT or flat-panel display controls the disel intensity by varying voltage. With printers, the choice is ink or no ink. Gray scales ("lightness") are achieved by controlling the distance between ink spots. As the distance increases, more white paper is seen and lightness increases. Since the ink spots are below the eye's resolution, the HVS blends the ink/no ink area into a uniform gray. Therefore, the number of dots per inch must be much greater than what the eye can resolve. Generally, 300 dpi produces "good" imagery. "Excellent" imagery requires more, with printers achieving 1200 dpi. Since perceptibility depends on the viewing distance, a 1200 dpi image can be viewed at any distance and still appear uniform. This is not true for CRTs or flat-panel displays. Thus, for equivalent "resolution," a printer needs more dots per inch than a monitor (disels per inch).

11.7. OBSERVER-BASED RESOLUTION

As with the MTF-based resolution metrics, the observer-based metrics assume that the scene contains all spatial frequencies. This may not be a reasonable assumption. We can always find a scene whose imagery is poor yet rated high on the observer-based rating scale. The observer-based metrics provide general design guidelines. Four metrics are compared (MTFA, SQF, SQRI, and TTP). Although the formulations are quite different, the results are similar to Schade's equivalent resolution: there is an optics-limited and a detector-limited region. As a first-order approximation, Schade's equivalent resolution can be used to estimate system performance. Hultgren et. al. provide[22] an overview of image quality evaluation methodologies applied to printed imagery.

11.7.1. MTFA

As the spatial frequency increases, the eye requires more contrast. Simultaneously, the system MTF is decreasing (Figure 11-13). The intersection of the $\boldsymbol{MTF_{SYS}}$ and $\boldsymbol{CTF_{EYE}}$ was called the "resolution" frequency. It is not unique because $\boldsymbol{u_{RES}}$ depends on the display size and viewing distance (visual angle). The area bounded by the system MTF and the eye's contrast threshold function is the modulation transfer function area (MTFA). It was originally created for film-based systems, where the CTF was called the demand modulation function. Because many measures were in support of aerial imagery, the intersection was also called the aerial image modulation (AIM). In one dimension

$$MTFA = \int_0^{u_{RES}} [MTF_{SYS}(u) - CTF_{EYE}(u)]\, du \qquad (10\text{-}9)$$

As the MTFA increases, the perceived image quality appears to increase. According to Snyder[23], the MTFA appears to correspond well with performance in military detection tasks where the targets are embedded in noise. A graph of MTFA as a function of $\boldsymbol{F\lambda/d}$ illustrates a detector-limited region and an optics-limited region with a transition around $\boldsymbol{F\lambda/d}=1$ (Figure 11-14).

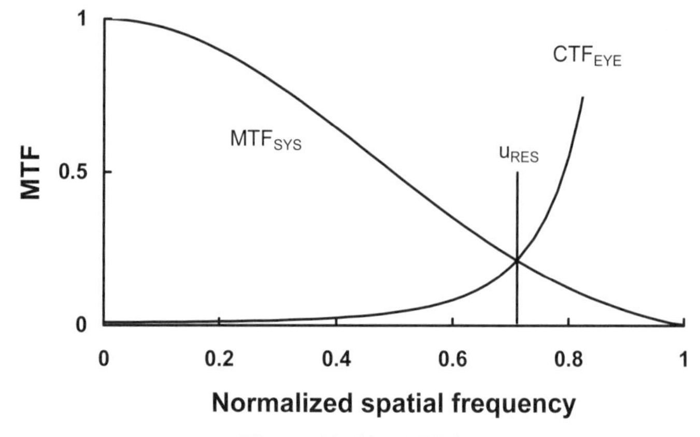

Figure 11-13. MTFA.

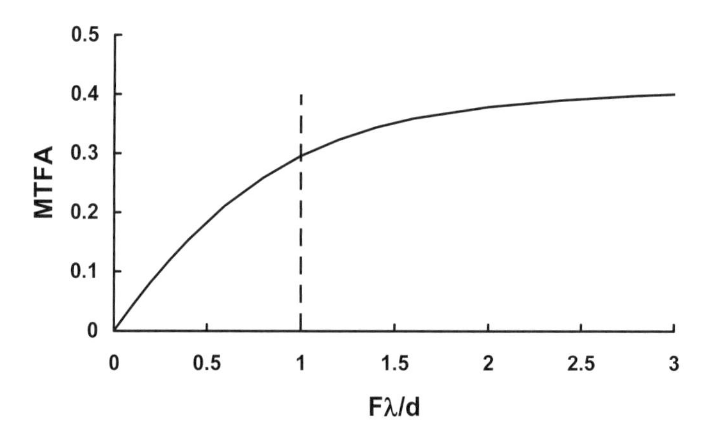

Figure 11-14. MTFA as a function of ***Fλ/d***. Detector-limited operation is to the left of the dashed line and optics-limited to the right. $MTF_{SYS} = MTF_{OPTICS} MTF_{DETECTOR} MTF_{FLAT\ PANEL}$.

11.7.2. SUBJECTIVE QUALITY FACTOR

According to Granger and Cupery,[2] the spatial frequency range important to image quality is in the region approximately 1/3 below the peak to three times above the peak sensitivity of the eye. The eye MTF is approximated by $MTF_{EYE} = 1/CTF_{EYE}$ normalized to one (Figure 9-49). The region is where the eye MTF is greater than 50% (Figure 11-15). In logarithmic units, the normalized subjective quality factor (SQF) is:

$$SQF = K \int_{\log(f_1)}^{\log(f_2)} MTF_{SYS}(f)\, d[\log(f)] \qquad (11\text{-}10)$$

where K is a normalization constant and $d[\log(f)] = df/f$. The spatial frequency presented to the eye depends on the image size on the display, the distance to the display, and electronic zoom. Table 11-9 provides Granger and Cupery's interpretation of the SQF. These results are based on many observers viewing noiseless photographs with known MTF degradation.

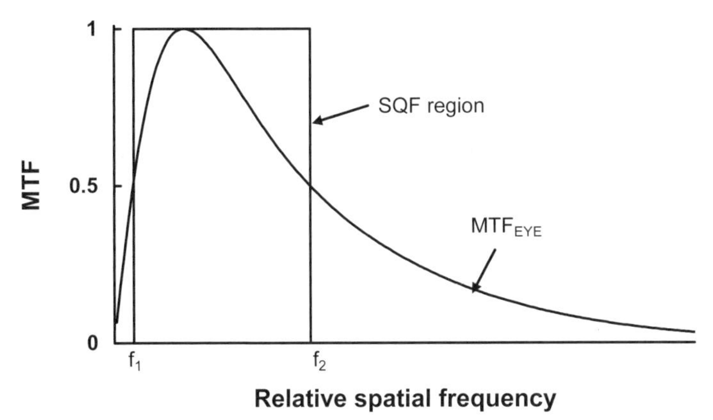

Figure 11-15. SQF region.

Table 11-9
SUBJECTIVE QUALITY FACTOR

SQF	SUBJECTIVE IMAGE QUALITY
0.92	Excellent
0.80	Good
0.75	Acceptable
0.50	Unsatisfactory
0.25	Unusable

The SQF is an adequate image quality metric for low-noise imagery when the illumination is fixed at a moderate level. As with the MTFA and Schade's equivalent pass band, the SQF is intended for general imagery only. It cannot be used for specific applications such as the legibility of alphanumeric characters. The SQR provides results similar to Schade's equivalent resolution: it has a detector-limited region and an optics-limited region (Figure 11-16). $MTF_{SYS} = MTF_{OPTICS} MTF_{DETECTOR} MTF_{FLAT\ PANEL}$.

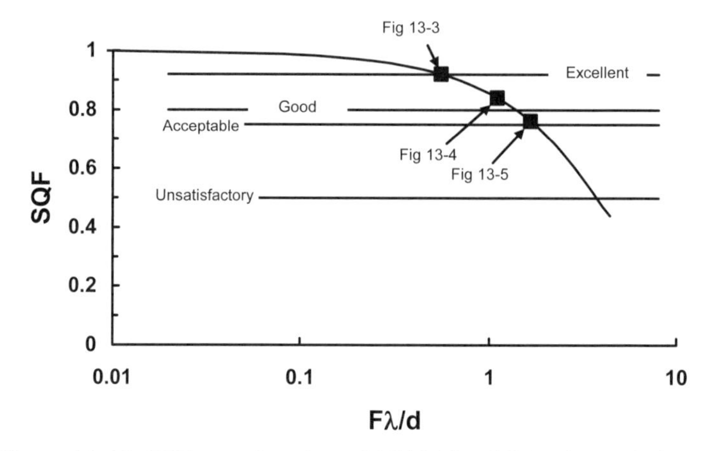

Figure 11-16. SQF as a function of $F\lambda/d$. The SQF values of Figures 11-3 through 11-5 are shown. The SQF does not include the sampling artifacts seen in those figures. Detector-limited operation occurs when $F\lambda/d < 1$ and optics-limited when $F\lambda/d > 1$.

SQF is currently used to evaluate lenses[24] using a modified rating scale (Table 11-10). It has been extended to imaging systems[25] where the lower limit is zero and the upper limit is the Nyquist frequency in Equation 11-10. While the Nyquist frequency would seem like a reasonable limit, there is no data to support this limit (i.e., observers rank order pixilated imagery). Although the viewing distance is usually sufficiently large so that pixilation is not perceived, the imagery may still contain significantly aliasing. Perceived image quality depends upon the scene content, $F\lambda/d$, image processing, and the reconstruction filter (See imagery in Section 11.9. *Image reconstruction*).

Table 11-10
LENS QUALITY ASSESSMENT

A+	A	B+	B	C+	C	D	F
94-100	89-94	84-89	79-84	69-79	59-69	49-59	<49

11.7.3. SQUARE-ROOT INTEGRAL

Barten[4-7] introduced the square-root integral (SQRI) as a measure of image quality. The SQRI is

$$SQRI = \frac{1}{\ln(2)} \int_{f_{MIN}}^{f_{MAX}} \sqrt{\frac{MTF_{SYS}(f)}{CTF_{EYE}(f)}} \frac{1}{f} df \qquad (11\text{-}11)$$

The factor 1/ln(2) allows SQRI to be expressed in just-noticeable-difference (JND) units. The multiplication overcomes the theoretical objection raised with the MTFA where the eye CTF was subtracted from the MTF. MTF_{SYS} includes all the subsystem MTFs up to and including the display.

This model includes the effects of various display parameters such as viewing ratio, luminance, resolution, contrast, addressability, and noise. Display viewing ratio and average luminance are incorporated in CTF_{EYE}. The SQRI versus $\boldsymbol{F\lambda/d}$ provides results similar to TTP (next section).

11.7.4. TARGETING TASK PERFORMANCE

Experiments by Vollmerhausen et. al.[26,27] demonstrated that military target acquisition follows a metric similar to Barten's which they called the targeting task performance (TTP). The formulation is identical with Barten's square root integral but the integration is performed over a linear scale:

$$TTP = \int_{f_{LOW}}^{f_{UPPER}} \sqrt{\frac{C_{RO}}{CTF_{SYS}(f)}}\, df \qquad (11\text{-}12)$$

The apparent target contrast, $\boldsymbol{C_{RO}}$, at the entrance aperture is the inherent contrast modified by atmospheric phenomena (discussed in Section 12.3. *Contrast transmittance*). The upper limit of integration, $\boldsymbol{f_{UPPER}}$, is the smaller of the intersection apparent target contrast $\boldsymbol{C_{TARGET}}$ with the $\boldsymbol{CTF_{SYS}}$ (labeled as $\boldsymbol{f_{HIGH}}$ in Figure 11-17) or the Nyquist frequency.

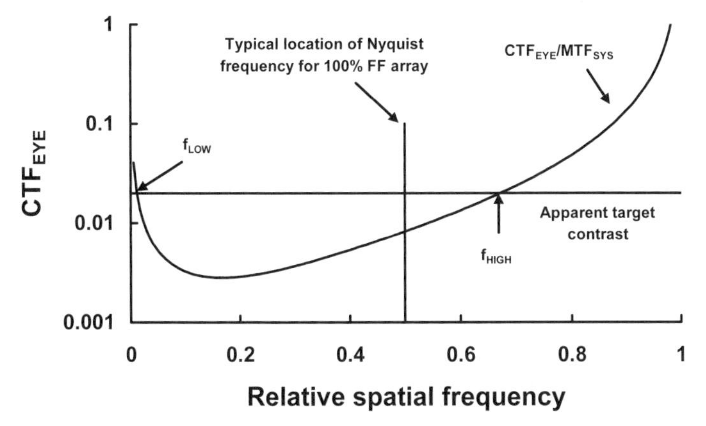

Figure 11-17. Definitions of the integration limits for the TTP metric.

As a contrast model, the TTP applies to all imaging systems

$$CTF_{SYS}(f) = \frac{CTF_{EYE}(f)}{M_{DISPLAY} MTF_{SYS}(f)} \sqrt{1 + \left(\frac{k_1 \sigma}{L}\right)^2} \tag{11-13}$$

The variable k_1 is a proportionality constant (model validation value) with units of √Hz, σ is the system noise filtered by the display and HVS in foot-lamberts/√Hz, and L is the display luminance with units of foot-lamberts. $MTF_{SYS} = MTF_{OPTICS} MTF_{DETECTOR} MTF_{FLAT\ PANEL}$. If there is glare from the monitor, the contrast is reduced by

$$M_{DISPLAY} = \frac{L - L_{GLARE}}{L + L_{GLARE}} \tag{11-14}$$

If the minimum luminance is zero (no glare), $M_{DISPLAY} = 1$. The normalized TTP (with the noise set to zero) is provided in Figure 11-18. Target acquisition range is proportional to the TTP. SSCamIP and IICamIP predict[9] target acquisition (discussed in Section 12.4. *Range predictions*).

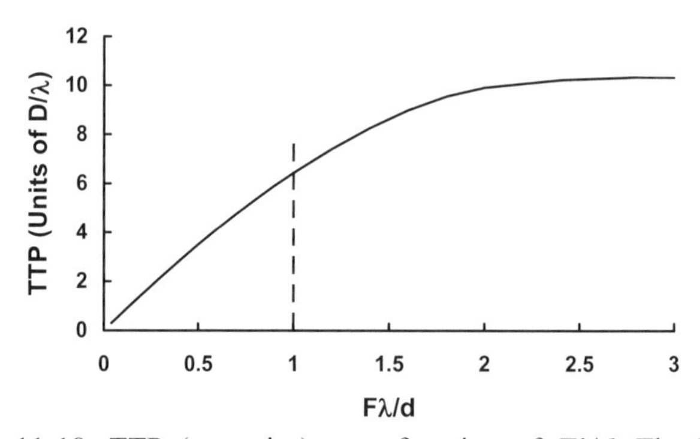

Figure 11-18. TTP (no noise) as a function of $F\lambda/d$. The TTP is normalized to D/λ. Detector-limited operation is to the left of the dashed line and optics-limited to the right. $M_{DISPLAY} = 1$.

11.8. VIEWING DISTANCE

For design purposes, perceptibility occurs when targets are larger than 1 arc minute. This corresponds to the normal visual acuity of 20/20. This design criterion is quite good. An observer can adjust his distance accordingly. An observer with poorer acuity (e.g., 20/40) will simply move closer to the screen whereas someone with better acuity will probably move further away. Most systems are observer-limited in that the observer cannot see all the detail because he is typically too far from the display.

For monochrome systems, the optimum viewing distance occurs when the raster pattern is barely perceptible. At closer distances, the raster pattern becomes visible and interferes with image interpretation. That is, it is considered annoying by most. For color CRTs, the observer must be far enough away so that the individual dots of the three primaries are imperceptible. Otherwise, the individual color spots become visible and the impression of full color is lost. Here, the eye's MTF attenuates the spatial frequencies associated with the disels thus producing a smooth, continuous image. This is how we view our computer monitor and TV. The individual red, green, and blue color spots cannot be seen unless we are only a few inches from the display.

Figure 11-19 illustrates the design viewing distance as a function of display diagonal size for monochrome displays. Each active line (raster line) subtends 1 arc minute (0.291 mrad). The viewing ratio is the viewing distance divided by the display height. The number of disels remains constant but the physical size of the display is variable. Therefore resolution and viewing distance is normalized to the picture height (Table 11-11). The typical viewing ratio for computer monitors is three.

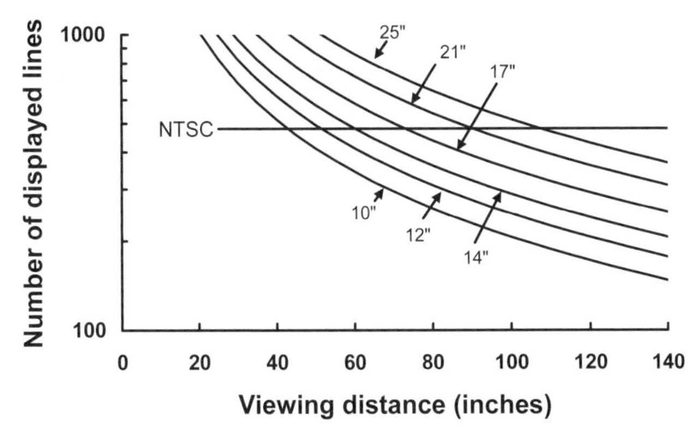

Figure 11-19. Viewing distance as a function of CRT diagonal picture size and number of displayed lines per picture height for monochrome displays. The aspect ratio is 4:3 so that the vertical extent is 60% of the diagonal. Actual viewing distance will probably be greater.

Table 11-11
TYPICAL VIEWING DISTANCES

Disels (H×V)	Viewing distance per PH	FOV (deg) (H×V)
640×480	7.2	10.7×8
1280×720	4.8	21.3×12
1920×1080	3.2	32×18

Printed imagery is considered "good" when there are more than 300 dpi. Photos look "excellent" when there are about 300 *pixels* per inch. To allow for half-toning, the printer should provide about 3 times more dots, or at least 900 dpi. The values in Table 11-12 assume 300 pixels/inch creates an "excellent" image. Larger images can be created. At some point, they will start to be blocky (you can see the individual pixels). That is, the sampling lattice is perceptible. However, blocky images are rarely seen because software interpolates the data to create a smooth image. Depending upon the viewing distance, this smoothed image may appear blurry. It has become fairly common practice to print a small image (usually of a person), then enlarge a portion (usually the eye) and claim the camera has good resolution. For example, a DSC4 image is printed on 4.3×3.4-inch paper and then enlarged to 10.2×6.8 inches to demonstrate its resolution.

Table 11-12
MAXIMUM DIGITAL STILL CAMERA PRINTS
(Assuming 300 dpi is acceptable)

Acronym	Pixels (H×V)	Print size (inches)
DSC1 "1.3 Megapixel"	1280×1024	4.3×3.4
DSC2 "2.0 Megapixel"	1600×1200	5.3×4.0
DSC3 "3.2 Megapixel"	1944×1672	6.5×5.6
DSC4 "6.3 Megapixel"	3072×2048	10.2×6.8

Whether the image is displayed on a CRT, flat-panel display, or printed, the viewing distance significantly affects perceived quality. For all the imagery in this book, it is suggested that you move several feet away to achieve the same visual angle that would normally exist. At the longer viewing distance, the HVS attenuates the amplitudes of the spatial frequencies associated with the blockiness, and the imagery appears continuous. In Figure 11-20, the image size is decreased for fixed viewing distance (reduces visual angle). The imagery was imported as a picture into Word and reduced in Word. It is not known how the decimation was performed nor the MTF associated with the decimation.

Figure 11-20. Lena with $F\lambda/d = 0.55$ (same as Figure 11-3). Original image in the upper left. Counter clockwise, the sizes are 80%, 60%, 40%, 20%, and 10%.

11.9. IMAGE RECONSTRUCTION

The quality of the viewed image depends upon the reconstruction filter used and display capability. Each CRT, flat-panel display, and printer has a different driver[28]. Video (PC), computer generated imagery, Macintosh, and Photoshop have different internal gammas resulting in different amounts of contrast enhancement. If using a CFA, the color quality depends on the manufacturer's personal theory on what color correction algorithm to use. Compression and decompression techniques affect image quality. It sometimes quite noticeable, and the degree of "corruption" depends upon the scene content. Decisions to incorporate a specific algorithm are often made on just a handfull of images. Finally, the image quality depends upon the viewing conditions (ambient lighting color temperature and visual angle).

MAVIISS[14] created the imagery shown throughout this book. The "quality" depends upon $\boldsymbol{F\lambda/d}$ (Figures 11-3 through 11-5). The imagery looked very good to the authors on their flat-panel computer display and laserjet printers. That imagery is different from what was created by MAVIIS in computer memory. The image quality printed in this book depends upon the paper quality and ink. Half-tone imagery will appear darker due to dot gain (ink bleeding into the paper). The printer uses his personal experience to compensate for dot gain. Although the printer supplies galley proofs, the printing method is different than that of the book, resulting in different tonal qualities. The only way to ensure the printing process is acceptable is to randomly view a few copies as they come off the printing press.

The largest difference in imagery is due to the reconstruction filter. Figures 11-21 through 11-23 illustrate three different scenes using a CRT and flat-panel display for reconstruction when $\boldsymbol{F\lambda/d} = 0.55$. Since your eye is a reconstruction filter, viewing distance has a dramatic impact on image quality. As you increase the viewing distance, the blockiness of the flat-panel display diminishes. At several feet, the two images will look comparable. Sampling artifacts are quite noticeable when viewing periodic targets.

Figure 11-21. Flat panel and CRT display imagery. See ***NOTE*** in Figure 11-3 caption.

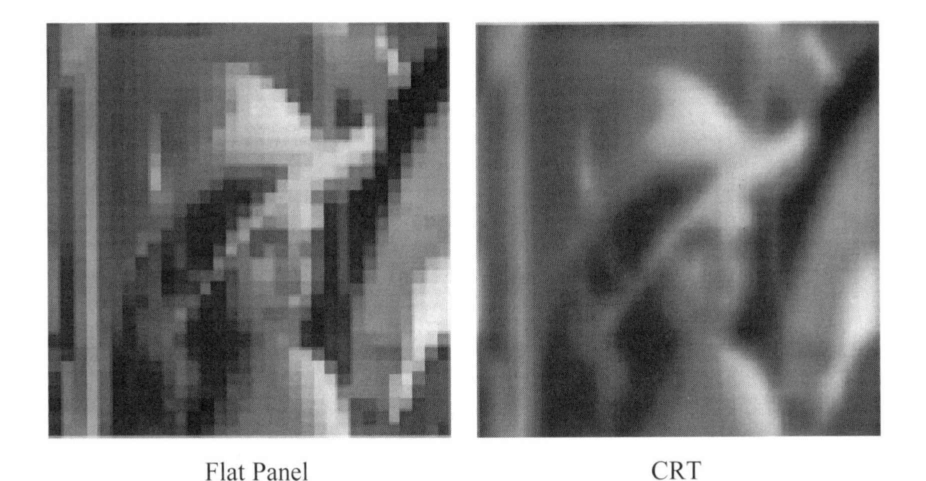

Figure 11-22. Flat panel and CRT display imagery. See ***NOTE*** in Figure 11-3 caption.

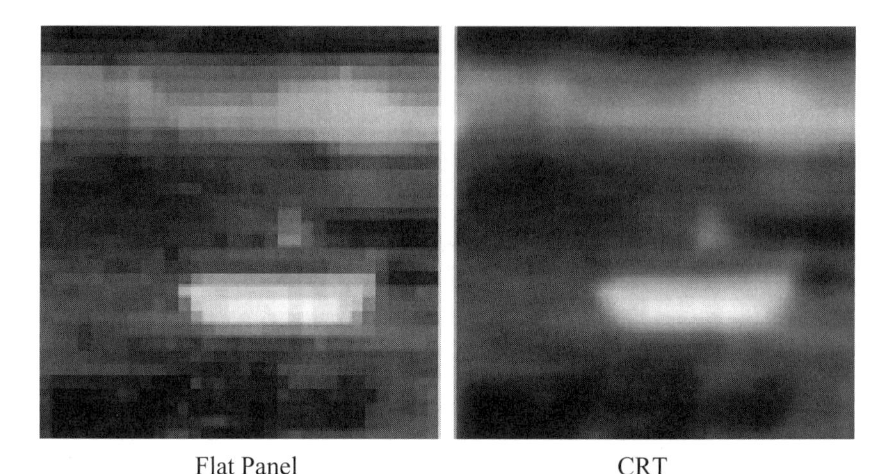

Figure 11-23. Flat panel and CRT display imagery. See ***NOTE*** in Figure 11-3 caption.

11.10. REFERENCES

1. H. Kusaka, "Consideration of Vision and Picture Quality - Psychological Effects Induced by Picture Sharpness," in *Human Vision, Visual Processing, and Digital Display*, B. E. Rogowitz, ed., SPIE Proceedings Vol. 1077, pp. 50-55 (1989).
2. E. M. Granger and K. N. Cupery, "An Optical Merit Function (SQF) which Correlates With Subjective Image Judgments," *Photographic Science and Engineering*, Vol. 16, pp. 221-230 (1972).
3. O. H. Schade, Sr., "Image Gradation, Graininess, and Sharpness in Television and Motion Picture Systems," published in four parts in *SMPTE Journal*: "Part I: Image Structure and Transfer Characteristics," Vol. 56(2), pp. 137-171 (1951); "Part II: The Grain Structure of Motion Pictures - An Analysis of Deviations and Fluctuations of the Sample Number," Vol. 58(2), pp. 181-222 (1952); "Part III: The Grain Structure of Television Images," Vol. 61(2), pp. 97-164 (1953); "Part IV: Image Analysis in Photographic and Television Systems," Vol. 64(11), pp. 593-617 (1955).
4. P. G. Barten, "Evaluation of Subjective Image Quality with the Square-root Integral Method," *Journal of the Optical Society of America. A*, Vol. 17(10), pp. 2024-2031 (1990).
5. P. G. Barten, "Evaluation of the Effect of Noise on Subjective Image Quality," in *Human Vision, Visual Processing, and Digital Display II*, J. P. Allenbach, M. H. Brill, and B. E. Rogowitz, eds., SPIE Proceedings Vol. 1453, pp. 2-15 (1991).
6. P. G. Barten, "Physical Model for the Contrast Sensitivity of the Human Eye," in *Human Vision, Visual Processing, and Digital Display III*, B. E. Rogowitz, SPIE Proceedings Vol. 1666, pp. 57-72 (1992).
7. P. G. Barten, *Contrast Sensitivity of the Human Eye and Its Effects on Image Quality*, SPIE Press Monograph PM72, Bellingham, WA (1999).
8. G. C. Holst, Electro-Optical Imaging System Performance, 4th edition, Chapter 12, JCD Publishing, Winter Park, FL (2006).

9. SSCamIP and IICamIP were developed by the U.S. Army Night Vision and Electronic Sensor Directorate. They can be obtained from SENSIAC (www.sensiac.gatech.edu/external/index.jsf)
10. L. M. Beyer, S. H. Cobb, and L. C. Clune, "Ensquared Power for Obscured Circular Pupils with Off-Center Imaging," *Applied Optics*, Vol. 30(25), pp. 3569-3574 (1991).
11. A. J. den Dekker and A. van den Bos, "Resolution: A Survey," *Journal of the Optical Society A*, Vol. 14(3), pp. 547-557 (1997).
12. See, for example, W. J. Smith, *Modern Optical Engineering*, 2nd Edition, McGraw-Hill, New York (1990).
13. R. D. Fiete, "Image quality and $\lambda FN/\rho$ for remote sensing systems," *Optical Engineering*, Vol, 38, pp. 1229-1240 (1999).
14. MAVIISS (MTF based Visual and Infrared System Simulator) is an interactive software program available from JCD Publishing at www.JCDPublishing.com.
15. M. Clampin, "Ultraviolet-Optical Charge-Coupled Devices for Space Instrumentation", *Optical Engineering*, Vol. 41, pp. 1185 – 1191 (2002).
16. E. Fossum, "What to do with Sub-Diffraction-Limit (SDL) Pixels? – A Proposal for a Gigapixel Digital Film Sensor (DFS)", *Proceedings of the 2005 IEEE Workshop on Charge-Coupled Devices and Advanced Image Sensors*, IEEE and IEEE Electron Devices Society, Nagano, Japan, pp. 214 – 217 (2005).
17. J. M. Lloyd, *Thermal Imaging*, p. 109, Plenum Press, New York (1975).
18. P. G. J. Barten, "Spot Size and Current Density Distributions of CRTs," in Proceedings of the SID, Vol. 25, pp. 155-159 (1984).
19. S. C. Hsu, "The Kell Factor: Past and Present," *SMPTE Journal*, Vol. 95, pp. 206-214 (1986).
20. C. Poynton, *Digital Video and HDTV: Algorithms and Interfaces*, pp. 67-68, Morgan Kaufmann (2003).
21. L. M. Biberman, "Image Quality," in *Perception of Displayed Information*, L. M. Biberman, ed., pp. 13-18, Plenum Press, New York (1973).
22. B. Hultgren, D. Hertel, J. Bullitt, "The influence of statistical variations on image quality," in Image Quality and System Performance III, L.C. Cui and Y. Miyake, eds., SPIE Proceeding Vol. 6059, paper 60590P (2006).
23. H. L. Snyder, "Image Quality and Observer Performance," in *Perception of Displayed Information*, L. M. Biberman, ed., pp. 87-118, Plenum Press, New York, NY (1973).
24. L. White, "Subjective Quality Factor: A New Way to Test Lenses" *Popular Photography and Imaging*, August 2003
25. Imatest software measures numerous image metrics. It is available at http://www.imatest.com
26. R. H. Vollmerhausen, E. Jacobs, and R. G. Driggers, "New Metric for Predicting Target Acquisition Performance." *Optical Engineering*, Vol. 43(11), pp. 1806-2818 (2004).
27. R. H. Vollmerhausen, E. Jacobs, J. Hixson, M. Friedman, "The Targeting Task Performance (TTP) Metric, A New Model for Predicting Target Acquisition Performance," NVESD Technical Report AMSEL-NV-TR-230, Fort Belvoir, VA (March 2006).
28. Comparative information is spread across numerous articles. See, for example, C. Poynton, *Digital Video and HDTV: Algorithms and Interfaces*, Morgan Kaufmann (2003).

12
RANGE PERFORMANCE

With high SNR, higher MTF cameras provide better image quality. While most resolution metrics are applicable to high-contrast targets, the relative rating of systems may change for low-contrast scenes. A high-MTF system with excessive noise will not permit detection of low-contrast targets.

The military is interested in detecting and recognizing targets at the greatest possible range. This is achieved by increasing the system gain until noise is apparent and the imagery is "snowy." The observer does not see the same contrast as that presented by the target to the camera. Both the displayed contrast and brightness can be adjusted such that the observed contrast can be significantly different from what is present at the entrance aperture of the imaging system.

Within the real world, there is a probability associated with every parameter. The target contrast is not one number, but a range of values that follow a diurnal cycle. The atmospheric transmittance is not fixed, but can change in minutes. There appears to be an overwhelming set of combinations and permutations. Therefore, only a few representative target contrasts and a few representative atmospheric conditions can be selected and range performance is calculated for these conditions. Because of this and other model uncertainties, range predictions cannot be placed on an absolute scale. *All analyses must be used only for comparative performance purposes.*

The symbols used in this book are summarized in the *Symbol List*, which appears after the *Table of Contents*.

12.1. ATMOSPHERIC TRANSMITTANCE

In the visible region of the spectrum, atmospheric transmittance is dominated by scattering. With a scattering coefficient of σ_{ATM}, the transmittance at range R is

$$T = e^{-\sigma_{ATM} R} \tag{12-1}$$

Meteorological range is defined quantitatively by the Koschmieder formula[1]

$$R_{MET} = \frac{1}{\sigma_{ATM}} \ln\left(\frac{1}{C_{TH}}\right) \tag{12-2}$$

where C_{TH} is the threshold contrast at which 50% of the observers can just detect the target. Koschmieder set C_{TH} to 0.02 and evaluated σ_{ATM} at λ = **0.555** μm. The transmittance (averaged over the eye's spectral response) is

$$T_{AVE}(R) = e^{-\sigma_{VIS} R} = \exp\left(-\frac{3.912}{R_{MET}} R \right) \qquad (12\text{-}3)$$

The subscript ***VIS*** is used to emphasize that σ_{VIS} is derived from the meteorological range. For wavelengths less than 3 μm, σ_{ATM} **(λ)** can be approximated by

$$\begin{aligned}
\sigma_{VIS}(\lambda) &\approx \left(\frac{3.912}{R_{MET}}\right)\left(\frac{0.555}{\lambda}\right)^{0.585(R_{MET})^{1/3}} & R_{MET} &< 6\,km \\
\sigma_{VIS}(\lambda) &\approx \left(\frac{3.912}{R_{MET}}\right)\left(\frac{0.555}{\lambda}\right)^{1.3} & 6 < R_{MET} &< 20\,km \qquad (12\text{-}4) \\
\sigma_{VIS}(\lambda) &\approx \left(\frac{3.912}{R_{MET}}\right)\left(\frac{0.555}{\lambda}\right)^{1.6} & R_{MET} &> 20\,km
\end{aligned}$$

and $\sigma_{ATM}(\lambda) = \sigma_{VIS}$ at λ = 0.555 μm. Table 12-1 provides the range of scattering coefficients associated with the international visibility code. The table should be used as guidance; the range is too coarse for scientific analyses.

Table 12-1
INTERNATIONAL VISIBILITY CODE

DESIGNATION	VISIBILITY	SCATTERING COEFFICIENT σ_{VIS} at λ = 0.555 μm.
Dense fog	0 - 50 m	> 78.2 km^{-1}
Thick fog	50 - 200 m	19.6 - 78.2 km^{-1}
Moderate fog	200 - 500 m	7.82 - 19.6 km^{-1}
Light fog	500 - 1 km	3.92 - 7.82 km^{-1}
Thin fog	1 - 2 km	1.96 - 3.92 km^{-1}
Haze	2 - 4 km	0.978 - 1.96 km^{-1}
Light haze	4 - 10 km	0.391 - 0.978 km^{-1}
Clear	10 - 20 km	0.196 - 0.391 km^{-1}
Very clear	20 - 50 km	0.0782 - 0.196 km^{-1}
Exceptionally clear	> 50 km	< 0.0782 km^{-1}

12.2. TARGET CONTRAST

The inherent target contrast is the target luminance minus the background luminance divided by the background luminance. Because both the target and background are assumed to be illuminated by the same source (sun, moon, etc.), the contrast depends on reflectances

$$C_O = \frac{|\Delta L|}{L_B} = \frac{|L_T - L_B|}{L_B} = \frac{|\rho_T(\lambda) - \rho_B(\lambda)|}{\rho_B(\lambda)} \tag{12-5}$$

Historically, the target was always considered darker than the background. However, the target can be brighter than the background and there does not appear to be any difference between the detectability of objects that are of negative or positive contrast.

12.3. CONTRAST TRANSMITTANCE

For most scenarios, the target and background are illuminated by the same source (sun, moon, etc.). The apparent contrast at the camera's entrance aperture, C_{RO}, is the inherent contrast modified by atmospheric phenomena. Dropping the wavelength notation for equation brevity,

$$\begin{aligned} C_{RO} &= \frac{\Delta L}{L} = \frac{e^{-\sigma_{ATM} R}(\rho_T - \rho_B) L_{SCENE}}{e^{-\sigma_{ATM} R} \rho_B L_{SCENE} + L_{PATH}} \\ &= C_O \frac{1}{1 + \dfrac{L_{PATH}}{\rho_B L_{SCENE}} e^{\sigma_{ATM} R}} \end{aligned} \tag{12-6}$$

The variable L_{SCENE} is the ambient illumination that irradiates the target and background. It is the sky, sun, or moon modified by the intervening atmospheric transmittance. The path radiance, L_{PATH}, is due to light scattered into the line-of-sight and the total amount received is an integral over the path length.[2] When only scattering is present (a reasonable approximation for the visible spectral band), $L_{PATH} \approx [1 - \exp(-\sigma_{ATM} R)] L_{SKY}$. The value L_{SKY} depends on the viewing direction and the location of the sun. The variable $L_{SKY}/\rho_B L_{SCENE}$ is called the sky-to-ground ratio (SGR)

$$C_{RO} = C_O \frac{1}{1 + SGR(e^{\sigma_{ATM} R} - 1)} \tag{12-7}$$

The SGR is approximately $0.2/\rho_B$ for a clear day ($L_{SKY}/L_{SCENE} \approx 0.2$) and $1/\rho_B$ for an overcast day ($L_{SKY}/L_{SCENE} \approx 1$). Table 12-2 provides several representative SGRs, with $SGR = 3$ considered to be an "average." Figure 12-1 illustrates the contrast transmittance for a variety of SGRs. When the SGR is one, the received contrast is simply the inherent contrast reduced by the atmospheric transmittance

$$C_{RO} \approx C_o e^{-\sigma_{ATM} R} \tag{12-8}$$

Table 12-2
TYPICAL SKY-TO-BACKGROUND RATIOS

Sky	Ground	SGR	Sky	Ground	SGR
Clear	Fresh snow	0.2	Overcast	Fresh snow	1
Clear	Desert	0.4	Overcast	Desert	2
Clear	Forest	5	Overcast	Forest	25

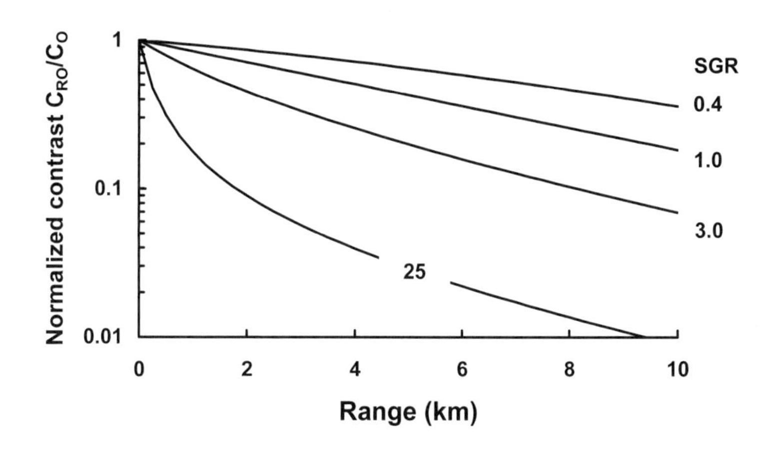

Figure 12-1. Normalized contrast as a function of range and SGR for a very clear day ($R_{MET} = 23$ km or $\sigma_{VIS} = 0.17$/km).

12.4. RANGE PREDICTIONS

Section 11.7.4 *Targeting task performance* described the image quality metric used in all NVESD contrast models[3] (e.g., the solid state camera model, SSCamIP, and the image intensified camera model, IICamIP). Range as a function of probability is determined as follows:

1. A range is selected. Using the assumed σ_{ATM} and SGR and the apparent contrast C_{RO} is determined.

2. Based upon C_{RO} the integration limits are determined. The TTP metric is calculated for the horizontal and vertical directions

$$TTP_H = \int_{u_{LOW}}^{u_{HIGH}} \sqrt{\frac{CTF_{RO}}{CTF_{SYS}(u)}}\, du$$
$$TTP_V = \int_{v_{LOW}}^{v_{HIGH}} \sqrt{\frac{CTF_{RO}}{CTF_{SYS}(v)}}\, dv \qquad (12\text{-}9)$$

to yield a two-dimensional TTP

$$TTP_{2D} = \sqrt{TTP_H TTP_V} \qquad (12\text{-}10)$$

3. The number of resolved cycles on a target of area A_{TARGET} is

$$N_{RESOLVED} = \frac{\sqrt{A_{TARGET}}}{R} TTP_{2D} \qquad (12\text{-}11)$$

4. The horizontal and vertical out-of-band spurious responses, SR_H and SR_V are determined. Sampling effects modify the number of effective cycles across the target. For recognition and identification, the number of *sampled* cycles is

$$N_{SAMPLED} = \sqrt{(1-0.58SR_H)(1-0.58SR_V)}\, N_{RESOLVED} \qquad (12\text{-}12)$$

For detection, the number of *sampled* cycles is identical to the number of resolved cycles

$$N_{SAMPLED} = N_{RESOLVED} \qquad (12\text{-}13)$$

5. The probability of acquisition is

$$P = \frac{\left(\frac{N_{SAMPLE}}{V_{50}}\right)^{E}}{1+\left(\frac{N_{SAMPLE}}{V_{50}}\right)^{E}} \tag{12-14}$$

where

$$\mathrm{E} = 1.51 + 0.24\frac{N_{SAMPLE}}{V_{50}} \tag{12-15}$$

The variable V_{50} is the 50% probability of correct choice for the task on hand. For detection, recognition, and identification, V_{50} = 2.7, 14.5, and 18.8, respectively. The number of cycles represents the average of an ensemble of similar-sized vehicles operating under a variety of conditions. For example, 80% probability of recognition means that ensemble of observers should correctly recognize (at P = 80%) the target when others similar-sized vehicles (confusers) are in the field of view. It does not mean that a specific individual will recognize a specific vehicle 80% of the time.

6. Repeat steps 1 through 5 for a variety of ranges to obtain the probability as a function of range (Figure 12-2).

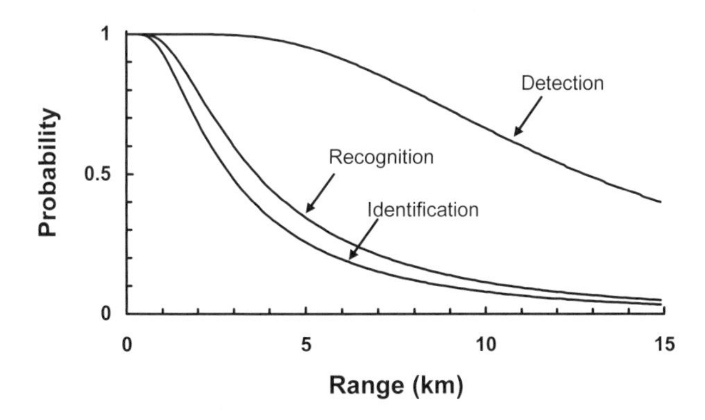

Figure 12-2. Representative range for a CCD camera.

The TTP

$$CTF_{SYS}(f) = \frac{CTF_{EYE}(f)}{M_{DISPLAY} MTF_{SYS}(f)} \sqrt{1 + \left(\frac{k\sigma}{L}\right)^2} \qquad (12\text{-}16)$$

can be separated into two components such that

$$CTF_{SYS}(f) = \sqrt{CTF^2_{EYE-SYS}(f) + CTF^2_{NOISE}(f)} \qquad (12\text{-}17)$$

where

$$CTF_{EYE-SYS}(f) = \frac{CTF_{EYE}(f)}{M_{DISPLAY} MTF_{SYS}(f)}$$
$$CTF_{NOISE}(f) = \frac{CTF_{EYE}(f)}{M_{DISPLAY} MTF_{SYS}(f)} \left(\frac{k\sigma}{L}\right) \qquad (12\text{-}18)$$

The performance models, SSCamIP and IICamIP, provide plots of $\boldsymbol{CTF_{EYE\text{-}SYS}}$ and $\boldsymbol{CTF_{NOISE}}$ (Figure 12-3). This allows the user to determine is his system is eye-limited or noise-limited. If eye-limited, a larger display, electronic zoom, or moving closer to the display will increase the range. If noise-limited, frame integration can be employed. In a well-designed camera, system noise is dominated by array noise (chapter 6).

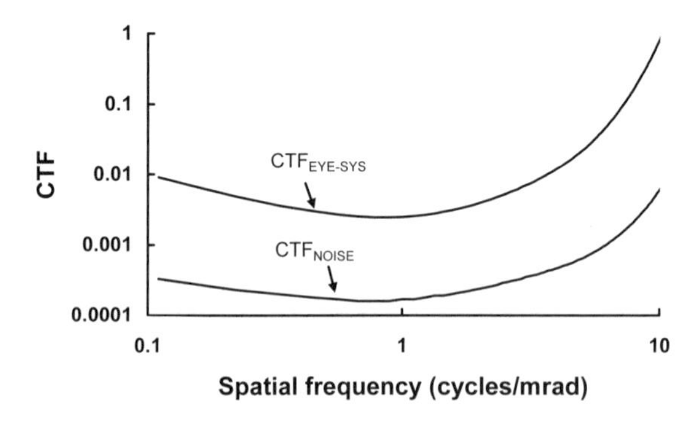

Figure 12-3. CTFs for a typical CCD camera. This system is eye-limited. Reducing camera noise will have little effect on range performance.

12.5. REFERENCES

1. W. E. K. Middleton, *Vision Through the Atmosphere*, University of Toronto Press (1958).
2. L. Levi, *Applied Optics*, pp. 118-124, Wiley and Sons (1980).
3. U.S. Army Night Vision and Electronic Sensor Directorate (Ft. Belvoir, VA) developed SSCamIP and IICamIP . The software is offered by SENSIAC (www.sensiac.gatech.edu/external/index.jsf).
4. R. H. Vollmerhausen, E. Jacobs, and R. G. Driggers, "New Metric for Predicting Target Acquisition Performance," *Optical Engineering*, Vol. 43(11), pp. 1806-2818 (2004).
5. R. H. Vollmerhasuen, E. Jacobs, J. Hixson, and M. Friedman, "The Targeting Task Performance (TTP) Metric, A New Model for Predicting Target Acquisition Performance," NVESD Technical Report AMSEL-NV-TR-230, Fort Belvoir, VA (March 2005).

APPENDIX

F-NUMBER

The radiometric equations (Chapter 3) were derived from plane geometry and paraxial ray approximations. For paraxial rays, the principal surfaces are assumed to be planes (Figure A-1). This is the representation shown in most introductory textbooks.

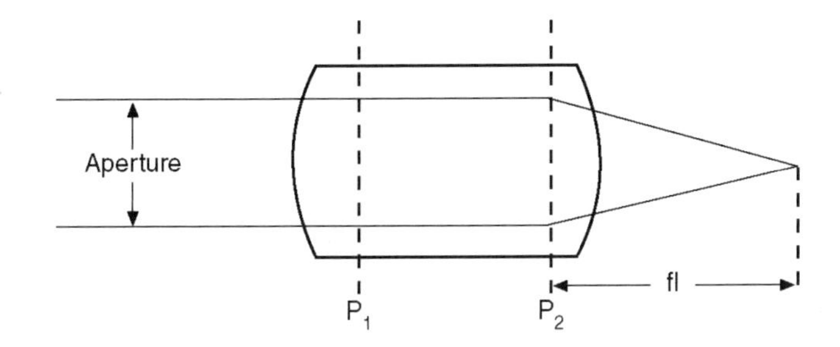

Figure A-1. The optical system can be considered as a single lens. P_1 and P_2 are the principal surfaces. The effective focal length is measured from the second principal plane. The clear aperture limits the amount of light reaching the detector.

Lens design theory[1] assumes that the principal surfaces are spherical: Every point on the surface is exactly a focal-length distance away (Figure A-2).

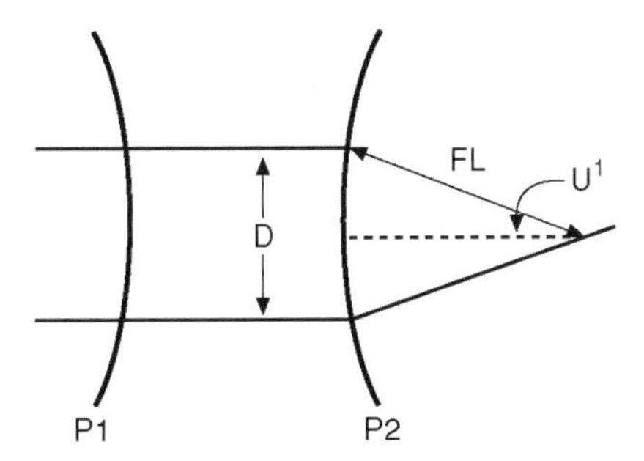

Figure A-2. Principal planes are typically spherical.

When using solid angles, the focal plane incidance is proportional to $\sin^2(\boldsymbol{U'})$ where $\boldsymbol{U'}$ is the maximum half-angle subtended by the lens. The numerical aperture is another measure of the energy collected by the optical system. When the image is in air (refractive index of unity) the numerical aperture is

$$NA = \sin U' = \frac{1}{2F^*} \tag{A-1}$$

Since the largest angle is π/2, the smallest theoretical value for F^* is ½. In the paraxial region the focal ratio is defined as

$$F = \frac{fl}{D} \tag{A-2}$$

Then

$$F^* = \sqrt{F^2 + \frac{1}{4}} \tag{A-3}$$

With the paraxial assumption, F can approach zero. The relationship between U' and F is plotted in Figure A-3 along with the error incurred with the paraxial approximation. The radiometric equations used throughout this text assumed paraxial validity. The analyst should insert F^* in the equation when $F < 3$.

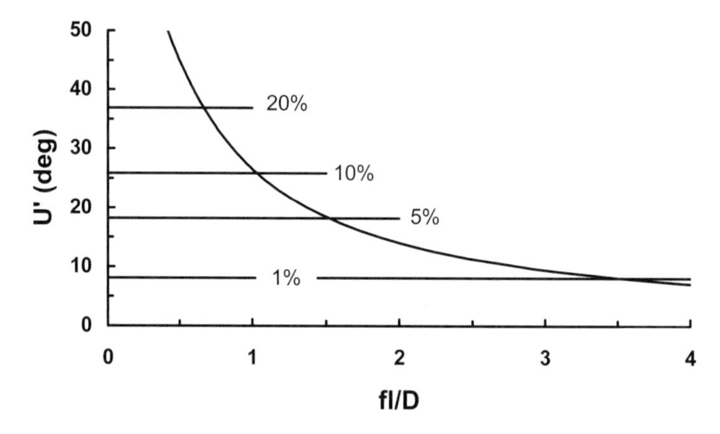

Figure A-3. Error associated with the paraxial approximation. The error is 20%, 10%, 5%, and 1% at F = 0.67, 1.03, 1.52, and 3.5, respectively.

Reference

1. W. J. Smith, *Modern Optical Engineering,* second edition, pp. 142-145, McGraw-Hill, New York (1990).

INDEX